Handbook of MTBE and Other Gasoline Oxygenates

CHEMICAL INDUSTRIES

A Series of Reference Books and Textbooks

Founding Editor

HEINZ HEINEMANN

1. *Fluid Catalytic Cracking with Zeolite Catalysts,* Paul B. Venuto and E. Thomas Habib, Jr.
2. *Ethylene: Keystone to the Petrochemical Industry,* Ludwig Kniel, Olaf Winter, and Karl Stork
3. *The Chemistry and Technology of Petroleum,* James G. Speight
4. *The Desulfurization of Heavy Oils and Residua,* James G. Speight
5. *Catalysis of Organic Reactions,* edited by William R. Moser
6. *Acetylene-Based Chemicals from Coal and Other Natural Resources,* Robert J. Tedeschi
7. *Chemically Resistant Masonry,* Walter Lee Sheppard, Jr.
8. *Compressors and Expanders: Selection and Application for the Process Industry,* Heinz P. Bloch, Joseph A. Cameron, Frank M. Danowski, Jr., Ralph James, Jr., Judson S. Swearingen, and Marilyn E. Weightman
9. *Metering Pumps: Selection and Application,* James P. Poynton
10. *Hydrocarbons from Methanol,* Clarence D. Chang
11. *Form Flotation: Theory and Applications,* Ann N. Clarke and David J. Wilson
12. *The Chemistry and Technology of Coal,* James G. Speight
13. *Pneumatic and Hydraulic Conveying of Solids,* O. A. Williams
14. *Catalyst Manufacture: Laboratory and Commercial Preparations,* Alvin B. Stiles
15. *Characterization of Heterogeneous Catalysts,* edited by Francis Delannay
16. *BASIC Programs for Chemical Engineering Design,* James H. Weber
17. *Catalyst Poisoning,* L. Louis Hegedus and Robert W. McCabe
18. *Catalysis of Organic Reactions,* edited by John R. Kosak
19. *Adsorption Technology: A Step-by-Step Approach to Process Evaluation and Application,* edited by Frank L. Slejko
20. *Deactivation and Poisoning of Catalysts,* edited by Jacques Oudar and Henry Wise
21. *Catalysis and Surface Science: Developments in Chemicals from Methanol, Hydrotreating of Hydrocarbons, Catalyst Preparation, Monomers and Polymers, Photocatalysis and Photovoltaics,* edited by Heinz Heinemann and Gabor A. Somorjai
22. *Catalysis of Organic Reactions,* edited by Robert L. Augustine

23. *Modern Control Techniques for the Processing Industries,* T. H. Tsai, J. W. Lane, and C. S. Lin
24. *Temperature-Programmed Reduction for Solid Materials Characterization,* Alan Jones and Brian McNichol
25. *Catalytic Cracking: Catalysts, Chemistry, and Kinetics,* Bohdan W. Wojciechowski and Avelino Corma
26. *Chemical Reaction and Reactor Engineering,* edited by J. J. Carberry and A. Varma
27. *Filtration: Principles and Practices, Second Edition,* edited by Michael J. Matteson and Clyde Orr
28. *Corrosion Mechanisms,* edited by Florian Mansfeld
29. *Catalysis and Surface Properties of Liquid Metals and Alloys,* Yoshisada Ogino
30. *Catalyst Deactivation,* edited by Eugene E. Petersen and Alexis T. Bell
31. *Hydrogen Effects in Catalysis: Fundamentals and Practical Applications,* edited by Zoltán Paál and P. G. Menon
32. *Flow Management for Engineers and Scientists,* Nicholas P. Cheremisinoff and Paul N. Cheremisinoff
33. *Catalysis of Organic Reactions,* edited by Paul N. Rylander, Harold Greenfield, and Robert L. Augustine
34. *Powder and Bulk Solids Handling Processes: Instrumentation and Control,* Koichi Iinoya, Hiroaki Masuda, and Kinnosuke Watanabe
35. *Reverse Osmosis Technology: Applications for High-Purity-Water Production,* edited by Bipin S. Parekh
36. *Shape Selective Catalysis in Industrial Applications,* N. Y. Chen, William E. Garwood, and Frank G. Dwyer
37. *Alpha Olefins Applications Handbook,* edited by George R. Lappin and Joseph L. Sauer
38. *Process Modeling and Control in Chemical Industries,* edited by Kaddour Najim
39. *Clathrate Hydrates of Natural Gases,* E. Dendy Sloan, Jr.
40. *Catalysis of Organic Reactions,* edited by Dale W. Blackburn
41. *Fuel Science and Technology Handbook,* edited by James G. Speight
42. *Octane-Enhancing Zeolitic FCC Catalysts,* Julius Scherzer
43. *Oxygen in Catalysis,* Adam Bielanski and Jerzy Haber
44. *The Chemistry and Technology of Petroleum: Second Edition, Revised and Expanded,* James G. Speight
45. *Industrial Drying Equipment: Selection and Application,* C. M. van't Land
46. *Novel Production Methods for Ethylene, Light Hydrocarbons, and Aromatics*, edited by Lyle F. Albright, Billy L. Crynes, and Siegfried Nowak
47. *Catalysis of Organic Reactions*, edited by William E. Pascoe
48. *Synthetic Lubricants and High-Performance Functional Fluids*, edited by Ronald L. Shubkin
49. *Acetic Acid and Its Derivatives*, edited by Victor H. Agreda and Joseph R. Zoeller
50. *Properties and Applications of Perovskite-Type Oxides*, edited by L. G. Tejuca and J. L. G. Fierro

51. *Computer-Aided Design of Catalysts*, edited by E. Robert Becker and Carmo J. Pereira
52. *Models for Thermodynamic and Phase Equilibria Calculations*, edited by Stanley I. Sandler
53. *Catalysis of Organic Reactions*, edited by John R. Kosak and Thomas A. Johnson
54. *Composition and Analysis of Heavy Petroleum Fractions*, Klaus H. Altgelt and Mieczyslaw M. Boduszynski
55. *NMR Techniques in Catalysis,* edited by Alexis T. Bell and Alexander Pines
56. *Upgrading Petroleum Residues and Heavy Oils*, Murray R. Gray
57. *Methanol Production and Use*, edited by Wu-Hsun Cheng and Harold H. Kung
58. *Catalytic Hydroprocessing of Petroleum and Distillates*, edited by Michael C. Oballah and Stuart S. Shih
59. *The Chemistry and Technology of Coal: Second Edition, Revised and Expanded,* James G. Speight
60. *Lubricant Base Oil and Wax Processing*, Avilino Sequeira, Jr.
61. *Catalytic Naphtha Reforming: Science and Technology*, edited by George J. Antos, Abdullah M. Aitani, and José M. Parera
62. *Catalysis of Organic Reactions*, edited by Mike G. Scaros and Michael L. Prunier
63. *Catalyst Manufacture,* Alvin B. Stiles and Theodore A. Koch
64. *Handbook of Grignard Reagents,* edited by Gary S. Silverman and Philip E. Rakita
65. *Shape Selective Catalysis in Industrial Applications: Second Edition, Revised and Expanded*, N. Y. Chen, William E. Garwood, and Francis G. Dwyer
66. *Hydrocracking Science and Technology*, Julius Scherzer and A. J. Gruia
67. *Hydrotreating Technology for Pollution Control: Catalysts, Catalysis, and Processes*, edited by Mario L. Occelli and Russell Chianelli
68. *Catalysis of Organic Reactions*, edited by Russell E. Malz, Jr.
69. *Synthesis of Porous Materials: Zeolites, Clays, and Nanostructures,* edited by Mario L. Occelli and Henri Kessler
70. *Methane and Its Derivatives*, Sunggyu Lee
71. *Structured Catalysts and Reactors*, edited by Andrzei Cybulski and Jacob Moulijn
72. *Industrial Gases in Petrochemical Processing*, Harold Gunardson
73. *Clathrate Hydrates of Natural Gases: Second Edition, Revised and Expanded*, E. Dendy Sloan, Jr.
74. *Fluid Cracking Catalysts*, edited by Mario L. Occelli and Paul O'Connor
75. *Catalysis of Organic Reactions*, edited by Frank E. Herkes
76. *The Chemistry and Technology of Petroleum, Third Edition, Revised and Expanded*, James G. Speight
77. *Synthetic Lubricants and High-Performance Functional Fluids, Second Edition: Revised and Expanded,* Leslie R. Rudnick and Ronald L. Shubkin

78. *The Desulfurization of Heavy Oils and Residua, Second Edition, Revised and Expanded,* James G. Speight
79. *Reaction Kinetics and Reactor Design: Second Edition, Revised and Expanded,* John B. Butt
80. *Regulatory Chemicals Handbook*, Jennifer M. Spero, Bella Devito, and Louis Theodore
81. *Applied Parameter Estimation for Chemical Engineers*, Peter Englezos and Nicolas Kalogerakis
82. *Catalysis of Organic Reactions,* edited by Michael E. Ford
83. *The Chemical Process Industries Infrastructure: Function and Economics*, James R. Couper, O. Thomas Beasley, and W. Roy Penney
84. *Transport Phenomena Fundamentals*, Joel L. Plawsky
85. *Petroleum Refining Processes,* James G. Speight and Baki Özüm
86. *Health, Safety, and Accident Management in the Chemical Process Industries,* Ann Marie Flynn and Louis Theodore
87. *Plantwide Dynamic Simulators in Chemical Processing and Control,* William L. Luyben
88. *Chemicial Reactor Design*, Peter Harriott
89. *Catalysis of Organic Reactions*, edited by Dennis Morrell
90. *Lubricant Additives: Chemistry and Applications*, edited by Leslie R. Rudnick
91. *Handbook of Fluidization and Fluid-Particle Systems*, edited by Wen-Ching Yang
92. *Conservation Equations and Modeling of Chemical and Biochemical Processes*, Said S. E. H. Elnashaie and Parag Garhyan
93. *Batch Fermentation: Modeling, Monitoring, and Control*, Ali Cinar, Satish J. Parulekar, Cenk Ündey, and Gülnur Birol
94. *Industrial Solvents Handbook: Second Edition*, Nicholas P. Cheremisinoff
95. *Petroleum and Gas Field Processing*, H. K. Abdel-Aal, Mohamed Aggour, and M. A. Fahim
96. *Chemical Process Engineering: Design and Economics*, Harry Silla
97. *Process Engineering Economics*, James R. Couper
98. *Re-Engineering the Chemical Processing Plant: Process Intensification*, Andrzej Stankiewicz and Jacob A. Moulijn
99. *Thermodynamic Cycles: Computer-Aided Design and Optimization*, Chih Wu
100. *Catalytic Naphtha Reforming, Second Edition, Revised and Expanded*, edited by George J. Antos and Abdullah M. Aitani
101. *Handbook of MTBE and Other Gasoline Oxygenates*, edited by Halim Hamid and Mohammad Ashraf Ali

ADDITIONAL VOLUMES IN PREPARATION

Handbook of MTBE and Other Gasoline Oxygenates

edited by
Halim Hamid
Mohammad Ashraf Ali
King Fahd University of Petroleum and Minerals
Dhahran, Saudi Arabia

CRC Press
Taylor & Francis Group
Boca Raton London New York

CRC Press is an imprint of the
Taylor & Francis Group, an **informa** business

CRC Press
Taylor & Francis Group
6000 Broken Sound Parkway NW, Suite 300
Boca Raton, FL 33487-2742

First issued in paperback 2019

ISBN-13: 978-0-8247-4058-0 (hbk)
ISBN-13: 978-0-367-39448-6 (pbk)

Visit the Taylor & Francis Web site at
http://www.taylorandfrancis.com

and the CRC Press Web site at
http://www.crcpress.com

Preface

This book was planned and written in view of the controversy associated with the usage of methyl *tertiary*-butyl ether (MTBE) in reformulated gasoline, generally in the United States and particularly in the state of California, which has sparked global discussions and arguments on the benefits and hazards associated with MTBE usage. The studies and arguments reflect both positive and negative aspects of MTBE. An effort has been made in this book to present studies, views, and reviews from all sides. It is now up to the readers and experts to conclude scientifically and technically the pros and cons of MTBE usage in gasoline.

Thousands of studies, reviews, websites, and web pages are available on this subject, each providing a particular point of view. With all this information available, it is surprising that a survey aimed at answering these questions and summarizing the preceding experiences is not readily found. This book provides information not currently available from a single literature source. All the editors and contributors of this book are experienced in the study of MTBE, and each of them would have employed such a survey in his own research program if it had been available.

The book comprises five main parts, each devoted to a specific aspect of MTBE and other oxygenate ethers. The introduction is devoted to the factors that influence the refining industry and that led to the development of this book. It also sheds light on the measures adopted by the U.S. Environmental Protection Agency with respect to the MTBE controversy and further actions taken to significantly reduce or eliminate MTBE, and to address prevention and remediation concerns. The introduction also presents key conclusions of an independent review of the University of California, Davis Report, by Stanford Research Institute (SRI). A global outlook of MTBE with respect to production, consumption, and future projection for demand and supply are also presented.

The first part is devoted to the properties of MTBE and other oxygenates and their blends with gasoline. The properties of oxygenates are discussed and tabulated. This part provides readers with a very clear understanding of the properties of oxygenates in pure and gasoline blended forms.

The second part deals with the synthesis and production of ethers using a number of catalysts and processes. A number of new catalytic systems are introduced and discussed. One of the chapters is dedicated to MTBE production using zeolite-based catalysts. The preparation of MTBE using fluorinated zeolites is also discussed. Another chapter focuses on the use of heteropolyacids for MTBE production. The use of catalytic distillation technology for ethers is discussed, as well as the production of ethanol from biomass. Another chapter is devoted to the commercial production of ethers.

The third part consists of chapters related to the kinetics and thermodynamics of MTBE synthesis. These chapters provide the detailed kinetics and thermodynamics of MTBE and other ethers production. These details are very helpful in reaction engineering studies. The fourth part discusses a number of technologies for remediation and the fate of MTBE in water. This part consists of three chapters contributed by researchers from and associated with the University of California. The technologies described include bioremediation, air stripping, oxidation, and adsorption processes for removing MTBE from water.

The fifth part contains a contribution on the impact of MTBE phaseout on the refining and petrochemical industries. The authors discuss the implications of MTBE phaseout on the world refining and petrochemical industries that are directly or indirectly associated or linked with MTBE production and utilization.

All the chapters are written by well-known authorities in the field encompassed by MTBE, starting with the process chemistry and focusing on the preparation and evaluation of catalysts as well as MTBE usage. Some attempt is made to predict the future of this fuel additive technology, a task made complicated by the conflicting demand for more octane and the resolution to reduce the effects resulting from their use. It has been our pleasure to work with these contributors. Their efforts in combining their own research with the recent literature is highly appreciated. The editors hope that industrial researchers, policymakers, chemists, chemical engineers, and graduate students will recognize this book as a valuable resource reference.

Halim Hamid
Mohammad Ashraf Ali

Contents

Contributors

Farid Aiouache, Ph.D. Assistant Professor, Research Center for Advanced Energy Conversion, Nagoya University, Nagoya, Japan

Mohammad Ashraf Ali, Ph.D. Research Scientist and Assistant Professor, Research Institute, King Fahd University of Petroleum and Minerals, Dhahran, Saudi Arabia

Adam Bielański, Ph.D., D.I.C. Professor, Department of Chemistry, Jagiellonian University, and Institute of Catalysis and Surface Chemistry, Polish Academy of Sciences, Krakow, Poland

Daniel P. Y. Chang, Ph.D. Ray B. Krone Professor of Environmental Engineering, Department of Civil and Environmental Engineering, University of California at Davis, Davis, California, U.S.A.

François Collignon, Ph.D. Center for Surface Chemistry and Catalysis, Katholieke Universiteit Leuven, Leuven, Belgium

Ravindra Datta, Ph.D. Professor and Head, Department of Chemical Engineering, Worcester Polytechnic Institute, Worcester, Massachusetts, U.S.A.

Blake T. Eskew, M.B.A. Vice President, Purvin & Gertz, Inc., Houston, Texas, U.S.A.

Christopher L. Geisler, B.S. Project Manager, Proprietary Services, Chemical Market Associates, Inc., Houston, Texas, U.S.A.

Shigeo Goto, Ph.D. Professor, Department of Chemical Engineering, Nagoya University, Nagoya, Japan

Halim Hamid, Ph.D. Professor, Department of Chemical Engineering, and Section Manager, Research Institute, King Fahd University of Petroleum and Minerals, Dhahran, Saudi Arabia

Harri Järvelin, Ph.D.* General Manager, Fortum Oil & Gas Oy, Edmonton, Alberta, Canada

Kyle L. Jensen, Ph.D. Department of Chemical Engineering, Worcester Polytechnic Institute, Worcester, Massachusetts, U.S.A.

Prakob Kitchaiya, Ph.D. Department of Chemical Engineering, Worcester Polytechnic Institute, Worcester, Massachusetts, U.S.A.

Aspi K. Kolah, Ph.D. Department of Physical and Chemical Process Engineering, Max Planck Institute for Dynamics of Complex Technical Systems, Magdeburg, Germany

Tuan Si Le, Ph.D. Research Associate, Departments of Chemistry and Biochemistry, Concordia University, Montreal, Quebec, Canada

Raymond Le Van Mao, Ph.D. Professor, Departments of Chemistry and Biochemistry, Concordia University, Montreal, Quebec, Canada

Sun Liang, Ph.D. Program Manager, Advanced Water Treatment Issues, Water Quality Section, Water Systems Operations, Metropolitan Water District of Southern California, La Verne, California, U.S.A.

Cris B. Liban, D.Env. Environmental Specialist III, Environmental Compliance and Services, Construction Division, Los Angeles County Metropolitan Transportation Authority, Los Angeles, California, U.S.A.

Anna Malecka-Lubańska, Ph.D. Institute of Catalysis and Surface Chemistry, Polish Academy of Sciences, Krakow, Poland

Anna Micek-Ilnicka, Ph.D. Institute of Catalysis and Surface Chemistry, Polish Academy of Sciences, Krakow, Poland

Cory B. Phillips, Ph.D. Staff Research Engineer, Department of Chemical Engineering, University of Michigan, Ann Arbor, Michigan, U.S.A.

Georges Poncelet, Ph.D. Professor, Department of Catalysis and Chemistry of Divided Materials, Université Catholique de Louvain, Louvain-la-Neuve, Belgium

**Current affiliation*: Alberta Envirofuels Inc, Edmonton, Alberta, Canada

Joanna Poźniczek, Ph.D. Institute of Catalysis and Surface Chemistry, Polish Academy of Sciences, Krakow, Poland

Liisa K. Rihko-Struckmann, D.Sc.(Tech) Department of Physical and Chemical Process Engineering, Max Planck Institute for Dynamics of Complex Technical Systems, Magdeburg, Germany

Tom C. Shih, D.Env., M.P.H. Los Angeles Regional Water Quality Control Board, State of California Environmental Protection Agency, Los Angeles, California, U.S.A.

I. H. (Mel) Suffet, Ph.D. Professor, Environmental Science and Engineering Program, Department of Environmental Health Sciences, UCLA School of Public Health, Los Angeles, California, U.S.A.

Kai Sundmacher, Ph.D. Director, Department of Physical and Chemical Process Engineering, Max Planck Institute for Dynamics of Complex Technical Systems, Magdeburg, Germany

Faisal H. Syed, Ph.D. Project Manager, Chemical Market Resources, Inc., Houston, Texas, U.S.A.

Andrzej Wyczesany, Ph.D. Institute of Organic Chemistry and Technology, University of Technology, Krakow, Poland

Tiejun Zhang, Ph.D. Department of Chemical Engineering, Worcester Polytechnic Institute, Worcester, Massachusetts, U.S.A.

Handbook of
MTBE
and Other Gasoline
Oxygenates

1
Introduction

Halim Hamid and Mohammad Ashraf Ali
Research Institute, King Fahd University of Petroleum and Minerals, Dhahran, Saudi Arabia

In the United States a debate is raging over the environmental consequences of the increased use of methyl *tertiary*-butyl ether (MTBE). Originally used as an antiknock agent for gasoline, this chemical is being used at concentrations of up to 10% (by weight) in the United States, and it is an oxygen source to improve gasoline combustion and hence reduce pollution from car exhausts. The controversy started with a report submitted in November 1998 by the University of California (UC Davis), on Health and Environmental Assessment of MTBE to the State of California. This study was authorized by the California Senate to assess a variety of issues and public concerns associated with the use of MTBE in gasoline. According to this report, MTBE and other oxygenates were found to have no significant effect on exhaust emissions from advanced-technology vehicles. There is no significant difference in the emissions reduction of benzene between oxygenated and nonoxygenated fuels. Thus, there is no significant additional air quality benefit to the use of oxygenates such as MTBE in reformulated gasoline (RFG), relative to alternative nonoxygenated formulations. There are significant risks and costs associated with water contamination due to the use of MTBE. MTBE is highly soluble in water and will transfer readily to groundwater from gasoline leaking from underground storage tanks, pipelines, and other components of the gasoline distribution system. In addition, the use of gasoline containing MTBE in motor boats results in the contamination of surface water reservoirs. It was stated in the report that the limited water resources in United States are at risk by using MTBE. If MTBE continues to be used at current levels and more sources become contaminated, the potential for regional degradation of water resources, especially groundwater basins, will increase. Severity of water shortages during drought years will be exacerbated. The UC Davis report recommended the phasing out of MTBE over an interval of several years, and that the refiners should be given flexibility to achieve air quality objectives by modifying the specifications to allow wide-scale production of nonoxygenated reformulated gasoline.

On the other hand, there are studies and arguments that say that the case of MTBE has been polluted by politics, media hype, and economic competitive interests, and that MTBE is a safe, beneficial, reliable, proven, and cost-effective component in today's clean-burning, beneficial gasoline. This has been proved by the extensively research and testing of MTBE for its performance, air quality improvement, health effects, and other benefits. In Europe, there continues to be support for MTBE to reduce air pollution, and there is a general belief that the real problem in the United States is leaking gasoline tanks, not the MTBE.

This book presents a forum to discuss the MTBE controversy in a broad spectrum of views, studies, and arguments in favor of and against MTBE. Conclusions may be drawn regarding the risks and benefits involved with the use of MTBE with respect to air, water, and human health.

I. WHAT IS METHYL *tertiary*-BUTYL ETHER?

Methyl *tertiary*-butyl ether (MTBE) is a volatile organic compound made as a by-product of petroleum refinery operations by combining methanol derived from natural gas and isobutylene. MTBE is a gasoline additive that has been used as an octane enhancer since the phase-out of lead in the late 1970s. MTBE is more soluble in water, has a smaller molecular size, and is less biodegradable than other components of gasoline. Consequently, MTBE is more mobile in groundwater than other gasoline constituents, and may often be detected when other components are not. MTBE has been used extensively around the country to reduce motor vehicle emissions. The passage of the U.S. 1990 Clean Air Act (CAA) resulted in increased use of MTBE in order to reduce carbon monoxide and hydrocarbon emissions. MTBE also reduces air toxics emissions and pollutants that form ground-level ozone. MTBE has been the additive most commonly used by gasoline suppliers throughout most of the country. It has been used because it is very cost-effective in meeting air quality and gasoline performance goals. Severe air quality problems in U.S. cities during the 1980s prompted increased use of MTBE in petroleum.

II. U.S. ENVIRONMENTAL PROTECTION AGENCY MEASURES

In response to the growing concerns about MTBE present in water, the U.S. Environmental Protection Agency (EPA) appointed an independent Blue Ribbon Panel of leading experts from the public health, environmental, and scientific communities, the fuels industry, water utilities, and local and state governments. The panel was tasked to investigate the air quality benefits and water quality concerns associated with oxygenates in gasoline, and to provide independent advice and recommendations on ways to maintain air quality while protecting water quality. The panel in July 1999 recommended the following:

- Remove the current congressional CAA requirement for 2% oxygen in RFG.
- Improve the nation's water protection programs, including over 20 specific actions to enhance underground storage tank, safe drinking water, and private well protection programs.
- Maintain current air quality benefits.
- Reduce the use of MTBE substantially nationwide.
- Accelerate research on MTBE and its substitutes.

The EPA has taken further actions to significantly reduce or eliminate MTBE, and to address prevention and remediation concerns. The agency has been working closely with the U.S. Congress, the individual states, and the regulated community to accomplish these efforts. The EPA is also assisting Congress toward a targeted legislative solution that addresses the panel's recommendations. The EPA released a legislative framework on March 20, 2000, to encourage immediate congressional action to reduce or eliminate MTBE and promote consideration of renewable fuels such as ethanol. On the same day, it announced the beginning of regulatory action under the Toxic Substances Control Act (TSCA) to significantly reduce or eliminate use of MTBE in gasoline while preserving clean air benefits. Meanwhile, the EPA is working with U.S. states to conduct an evaluation of underground storage tank (UST) systems performance to verify and validate how effectively leak detection and other UST systems are working. The EPA-released air quality trends report shows that cleaner cars and fuels accounted for almost two-thirds of total national emission reductions between 1970 and 1998. This is the period in which MTBE has had its maximum use in U.S. gasoline.

III. REVIEW AND EVALUATION OF THE UC DAVIS REPORT BY INDEPENDENT BODIES (SRIC AND SRI)

The report that follows presents key conclusions of an independent review of the UC Davis report mentioned above. The review was conducted by SRI Consulting (SRIC) and SRI International (SRI), having funding from the Oxygenated Fuels Association.

The UC Davis report does not adequately appreciate and quantify the air quality benefits of the use of MTBE and reformulated gasoline (RFG) in general. Much of the report's air evaluation focuses on emissions of MTBE and its combustion products, and on distinguishing MTBE benefits from California Cleaner Burning Gasoline (CBG). The benefits of reduced air toxics and other pollutants have been well documented in numerous reports by the U.S. EPA, the California Air Resources Board (CARB), and other regulatory agencies.

The UC Davis conclusion that there is no significant air quality benefit to the use of oxygenates is not supported by research. The Davis report from which this statement is extrapolated is extremely limited. It does not represent the actual vehicle fleet or commercial gasoline formulations. Therefore, it is not an adequate foundation for major air quality policy decisions.

The UC Davis report demonstrates that exposure to high levels of MTBE poses risks, but a variety of national and international organizations have concluded that continued use of MTBE in gasoline, which involves much lower levels of exposure, is safe. Scientific scrutiny of the potential health effects of exposure to MTBE is justified, but it is essential for the investigators to present the available research as complete as possible. The UC Davis report places unwarranted emphasis on the uncertainties in the research rather than making judgments based on the overall evidence and the intended use of the product.

The conclusions and cost–benefit analysis in the UC Davis report do not focus on forward-looking policy issues. The decision should be based on the future benefits and risks to the state, not on events that have already occurred. The value and necessity of analyzing historical information is fully understood. A number of extrapolations made in the UC Davis report are not valid.

A number of errors have been made in the UC Davis report regarding water contaminated by MTBE and the remediation cost calculations, the most significant of which is the inclusion of up to $1 billion in "sunk costs" for correcting contamination that has already occurred. Regardless of the ultimate magnitude of these costs, they will be incurred equally under all future MTBE use and MTBE ban options. Therefore, sunk costs are not relevant to policy decisions about the future use of MTBE. Their incorporation with other, legitimate costs makes it difficult to determine the appropriate economics.

The fuel analysis economics is another major component of the cost–benefit analysis that does not accurately reflect industry practices, commercial gasoline blends, or real world economics. After the UC Davis report was submitted, the California Energy Commission (CEC) published its report, *Supply and Cost of Alternatives to MTBE in Gasoline.* That report, which is based on a much more detailed analysis of gasoline formulations and industry conditions than the UC Davis report, concludes that ethanol and nonoxygenated gasoline blends will both be more expensive as compared to CBG with MTBE. It is believed that the CEC analysis provides a more accurate projection, but it still understates the cost of switching away from MTBE.

Because of the specific errors cited above, the overall cost–benefit analysis in the UC Davis report leads to the wrong conclusions. The analysis shows that CBG with MTBE is the least costly of the three fuel options considered in the UC Davis study, and not the most expensive.

From the review and evaluation of the UC Davis report by independent bodies, it can be understood that there is not a clear basis, either economically or technically, for a regulatory ban or phase-out of MTBE. In fact, there are certainly considerable costs involved in switching to a new fuel, particularly if the switch is done hastily and without a complete assessment. SRIC concluded that MTBE remains a viable and economical choice for reducing automotive emissions in California, and therefore its use should not be banned.

IV. RECENT REPORT OF THE COMMISSION OF EUROPEAN COMMUNITIES

The scheduled phaseout of MTBE in California now appears both unnecessary and economically risky, due to an important new European study. A risk assessment by the European Union (EU) of MTBE has concluded that it does not pose a danger to human health, but tighter controls on the handling and storage of the chemical are required. The Commission of European Communities, Europe's official scientific investigative body, released findings from a comprehensive study of MTBE health effects. The commission concluded that MTBE poses very limited risks that can be essentially mitigated by existing control mechanisms such as sound fuel tank management and code enforcement. The commission found no compelling reasons to limit use of MTBE in motor fuel. The European Commission's findings are significant because they validate the findings of the World Health Organization's International Agency for Research on Cancer, the U.S. National Toxicology Panel, California's own Science Advisory Board for Proposition 65, and several other studies that all agree there is no compelling evidence that MTBE causes cancer in human beings. In fact, there is not a single peer-reviewed study that concludes that MTBE causes cancer in humans. The European Commission is yet another credible scientific body that has declared MTBE safe, given rigorous enforcement of the underground gasoline storage tank program.

V. STATISTICS FROM THE CALIFORNIA DEPARTMENT OF HEALTH SERVICES

Updated, cumulative statistics from the California Department of Health Services indicate that the actual detections of MTBE in drinking water have been extremely small, less than 1% of the water sources tested, and the trends have declined to very low levels. This marked improvement is due to advances made in upgrading underground gasoline storage tanks and better tank program enforcement throughout the state. Governor Gray Davis's decree that MTBE be phased out from California gasoline by 2003 was based largely on two projections that have proven to be erroneous: that MTBE is a pervasive groundwater contaminant throughout the state, and that MTBE poses a health threat to Californians. The governor has extended the deadline for the MTBE ban due to supply and transportation problems associated with importing ethanol (the primary alternative to MTBE) to California. The California Energy Commission has concluded that a switch from MTBE to ethanol in California gasoline would result in significant gasoline price spikes. Governor Davis has stated publicly that ethanol-related price spikes at the gas pump could be as high as 50 cents per gallon. Better enforcement in California is now preventing gasoline leaks into groundwater. The phaseout becomes even more questionable when considering new statistics compiled by the State of California indicating that MTBE detections in drinking water have largely been eradicated.

The health benefits associated with the use of MTBE in gasoline outweigh the health risks posed by its detection in groundwater. By featuring reduced concentrations of known cancer-causing compounds such as benzene, cleaner burning gasolines containing MTBE have been demonstrated to reduce risk of cancer from toxics in automotive exhaust. The annual reduction from baseline fuel is somewhere between 10% and 50%, depending on other environmental factors. As a result of this air quality improvement, a large public health benefit accrues over the long run. While MTBE may render water unpalatable at very low concentrations, it does not pose a significant health risk when used in gasoline. MTBE does not accumulate in the body. It does not have pathological effects, nor does it injure developing offspring or impede reproductive functions. Its effects occur only at high doses not encountered by humans, and, despite extensive testing, there exists no scientific consensus on whether it can cause cancer in humans. In 1999, National Toxicology Program (NTP) leaders excluded MTBE from its recently revised list of human carcinogens, which has been published in the NTP Ninth Edition report on carcinogens. Panel members have concluded that while there is some evidence of MTBE carcinogenicity in test animals, the data are not strong enough to list MTBE as a human carcinogen.

VI. ECONOMIC LOSS AND BURDEN

According to a report by the U.S. Department of Energy, an MTBE ban in the United States would be equivalent to a loss of 300,000 barrels per day of premium blend stock and it would need to be compensated by crude processing capacity equivalent to five average U.S. refineries. An MTBE ban will invite a series of problems which are as follows: there will be an increased in gasoline production cost, increased reliance on gasoline imports, increased refinery investment, and increased pollutants emissions.

VII. CALIFORNIA DELAYED MTBE BAN DEADLINE

On Friday, March 15, 2002, California Governor Gray Davis postponed, by one year, the deadline to ban MTBE as a fuel additive across California. Through an executive order, he extended the MTBE ban from January 1, 2003, until January 1, 2004. This has provided refiners more time to overhaul their plants. The replacement of MTBE in California's reformulated gasoline program is expected to require an additional 675 million gallons of ethanol per year. Since March 1999, when Governor Davis issued an Executive Order requiring the elimination of MTBE within two-and-a-half years, 10 new ethanol plants have opened and many older ones have expanded, increasing the nation's ethanol capacity by more than 550 million gallons. Eighteen additional plants are under construction and slated to begin production before California's deadline for phasing out MTBE. These plants will add yet another 470 million gallons of capacity. There are also scores of new plants in various stages of development. The ethanol industry is alarmed by persistent rumors suggesting the state may yet delay the implementation of the MTBE phase-out because of concerns about the potential impact of transitioning from MTBE to ethanol.

Some refiners in California have not done anything to make the changes needed to adopt ethanol. The governor's 1999 order to ban MTBE by the end of 2002 included a request to the Environmental Protection Agency to waive the oxygenate requirement. The EPA did not rule until the summer of 2001, and it ruled against waiving the oxygenate requirement. Many of the state's refiners would not act until the EPA ruled on the oxygenate requirement. Now that refiners know the rules, they need sufficient time to get permits to retrofit their plants and logistics systems to comply with state guidelines. Permitting agencies in California are reluctant to issue permits for the required modifications. In addition, it takes only about half as much ethanol as MTBE to meet the oxygen requirement. This still leaves a void of approximately 50,000 bbl/day of gasoline. Most California refineries are already at maximum rates. Alkylate or isooctane will be needed to make up the difference, and it is not available and likely will not be in the near future. An extension of the ban is the only course that makes sense. It has been stated by Gray Davis that California's gasoline specifications are hard to meet, and the state must be careful to avoid a gasoline shortage or backsliding on air quality.

VIII. MTBE OUTLOOK FOR EUROPE

When Europe started to phase down lead octane additives in petrol (gasoline) in the 1980s, many refiners replaced them with aromatics, which were the lowest-cost alternative at that time. Toward the end of the 1990s, new environmental regulations limited the aromatic content of gasoline. Refiners seeking alternative blending components came to rely more on fuel oxygenates—oxygen-rich, cost-effective compounds that act as octane enhancers, with the additional benefit of making gasoline burn more completely, thereby significantly reducing toxic exhaust emissions. MTBE is currently the most commonly used oxygenate in Europe.

In Europe, the demand is approximately equal to the production capacity. Stricter gasoline quality requirements have increased the demand for high-octane blending components such as MTBE, and its consumption is expected to remain fairly stable in Europe over the next few years. Unlike the United States, there is no mandate on the use of MTBE in Europe. In Europe, gasoline is delivered under suction, with similar specifications for all EU states, whereas in the United States gasoline is delivered with pressure. In 1997, MTBE was included in the third Priority List of substances selected for risk assessment under the EU Existing Substances Regulation. Finland was chosen as the Member State responsible for progressing the risk assessment on MTBE on behalf of the European Commission's working group. In November 2000, the EU Working Group on the Classification and Labeling of Dangerous Substances examined the status of MTBE in a meeting of the relevant Competent Authorities of the 15 Member States held in Italy. This meeting of experts resulted in the EU deciding that MTBE will not be classified as a carcinogen, mutagen, or reproductive toxin.

The result of the EU risk assessment carried out in Finland was presented in January 2001. It concluded that MTBE was not a toxic threat to health, but can leave a bad taste in drinking water. As a result, in the draft directive for new fuel quality laws called Auto Oil II, published May 11, 2001, the EU set no limitations on the use of MTBE in fuel after 2005. Using a similar approach as the official EU Risk Assessment process, the European Centre for Eco-toxicology and Toxicology of Chemicals (ECETOC) recently concluded, "the risk characterization for MTBE does not indicate concern for human health with regard to current occupations and consumer exposures." A ban on the use of MTBE is not expected in Europe, as widespread contamination on the same scale as in the United States is unlikely. The EU has, however, recommended that MTBE be prevented from seeping into groundwater through storage tanks at service stations. One example showing the EU's initiative was the September 28, 2001, approval of a Danish tax initiative implemented to improve the quality of storage tanks. Denmark, one of the countries in Europe that is most concerned about the effects of MTBE on the environment, had proposed tax breaks of Euro 0.02 per liter of gasoline sold at service stations fitted with leak-resistant underground tanks.

In Europe, there continues to be support for MTBE's ability to reduce air pollution and a general belief that the real problem in the United States is leaking gasoline tanks. There is also surprise that the United States has taken so long to address the leaking tank problem. In response to European directives, countries such as Germany and Holland have some of the tightest environmental legislation in the world. The European Commission briefed the Strasbourg plenary session of the European Parliament that Europe is fundamentally different from the United States and for that reason MTBE was not considered a risk. Europe will phase down both its exports of straight MTBE and that of MTBE in RFG to the North American market, mainly as the product will be required locally but also as the United States presently looks to be moving away from an oxygen mandate. Eastern Europe is presently experiencing some solid growth and is demanding supply from West European producers.

Nonetheless, MTBE producers expect that the results of the assessment mean that this chemical will not face a crisis of public confidence like that in the United States due to leakages from storage facilities. The report provides a balanced and objective overview of the risks associated with MTBE. The conclusions confirm that MTBE has a key role to play in permitting Europe to reach better air quality. The assessment was carried out after MTBE was placed on an EU priority list of chemicals for risk investigation. It was conducted by a group of environmental and safety institutes in Finland, where average MTBE content in gasoline of 8–10% is the highest in the EU. The European Commission is likely to draw up legislation for tighter storage standards following approval of the assessment by competent authorities in the EU member states. The Finnish assessors recommended measures such as the use of double-walled tanks, embankments around tanks, and leak detection equipment. Unlike the United States, the EU has had few problems with leakages of gasoline containing MTBE. Gasoline retailers have been willing to invest in secure storage and distribution facilities at service stations. The storage equipment is of a relatively high standard because gasoline is much more expensive in Europe than in the United States. In the wake of the risk assessment, MTBE is now poised to take full advantage of a predicted rising demand for oxygenates because of tighter EU regulations on fuel content. MTBE may face limited competition from ethyl *tertiary*-butyl ether (ETBE), the production of which could increase as a result of EU measures to stimulate the production of biofuels. Now that the health and environmental risks of MTBE have been properly assessed, there is a strong argument in favor of raising its average content in EU gasoline. Analysts projected that MTBE will continue to account for the vast majority of oxygenates consumed in the EU market. The higher demand for MTBE could even lead to a major increase in imports, because of shortages of capacity in Europe. Adding to these points is the fact that the Asia/Pacific area is using more and more MTBE as nations in that part of the world switch to nonleaded and cleaner-burning motor fuels, and it can be concluded that all the MTBE produced will be consumed in the near future and not available for export.

The German Federal Environmental Agency (UBA) has concluded their study into MTBE, which was initiated after the situation in California. They have advised the Ministry for the Environment that MTBE does not constitute a threat for the German environment at this time, as the general condition of the underground storage tanks is considered to be very good and the potential health risks from MTBE are negligible. One of the main reasons for reaching these conclusions is the fact that since 1992 there has been a German directive on the quality of underground storage tanks. Retailers selling more than 5000 metric tons per annum had a deadline of 3 years to become compliant, whereas those that sold lesser volumes had extended deadlines of 6 and 9 years. At the beginning of 1998, 70% of all stations were completed, covering more than 95% of the retail stations that sold over 5000 metric tons per annum. The water supplies in Germany are regulated by federal agencies. Evidence suggests that MTBE is found in rivers and in surface water at levels near to 100 ng. MTBE is also not of concern based on its current usage rates in gasoline, approximately 11% in 98 RON Super Premium (only 6% of the total market), and 1.3% to 1.6% in 95 RON Eurograde. With the new European specifications, it is probable that the use of MTBE and oxygenates will increase as total aromatics and benzene levels are reduced. However, with the right equipment in place, MTBE is unlikely to become a concern in the future.

The European MTBE market appears to be relaxed in the face of California's proposed ban on the oxygenate, but market comment indicates that this calm exterior may conceal a degree of anxiety about the future of the blendstock. These concerns were said to be fuelled by unresolved or unknown issues relating to the likely benefits of any alternative to MTBE, and worries about the alternative destination for the octane booster if the planned ban in California was extended to other regions. Opinions from European MTBE producers indicate that none was simply watching and waiting. They were pro-active and had been preparing for some time for a change in production and consumption dynamics that could be triggered by the pending California ban. Currently, the European market appears to consider MTBE the blendstock of choice in terms of cost and efficiency, but it could lose its prime spot in Europe if California presents a glowing report on MTBE's successor. If the replacement turns out to be better environmentally, and the costs to industry and consumers are acceptable, Europe could be forced to review its use of MTBE.

According to MTBE industry sources, expansion in Eastern Europe had long been the hope for MTBE sellers. The aim was that economic growth in Eastern Europe would promote an increase in demand for private car ownership, and this in turn would help to offset the impact of the removal of California's market from the equation. However, global economic and political instability has put a dampener on the previously positive growth forecasts for Eastern Europe. Another industry source warned of the negative political impact of proposing the use of MTBE in Eastern Europe at a time when a ban was being applied in the United States on health and environmental grounds. Broader uncertainty about the timetable for the California ban was also voiced by a number of producers and refiners. It might not be practical for the ban to apply to the whole of California in one shot, and it is likely the state will have to implement a gradual reduction in the quantity of MTBE in gasoline to around 5% over a period of time. California will have to allow time for MTBE to be flushed through the pipelines, because it is not a simple case of shutting off supply of MTBE and replacing it with another oxygenate overnight.

IX. MTBE OUTLOOK FOR THE UNITED STATES

Even with the controversy MTBE has been encountering in the United States, demand for MTBE continues at a high level. The publicity machines continue to hammer away at MTBE. California has finalized new pump-labeling rules for gasoline containing MTBE, stating that the State has determined that it represents a significant environmental risk. Despite all the negative publicity, the demand for MTBE in the United States continues at a high level. Production and the operating percentage of the plants have been rising steadily for the past few years. Considering all these impacts on the market, we anticipate that MTBE use in the United States will continue more or less at its present level until about the year 2004.

MTBE production began two decades ago to produce cleaner-burning gasoline. Between 1990 and 1994, MTBE production nearly doubled, from 83,000 bbl/day to 161,000 bbl/day, according to the U.S. Energy Information Administration (EIA). Because of the reformulated gasoline program, MTBE production was increased to meet oxygenated fuel demands. In September 1999, U.S. MTBE plants produced 231,000 bbl/day of MTBE, an all-time record, representing more than 90% of the capacity of the industry. California and the New England region are the biggest users of MTBE in reformulated gasoline.

According to a 2000 EIA analysis, if MTBE is banned in California, reformulated gasoline prices could go up by 2 cents/gal in the near term (in about three years), and go down by 3 cents/gal over the long term (about six years). By EIA estimates, if MTBE or all ethers are banned national RFG would increase by 2.8 cents/gal in the near term. A 1999 Chevron/Tosco MathPro analysis showed that a California ether ban in gasoline would increase prices by 2.7 cents/gal in the near term and 1.2 cents/gal in the long term. The U.S. Department of Energy (DOE) estimated a nationwide MTBE and ether ban with no oxygenate requirement would increase RFG production cost on the East Coast by 1.9 cents/gal over four years.

The political nature of the current MTBE situation in the United States makes it exceptionally difficult to predict the future course of events with any certainty. According to a low-MTBE-demand case, California phases out MTBE by the end of 2003, the U.S. East Coast reduces MTBE in reformulated gasoline from 95% to about 47% in 2005, and the East Coast eliminates the oxygenate by 2010. Under this low-MTBE-demand scenario, it was predicted that U.S. MTBE demand would fall from around 300,000 bbl/day to around 80,000 bbl/day in 2005, which would cause U.S. methanol demand to decline by 3.5 million metric tons per year. In a medium-MTBE-demand case, California delays its MTBE ban until the end of 2005, and the U.S. East Coast market reduces MTBE in reformulated gasoline from 95% to 71% in 2005 and by 2010 reduces MTBE to 46% in RFG. About one-third of methanol demand is for MTBE production. If an MTBE ban goes into effect over the next four to six years, the methanol industry would have to compensate for the drop in demand. The methanol industry is looking at several initiatives to fill demand, such as methanol as a hydrogen carrier in fuel cells, as fuel for turbine generators, and for use in wastewater denitrogenation.

X. MTBE OUTLOOK FOR ASIA

Asia and Australia currently consume about 29.1 billion bbl/day of motor gasoline, which is about 17% of the world's gasoline supply, according to the U.S. Energy Information Agency (EIA). Of this amount, China and Japan make up about 50% of the region's fuel demand. The vehicle population in this region is expected to increase to about 175 million by the year 2020, fueled by a projected increase in wealth. This in turn would cause an increase in demand for motor gasoline. With this projected increase in vehicle population and gasoline consumption, Asian governments are under pressure to promote the use of clean fuels to combat urban air pollution and improve air quality.

Most Asian countries, such as South Korea, Japan, Hong Kong, Taiwan, China, Malaysia, Singapore, the Philippines, and Thailand, have already phased lead out of their gasoline pool and are replacing it with oxygenates such as MTBE. Due to the relative ease in blending of MTBE into gasoline, easy transportation and storage, as well as relatively cheap and abundant supply, MTBE is the most widely used oxygenate in Asia.

Demand for MTBE is expected to be marginally firmer in the near future, as more Asian countries such as Indonesia and India are working to totally phase out lead from their gasoline pool. Supply, on the other hand, is expected to remain abundant, as Asia is able to produce about 3 million metric tons per year of MTBE for its captive consumption. In addition, Asia attracts a regular supply of about 500,000 metric tons per year of MTBE from Middle Eastern and European sources. Differing from the rest of Asia, refiners in Japan have limited the use of MTBE for enhancing the octane value of its gasoline due to poor blending margins. Refiners have drastically reduced MTBE plant operating rates. Japan has no oxygenate mandate and requires a maximum of only 6% MTBE in its gasoline pool, so a number of Japanese refiners have reduced MTBE use. Relatively high MTBE production costs brought about by high domestic methanol feedstock prices have prompted Japanese refiners to look to other blending components such as alkylates to reduce production costs. Refiners also cited high transportation costs for shipping small cargos to local blenders as another factor for limiting MTBE use.

The economic situation in Indonesia, Thailand, and Korea is improving, and it is estimated that the demand for MTBE in this region will increase. The demand forecast for 2000 was 1855 kilotons, rising to 2588 and 3450 kilotons in 2005 and 2010, respectively. In other countries in Asia where leaded gasoline is being used, the efforts are being put forward to phase out lead and it will be largely eliminated in a decade. This trend and the need to improve air pollution problems in many major urban areas have created a major international market for MTBE.

With respect to California's MTBE situation, Asian trading sources have said there is little possibility of MTBE being phased out in Asia. There is no issue of groundwater contamination or any other health hazards. MTBE is currently the most cost-effective alternative to lead in gasoline and there is currently no reason for MTBE to be replaced. It has been observed that economics rather than politics were the determining factor for MTBE survival in Asia. There are a lot of issues that Asian governments need to address if they want to get rid of MTBE, such as the cost effectiveness to build new infrastructure such as new processing plants and additional road network, ensuring adequate supply of MTBE alternatives, and how much would gasoline cost in the end.

XI. ALTERNATIVES TO MTBE

Most refiners use MTBE over other oxygenates primarily for its blending characteristics as well as for economic reasons. The U.S. EPA in 1999 recommended the amount of MTBE in gasoline be reduced because of the hazards this additive poses to drinking water supplies nationwide. Similar to MTBE, aromatics such as toluene serve to boost octane levels but can only be used to replace MTBE in limited amounts, as they increase toxic emissions. There are a few other nonaromatic octane boosters, such as alkylates, isomerates, and ethanol, but they are very limited in terms of both octane contribution and availability. Each of the various alternatives has several advantages and disadvantages. While the debate about whether MTBE is the best additive for cleaner gasoline continues, ethanol has emerged as a strong contender. Some lobbyists have stated that the two most viable alternatives to MTBE are ethanol and no oxygen requirement in gasoline. However, no oxygenates would require a change in the CAA or a waiver from the oxygen requirement in the CAA. Ethanol or ethyl alcohol is produced chemically from ethylene or biologically from the fermentation of corn and other agricultural products. Ethanol, used as a gasoline octane enhancer and oxygenate, increases octane numbers by 2.5 to 3.0 at 10% concentration. Ethanol can also be used in higher concentration as E85 (85% ethanol and 15% gasoline) in vehicles optimized for its use. As an alternative to MTBE, the MTBE plants could be converted to produce ETBE, isobutane, or ethanol. There are plans to shut down and convert a number of MTBE plants for isooctane production. However, the cost of conversion contains unknown economics of producing a product for reformulated gasoline, and the price fluctuation is a totally uncertain and unexpected phenomenon. The production cost of an alternative additive would have to be cost-effective given the fluctuation of gasoline prices.

XII. CONCLUSIONS

There are several factors affecting the debate on the efficacy and economics of MTBE in the U.S. gasoline pool. MTBE has for many years been a vital component in the U.S. refiner's arsenal to produce high-quality, cleaner-burning motor fuels. It will be some time before this situation is resolved. Following are the key points regarding the MTBE controversy.

- The UC Davis report and recommendations were made in a very hurried manner, and enough time should have been given to study MTBE and to take into account the consideration of other organizations to deal with the matter. It seems that the report was partially influenced politically by the ethanol lobby. This opinion is also strengthened by the fact that now the MTBE ban deadline has been delayed for one year. A balance and thoughtful decision would be the best approach to the problem.

- The independent review of the UC Davis report conducted by SRI Consulting (SRIC) and SRI International (SRI) concluded that the report does not adequately recognize and quantify the air quality benefits from the use of MTBE and reformulated gasoline in general. Research demonstrates that exposure to high levels of MTBE poses risks, but a variety of national and international organizations have concluded that continued use of MTBE in gasoline, which involves much lower levels of exposure, is safe. The conclusions and cost–benefit analysis in the UC Davis report do not focus on forward-looking policy issues. A number of errors were made in UC Davis report with respect to MTBE in water and remediation cost calculations. The fuel analysis economics, another major component of the cost–benefit analysis, do not accurately reflect industry practices, commercial gasoline blends, or "real world" economics. Because of the specific errors cited above, the overall cost–benefit analysis in the UC Davis report leads to the wrong conclusions. The analyses by SRIC and SRI show that gasoline with MTBE is the least costly of the three fuel options considered in the UC study, not the most expensive.
- MTBE demand has been continuing at a high level despite the controversy.
- The political nature of the current MTBE situation in the United States makes it exceptionally difficult to predict the future course of events with any certainty.
- Automakers in the United states are concerned that replacing MTBE with something else may cause serious problems with gasoline quality.
- A recently published study on risk assessment of MTBE by the European Union has concluded that MTBE does not pose a danger to human health, but tighter controls on the handling and storage of the chemical are required.
- In Europe, there continues to be support for MTBE's ability to reduce air pollution and a general belief that the real problem in the United States is leaking gasoline tanks. With the new European specifications, it is probable that the use of MTBE/oxygenates will increase as total aromatics and benzene levels are reduced;

however, with the right equipment in place, MTBE is unlikely to become a concern in the future.

- Demand for MTBE is expected to be marginally firmer in the near future, as more Asian countries such as Indonesia and India are working to totally phase out lead from their gasoline pool.
- According to a presentation by the U.S. Department of Energy, an MTBE ban in the United States would be equivalent to loss of 300,000 bbl/day of premium blendstock, and it would need to be compensated by crude processing capacity equivalent to five average U.S. refineries. An MTBE ban will invite a series of problems which include increases in gasoline production cost, increased reliance on gasoline imports, increased refinery investment, and increased pollutants emissions.

ACKNOWLEDGMENT

The authors wish to acknowledge the support of the Research Institute of the King Fahd University of Petroleum and Minerals, Dhahran, Saudi Arabia.

2
Properties of MTBE and Other Oxygenates

Mohammad Ashraf Ali and Halim Hamid
Research Institute, King Fahd University of Petroleum and Minerals, Dhahran, Saudi Arabia

I. INTRODUCTION

Ethers and alcohols are being blended with gasoline to increase octane number and to reduce air pollution problems associated with leaded gasoline. These oxygenates have replaced lead alkyl and other metal-containing compounds in gasoline because the use of compounds such as tetraethyl lead (TEL), tetramethyl lead (TML), and methylcyclopentadienyl manganese tricarbonyl (MMT) in gasoline has created air pollution problems. The emission of their combustion products from vehicle exhausts creates atmospheric pollution causing serious health hazards. These oxygenates are methyl *tertiary*-butyl ether (MTBE), ethyl *tertiary*-butyl ether (ETBE), *tertiary*-amyl methyl ether (TAME), *tertiary*-amyl ethyl ether (TAEE), diisopropyl ether (DIPE), methyl alcohol, ethyl alcohol, and *tertiary*-butyl alcohol (TBA). Among all these oxygenates, MTBE appears to be the most effective choice because its physical, chemical, and thermal properties are compatible with that of gasoline, especially in the boiling range where gasoline typically shows lowest antiknock characteristics. In this chapter, the properties of ethers and alcohol oxygenates are presented.

II. PROPERTIES OF MTBE

Methyl *tertiary*-butyl ether is a colorless liquid of low viscosity with a distinct odor having a boiling point of 55°C and a density of 0.74 g/mL. MTBE belongs to the ether class of organic compounds and it is combustible. Its molecular structure is shown below.

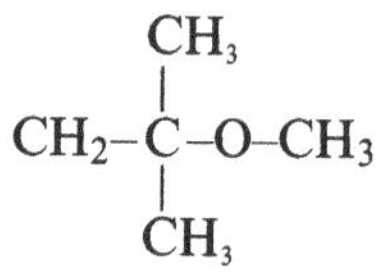

MTBE is made by reacting methanol, made from natural gas, with isobutene (2-methyl-1-propene) in the liquid state, using an acidic catalyst at 100°C. The physical, chemical, and thermal properties of MTBE are given in Table 1 [1–37].

III. PROPERTIES OF ETBE

Ethyl *tertiary*-butyl ether (ethyl *t*-butyl ether, ETBE) is a combustible colorless liquid boiling at 77°C and belongs to the ether class of chemicals. Its molecular structure is shown below.

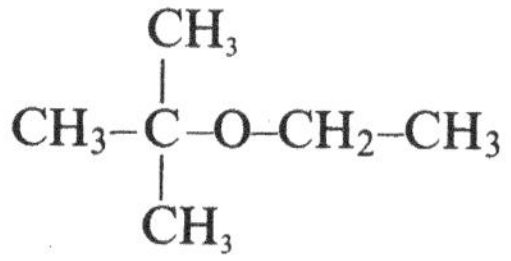

ETBE is made by reacting ethanol with isobutene (2-methyl-1-propene) in the liquid state, using an acidic catalyst. It has a higher octane value than MTBE. Also, the vapor pressure of ETBE is lower, providing advantages in meeting the seasonal volatility standards and additions of butanes, which have high octane number and vapor pressures. The typical properties of ETBE, as reported in various studies [6,10,14,16,25,38], are listed in Table 2.

IV. PROPERTIES OF TAME

Tertiary-amyl methyl ether (TAME) is a higher analog to MTBE, produced by the reaction of a branched C_5 olefin called *tertiary*-amylene with

Table 1 Physical, Chemical, and Thermal Properties of MTBE

Molecular formula	$C_5H_{12}O$
Molecular weight	88.15
Elemental analysis	
Carbon content, wt%	68.1
Hydrogen content, wt%	13.7
Oxygen content, wt%	18.2
C/H ratio	5.0
Density, g/cm^3	
at 15/4°C	0.7456
at 20/4°C	0.7404
at 25/4°C	0.7352
at 30/4°C	0.7299
Reid vapor pressure at 25°C, psi	4.7
Reid vapor pressure at 37.8°C, psi	7.8
Blending Reid vapor pressure, psi	8.0
Boiling point, °C	55.0
Freezing point, °C	−108.6
Vapor density, calculated (air = 1), g/cm^3	3.1
Solubility of MTBE in water at 25°C, wt%	5.0
Solubility of water in MTBE at 25°C, wt%	1.5
Viscosity at 37.8°C, cSt	0.31
Stoichiometric air/fuel ratio	11.7
Refractive index at 20°C	1.3694
Surface tension, din/cm^2	19.4
Latent heat of vaporization at 25°C, Cal/g	81.7
Specific heat at 25°C, Cal/g-°C	0.51
Lower heating value, Cal/g	8,400
Flammability limits in air	
Lower limit, vol%	1.5
Upper limit, vol%	8.5
Auto ignition temperature, °C	425
Flash point, closed cup, °C	−30
Blending octane number[a]	
RON	117
MON	101
(RON + MON)/2	110

[a] Obtained by adding 10 vol% MTBE to a base gasoline having RON clear = 94.3 and MON clear = 84.3 [6]. Laboratory RON and MON rating procedures are not suitable for use with pure oxygenates.

Table 2 Physical, Chemical, and Thermal Properties of ETBE

Molecular formula	$C_6H_{14}O$
Molecular weight	102
Elemental analysis	
Carbon content, wt%	70.6
Hydrogen content, wt%	13.7
Oxygen content, wt%	15.7
C/H ratio	4.5
Density, g/cm3	
at 15/4°C	0.7456
at 20/4°C	0.7404
at 25/4°C	0.7353
at 30/4°C	0.7300
Reid vapor pressure (RVP), bar (psi)	0.30 (4.4)
Blending RVP, bar (psi)[a]	0.32 (4.7)
Boiling point at 760 mmHg, °C	72
Freezing point, °C	−94
Viscosity at 40°C, cSt	0.528
Solubility at 20°C	
ETBE in water, wt%	1.2
Water in ETBE, wt%	0.5
Refractive index at 20°C	1.376
Surface tension at 24°C, din/cm^2	19.8
Latent heat of vaporization, kcal/kg	74.3
Specific heat at 25°C, Cal/g-°C	0.51
Lower heating value, kcal/kg	8600[a]
Stoichiometric air/fuel ratio	12.17
Blending octane number[b]	
RON	118
MON	101
(RON + MON)/2	111

[a]An average from six series of fuels [38].
[b]Obtained by adding 10 vol% ETBE to a base gasoline having RON clear = 94.3 and MON clear = 84.3 [6].

methanol. TAME has a slightly higher boiling point and a somewhat lower octane value than MTBE, and is fully compatible with gasoline hydrocarbon blends. It has the molecular structure given below.

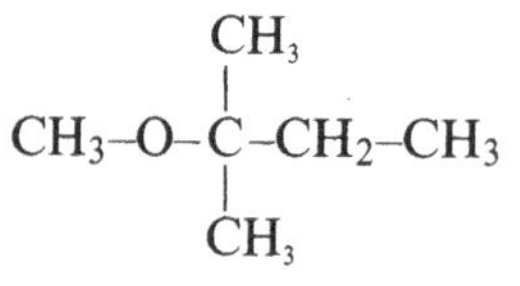

Table 3 Physical, Chemical, and Thermal Properties of TAME

Molecular formula	$C_6H_{14}O$
Molecular weight	102
Elemental analysis	
Carbon content, wt%	70.6
Hydrogen content, wt%	13.7
Oxygen content, wt%	15.7
C/H ratio	5.3
Density, g/cm^3	
at 15/4°C	0.7750
at 20/4°C	0.7703
at 25/4°C	0.7636
at 30/4°C	0.7607
Reid vapor pressure (RVP), bar (psi)	0.10 (1.5)
Boiling point at 760 mmHg, °C	86
Solubility at 20°C	
TAME in water, wt%	1.15
Water in TAME, wt%	0.6
Refractive index at 20°C	1.3888
Surface tension at 24°C, din/cm^2	22.6
Stoichiometric air/fuel ratio	12.1
Latent heat of vaporization, kcal/kg (Btu/lb)	78.0 (140)
Specific heat at 25°C, cal/g-°C	0.52
Lower heating value, kcal/kg	8600[a]
Flash point, °C	−7
Blending octane number[b]	
RON	112
MON	98
(RON + MON)/2	105

[a]Estimated [6].
[b]Obtained by adding 10 vol% TAME to a base gasoline having RON clear = 94.3 and MON clear = 84.3.

Typical properties of TAME, as reported by various investigators [6,10,11,14,16,25,26,32,36] are listed in Table 3. The results of vapor pressure measurements are given in Table 4. The calculated values used in the third column were obtained by the Antoine equation. A plot of vapor pressure as a function of temperature in given in Figure 1. The vapor pressures of TAME in the range 21–86°C were measured by comparative ebulliometry. Pure TAME is prepared from the technical product by distillation, rectification, and drying [39].

Table 4 Vapor Pressures of TAME at Various Temperatures

t (°C)	*P* (kPa)	$P_{calc} - P$ (kPa)
21.109	8.332	−0.002
21.128	8.331	0.006
27.070	11.075	−0.018
27.096	11.073	−0.002
32.140	13.922	−0.002
32.148	13.923	0.002
36.451	16.810	0.000
36.449	16.807	0.001
40.928	20.311	0.000
40.912	20.299	−0.002
45.596	24.571	0.000
45.601	24.578	−0.002
50.137	29.387	−0.002
50.142	29.395	−0.005
54.591	34.814	0.005
54.587	34.815	−0.001
58.857	40.754	−0.001
58.852	40.751	−0.004
63.219	47.627	0.006
63.207	47.612	−0.008
70.048	60.227	0.007
70.057	60.245	0.009
76.615	74.710	0.007
76.616	74.707	0.013
83.738	93.395	−0.004
83.748	93.405	0.014
85.386	98.205	−0.015
86.027	100.126	−0.018

V. PROPERTIES OF TAEE

Tertiary-amyl ethyl ether (TAEE) is a higher analog to TAME, produced by reaction of a branched C_5 olefin called *tertiary*-amylene with ethanol. TAEE has a higher boiling point, a somewhat lower octane value than MTBE, and is fully compatible with gasoline hydrocarbon blends. Typical properties of TAEE are listed in Table 5, and it has the molecular structure given below.

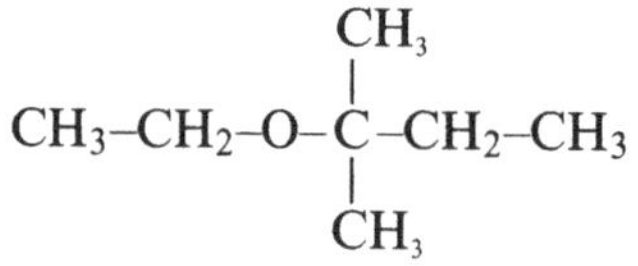

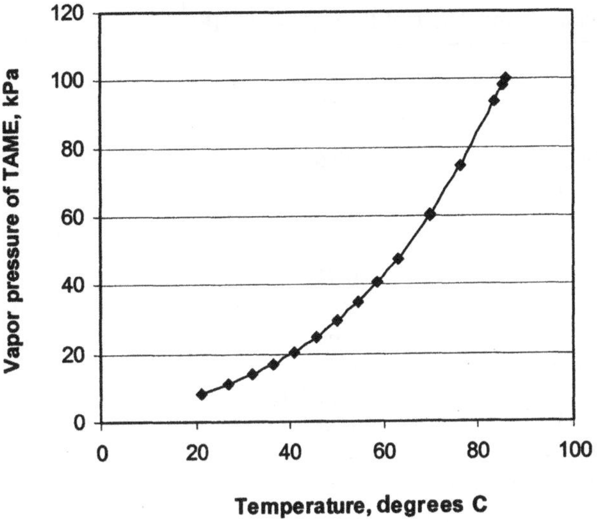

Figure 1 Vapor pressures of TAME as a function of temperature.

VI. PROPERTIES OF DIPE

Diisopropyl ether (DIPE) is a colorless liquid with a characteristic odor. It is produced by the reaction of propylene with water to form isopropyl alcohol, which then reacts with water to produce diisopropyl ether. The vapor is

Table 5 Physical, Chemical, and Thermal Properties of TAEE

Molecular formula	$C_7H_{16}O$
Molecular weight	116
Elemental analysis	
Carbon content, wt%	72.4
Hydrogen content, wt%	13.8
Oxygen content, wt%	13.8
C/H ratio	4.5
Specific gravity, g/cm^3	0.70
Reid vapor pressure at 25°C, mmHg	1.2
Boiling point at 760 mmHg, °C	101
Solubility of TAME in water at 20°C, wt%	0.4
Blending octane number[a]	
RON	105
MON	95
(RON + MON)/2	100

[a]Obtained by adding 10 vol% TAEE to a base gasoline having RON clear = 94.3 and MON clear = 84.3.

Table 6 Physical, Chemical, and Thermal Properties of DIPE

Molecular formula	$C_6H_{14}O$
Molecular weight	102
Elemental analysis	
Carbon content, wt%	70.6
Hydrogen content, wt%	13.7
Oxygen content, wt%	15.7
C/H ratio	5.3
Specific gravity, g/cm^3	0.75
Boiling point, °C	69
Melting point, °C	−60
Relative density (water = 1)	0.7
Solubility in water	very little
Vapor pressure at 20°C, kPa	15.9
Relative vapor density (air = 1)	3.5
Relative density of the vapor/air mixture at 20°C (air = 1)	1.5
Flash point, °C	−28
Autoignition temperature, °C	443
Explosive limits, vol% in air	1.4–7.9
Blending octane number[a]	
RON	105
MON	95
(RON + MON)/2	107
Occupational exposure limits	
TLV	250 ppm; 1040 mg/m^3
STEL	310 ppm; 1300 mg/m^3

[a] Obtained by adding 10 vol% DIPE to a base gasoline having RON clear = 94.3 and MON clear = 84.3.

heavier than air and may travel along the ground, and thus distant ignition is possible. Harmful contamination of the air can be reached rather quickly upon evaporation of this substance at 20°C. DIPE can be absorbed into the body by inhalation of its vapors, and can readily form explosive peroxides if unstabilized and may explode upon shaking. DIPE irritates the eyes, the skin, and the respiratory tract, and may cause effects on the central nervous system. Exposure above the occupational exposure limit (OEL) could cause lowering of consciousness. Repeated or prolonged contact with skin may cause dermatitis. Typical properties of DIPE are listed in Table 6, it has the molecular structure given below.

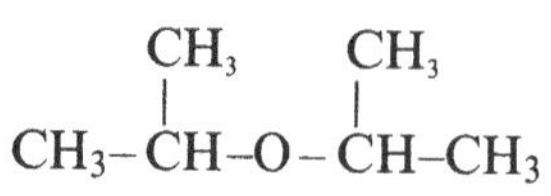

VII. PROPERTIES OF METHANOL (METHYL ALCOHOL)

Typical properties of methanol, as reported by several investigators in the literature, are listed in Table 7 [1,3,6,10,11,13,14,20,24,25,32,36,40–46]. The current approach to the use of methanol in gasoline engines is to blend it in low amounts with gasoline. Methanol blends lead to phase-separation problems because of its miscibility with water. For this reason, co-solvents such as ethyl, propyl, butyl, and higher alcohols (octyl alcohols) are needed to produce methanol blends. Atlantic Richfield Co. (ARCO) uses methanol blends with a co-solvent gasoline-grade *tertiary*-butyl alcohol (GTBA). The molecular structure of methanol is between that of water and hydrocarbons. Since it contains a hydroxyl group, like water, it has solvent properties like water and a strong affinity for water. There is no way to reclaim methanol when it settles to the bottom of the tank with water. However, water separates from gasoline or MTBE during draining water bottoms from tanks. Water–methanol mixture is a hazardous waste that requires proper disposal [47].

Due to its high oxygen content (49.9 wt%), pure methanol has a markedly different stoichiometric air/fuel ratio than gasoline (6.4 versus 14.2–15.1). When alcohol is added to gasoline, the physical qualities of the blend do not change significantly for up to 10% volume [1]. A carburetor will meter the blend the same way as it would meter straight gasoline. The blend with no other changes to the engine is automatically carbureted at leaner equivalence ratios. The phenomenon is called the blend leaning effect.

The heating value of methanol is lower than that of other oxygenates, due to its low carbon and hydrogen contents. For this reason methanol gasoline blends show lower fuel economy than gasoline, because of its slightly reduced energy content. Some of the physical constants of pure methanol are reported in Table 7. The vapor pressures of methanol reported in the literature in the temperature range 26–64°C are given in Table 8, and a plot showing vapor pressure as a function of temperature in given in Figure 2. The values were determined by comparative ebulliometry, using a standard ebulliometer connected in parallel with a second ebulliometer filled with water to a buffer reservoir of pressure [37,39]. The calculated values were obtained using the Antoine equation.

VIII. PROPERTIES OF ETHANOL (ETHYL ALCOHOL)

Ethanol, known as ethyl alcohol, alcohol, grain spirit, or neutral spirit, is a clear, colorless, flammable oxygenated fuel. Ethanol is usually

Table 7 Physical, Chemical, and Thermal Properties of Methanol

Molecular formula	CH_3OH
Molecular weight	32.0
Elemental analysis	
Carbon content, wt%	37.5
Hydrogen content, wt%	12.6
Oxygen content, wt%	49.9
C/H ratio	3.0
Specific gravity at 15°C	0.796
Reid vapor pressure (RVP)	
at 37.8°C, bar (psi)	0.32 (4.6)
at 25°C, bar (psi)	0.16 (2.3)
Blending RVP, bar (psi)	4.14 (60.0)
Boiling point, °C	65.0
Freezing point, °C	(−93.0)
Specific heat, kJ/kg-°C (Btu/lb-°F)	2.51 (0.60)
Solubility at 21°C (70°F)	
Methanol in water, wt%	100
Water in methanol, wt%	100
Viscosity at 40°C, cSt (cP)	0.46 (0.58)
Stoichiometric air/fuel ratio, wt	6.4
Dielectric constant	32.6
Latent heat of vaporization, kcal/kg (Btu/lb)	260 (506)
Lower heating value, kcal/kg (Btu/lb)	4650 (8600)
Flammability limits in air	
Lower limit, vol%	7.0
Upper limit, vol%	36.0
Autoignition temperature, °C	470
Flash point, closed cup, °C	11
Refractive index at 20°C and 760 mmHg[a]	64.51–64.70
Blending octane number[b]	
RON	123
MON	91
(RON + MON)/2	107

[a]Values reported [37,39] in which methanol has density values in the range 0.7910–0.7915 g/cm^3 and a boiling-point range of 64.515–64.527.
[b]Obtained by adding 10 vol% methanol to a base gasoline having RON clear = 94.3 and MON clear = 84.3 [6].

blended with gasoline to create what is sometimes known as gasohol. Typical properties of ethanol, as reported in various studies [1,6,10,11,14,20,25,32,36,41,45,46,67,68] are listed in Table 9. The property differences of ethanol and gasoline are directionally the same as methanol

Table 8 Vapor Pressures of Methanol

t (°C)	P (kPa)	$10^3 \times (P_{calc} - P)$ (kPa)
26.028	17.900	−1
29.431	21.286	0
33.316	25.736	0
36.316	29.830	2
39.963	35.429	3
43.371	41.441	5
43.392	41.494	−8
47.402	49.644	6
47.425	49.708	−7
52.229	61.245	−11
52.230	61.228	8
57.705	77.032	8
57.737	77.151	−10
57.857	77.512	11
63.855	98.729	−5
64.497	101.265	−10
64.528	101.367	11

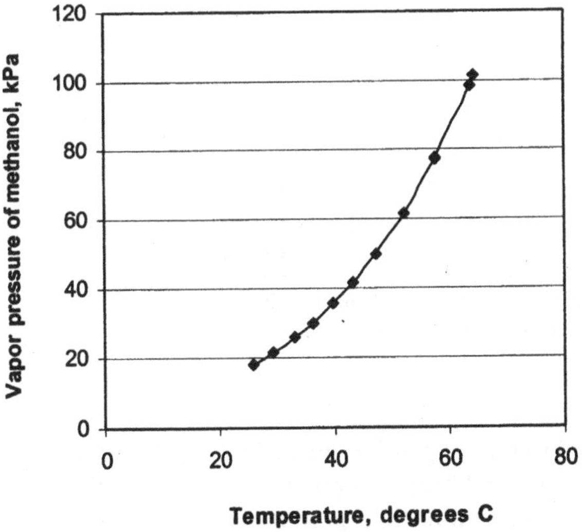

Figure 2 Vapor pressures of methanol as a function of temperature.

Table 9 Physical, Chemical, and Thermal Properties of Ethanol

Molecular formula	C_2H_5OH
Molecular weight	46.1
Elemental analysis	
Carbon content, wt%	52.1
Hydrogen content, wt%	13.1
Oxygen content, wt%	34.7
C/H ratio	4.0
Specific gravity at 15°C	0.794
Reid vapor pressure (RVP)	
at 37.8°C, bar (psi)	0.2 (2.5)
at 25°C, bar (psi)	0.06 (0.85)
Boiling point, °C	78.5
Freezing point, °C	−114
Specific heat, kJ/kg-°C (Btu/lb-°F)	2.51 (0.60)
Solubility at 21°C	
Ethanol in water, wt%	100
Water in ethanol, wt%	100
Viscosity at 40°C, cSt (cP)	0.83 (1.1)
Stoichiometric air/fuel ratio	9.0
Dielectric constant	24.3
Latent heat of vaporization, kcal/kg (Btu/lb)	200 (396)
Lower heating value, kcal/kg (Btu/lb)	6,380 (11,500)
Flammability limits in air	
Lower limit, vol%	4.3
Upper limit, vol%	19.0
Autoignition temperature, °C	363
Flash point, closed cup, °C	13
Blending octane number[a]	
RON	123
MON	96
(RON + MON)/2	109.5

[a]When adding 10 vol% ethanol to a base gasoline having the following octane properties: RON clear = 94.3, MON clear = 84.3 [6].

but less severe. Compared to typical gasoline air/fuel ratio (14.2–15.1), lower heating value (18,900 Btu/lb), and heat of vaporization (150 Btu/lb), ethanol requires 60% as much air for combustion, produces 65% as much energy, and requires 2.6 times as much heat for vaporization. The corresponding values for methanol are 45% as much air, 50% as much energy, and 3.7 times as much heat for vaporization. The ratios of heat of vaporization to heat of combustion show that for a given energy output, methanol requires

7.6 and ethanol 4.0 times as much heat to vaporize the fuel [45]. Typically alcohols are blended 5–10 vol% with gasoline for use in existing vehicles with no modifications. Conventional cars cannot operate on alcohol-rich fuels without drastic modifications.

IX. PROPERTIES OF TBA

The typical properties of *tertiary*-butyl alcohol (TBA), as reported in various studies [6,10,11,13,18,20,32,36], are listed in Table 10. Among all the alcohols, TBA has properties that are most similar to those of ethers [10]. Its gasoline blending value is relatively lower than that of other oxygenates. TBA minimizes or eliminates many of the undesirable physical characteristics often associated with methanol gasoline blends [48]. The GTBA/methanol mixtures called Oxinol blending component is the trademark of the Atlantic Richfield Co. (ARCO) (specifically GTBA, gasoline-grade tertiary butyl alcohol). TBA is added to methanol as a co-solvent to improve water tolerance and drivability of the performance of oxygenates. GTBA is marketed under ARCO trademark as Arconol. In 1979, the U.S. Environmental Protection Agency (EPA) approved an ARCO application allowing up to 7 vol% Arconol to be added to unleaded gasoline. Another waiver was granted to ARCO in 1981 for GTBA/methanol product, marketed as Oxinol blending component. This waiver allows up to 3.5 wt% oxygen to be added to unleaded gasoline and limits the methanol content to no more than the equal volume of GTBA [49]. For example, with Oxinol 50, which is a 50/50 blend of methanol and GTBA, up to about 9.6 vol% alcohol can be blended into unleaded gasoline. In order to solve the problems associated with alcohols, the use of co-solvents such as TBA is necessary.

X. A COMPARISON OF PROPERTIES OF OXYGENATES

Table 11 compares the physical and blending data on oxygenates which have been considered to be blended with U.S. gasoline, based on manufacturing capabilities and availability of technology. Among these oxygenates, ethers have low blending Reid vapor pressure (RVP) as compared to alcohols. The blending RVP of most alcohols is much higher than their true vapor pressures. This nonideal blending effect is due to the unfavorable interaction between the highly polar hydroxyl group of alcohols and nonpolar hydrocarbons of gasoline. This nonideal behavior increases as the oxygen content of alcohol increases (polarity increases). Ethers

Table 10 Physical, Chemical, and Thermal Properties of TBA

Molecular formula	$(CH_3)_3COH$
Molecular weight	74.12
Elemental analysis	
Carbon content, wt%	64.8
Hydrogen content, wt%	13.6
Oxygen content, wt%	21.6
C/H ratio	4.8
Specific gravity at 15°C	0.791
Reid vapor pressure at 37.8°C, psi	1.8
Blending RVP, psi	12
Boiling point, °C	83
Freezing point, °C	25
Solubility at 21°C	
TBA in water, wt%	100
Water in TBA, wt%	100
Viscosity at 25.5°C, cP	4.2
at −20°C, cP	Solid
Stoichiometric air/fuel ratio	11.1
Latent heat of vaporization, kcal/kg (Btu/lb)	130 (258)
Lower heating value, kcal/kg (Btu/lb)	7,850 (14,280)
Flammability limits in air	
Lower limit, vol%	2.4
Upper limit, vol%	8.0
Autoignition temperature, °C	478
Flash point, closed cup, °C	11
Blending octane number[a]	
RON	106
MON	89
(RON + MON)/2	97.5

[a]When adding 10 vol% TBA to a base gasoline having the following octane properties: RON clear = 94.3, MON clear = 84.3 [6]. Laboratory RON and MON rating procedures are not suitable for use with pure oxygenates.

have low boiling points as compared to alcohols and thus require less heat of vaporization. Compounds which require more heat for vaporization are more difficult to vaporize during cold engine operation. Incomplete vaporization leads to poor fuel–air mixing and thus contributes to incomplete combustion and higher hydrocarbons emissions. The alcohols also pose phase-separation problems in the presence of trace amounts of water as well as corrosion in the engine. This comparison makes it clear that ethers are clearly superior to alcohols in achieving reformulated gasoline

Table 11 Physical and Blending Properties of Oxygenates and Their Blending Limits

Oxygenate	Specific gravity	Boiling point (°C)	Average octane (R + M)/2	Oxygen contents (wt%)	Water tolerance
Ethers					
MTBE	0.744	55.2	110	18.2	Excellent
ETBE	0.747	67.0	111	15.7	Excellent
TAME	0.770	86.0	105	15.7	Excellent
TAEE	0.750	101	100	13.8	Excellent
DIPE	0.736	68.0	107	15.7	Excellent
Alcohols					
Methanol	0.796	149	120	50	Poor
Ethanol	0.794	172	115	35	Very poor
Methanol/ GTBA blend[a]	0.793	145/180	108	35	Poor
GTBA[b]	0.791	181	100	21	Poor

[a] Methanol must be used with an appropriate co-solvent.
[b] Gasoline-grade *t*-butyl alcohol.

goals, and they are oxygenates of choice. They contribute more favorable properties to gasoline such as low heat of vaporization, low RVP, high front-end octane number, and low flame temperature [9].

REFERENCES

1. P. W. Mccallum, T. J. Timbario, R. L. Bechtold. Alcohol fuels for high way vehicles. Chem Eng Prog, Aug. 1982, pp 52–59.
2. R. W. Reynolds, J. S. Smith, I. Steinmetz. Methyl ether (MTBE) scores well as high-octane gasoline component. Oil Gas J, June 16, 1975, pp 50–52.
3. M. F. Ali, A. Bukhari, A. Amer. Evaluation of MTBE as a high octane blending component for gasoline. Arabian Journal of Science and Engineering, 12(4): 451–458, 1987.
4. R. G. Clark, R. B. Morris, D. C. Spence, E. Lee Tucci. Unleaded octanes from RVP changes: MTBE from butane. Energy Prog, 7(3): 164–169, 1987.
5. R. Csikos, I. Pallay, J. Laky. Practical use of MTBE produced from C_4 fraction. 10th World Petroleum Congress, Bucharest, 1979, pp. 167–175.
6. G. Marceglia, G. Oriani. MTBE as alternative motor fuel. Chem Econ Eng Rev, 14(4): 39–45, 1982.

7. I. S. Al-Mutaz. Saudi MTBE plant and its role in the lead phasedown in the country. Energy Prog, 7(1): 18–22, 1987.
8. G. H. Unzelman. Reformulated gasolines will challenge product-quality maintenance. Oil Gas J, April 9, 1990, pp 43–48.
9. G. H. Unzelman. Ethers have good gasoline-blending attributes. Oil Gas J, April 10, 1989, pp 33–44.
10. W. J. Piel, R. X. Thomas. Oxygenates for reformulated gasolin. Hydrocarbon Proc, July 1990, pp 68–72.
11. M. Prezeij. Pool octanes via oxygenates. Hydrocarbon Proc, Sept., 1987, pp 68–70.
12. R. Csikos, Low-lead fuel with MTBE and C_4 alcohols. Hydrocarbon Proc, 55: 121–125, July 1976.
13. G. Pecci, T. Floris. Ether ups antiknock of gasoline. Hydrocarbon Proc, 56(12): 98–102, Dec. 1977.
14. B. V. Vora. Ethers for gasoline blending. UOP Document, 1990, pp 1–15 .
15. S. C. Stinson. New plants, processing set for octane booster. Chem Eng News, June 25, 1979, pp 35–36.
16. T. W. Evans, K. R. Eklund. Tertiary alkyl ethers, preparation and properties. Ind Eng Chem, 28(10): 1186–1188, 1963.
17. J. J. McKetta and W. A. Cunningham. Octane options. Encyclopaedia of Chemical Technology, Vol. 31, pp 436–450, 1990.
18. SRI. MTBE and TBA. SRI Report No 131, SRI International, Menlo Park, CA, Aug. 1979.
19. H. L. Hoffman. Components for unleaded gasoline. Hydrocarbon Proc, 59(2): 57–59, Feb. 1980.
20. M. F. Ali, M. M. Hassan. A. Amer. The effect of oxygenates addition on gasoline quality. Arabian Journal of Science and Engineering, 9(3): 221–226, 1984.
21. G. H. Unzelman. Ethers will play larger role in octane, enviromental specs for gasoline blends. Oil Gas J, April 17, 1989, pp 44–49.
22. ARCO. ARCO to use MTBE to improve gasoline octane. Oil Gas J, June 26, 1978, p 62.
23. R. T. Johnson, B. Y. Taniguchi. Methyl tertiary-butyl ether, evaluation as a high octane blending component for unleaded gasoline. Symposium on Octane in the 1980's, ACS Miami Beach Meeting, Sept 10–15, 1978.
24. P. L. Dartnell, K. Campbell. Other aspects of MTBE/methanol use. Oil Gas J, Nov. 13, 1978, pp 205–212.
25. B. Davenport, R. Gubler, M. Yoneyama. Gasoline Octane Improvers, Chemical Economics Handbook Report, SRI Consulting, SRI, Menlo Park, CA, May 2002.
26. Ch. Thiel, K. Sundmacher, U. Hoffmann. Residue curve maps for heterogeneously catalyzed reactive distillation of fuel ethers MTBE and TAME. Chemical Engineering Science, 52, pp 993–1005, 1997.
27. ARCO. MTBE octane enhancer. Tech Bull, ARCO Chemical Company, 1985.

28. F. Obenaus, W. Droste. Huls process: methyl tertiary butyl ether. Erdol und Kohle- Erdgas, 33(6): 271–275, 1980.
29. Huls. Huls Data Sheet, Technical Data Sheet No. 2148, Chemische Werke Huls AG, 1978.
30. Huls. The Huls MTBE process. Technical Brochure, Chemische Werke Huls AG, Oct. 1980.
31. Phillips. Phillips methyl tertiary butyl ether process. Technical Brochure, Phillips Petroleum Company, 1981.
32. SRI. SRI Report No. 158, SRI international, Menlo Park, CA, 1983.
33. MAFKI. Lead-free or low leaded fuel composition MTBE production technology. Technical Brochure, MAFKI, Hungary, 1980.
34. W. J. Piel. The role of MTBE in future gasoline production, Paper 47d. Spring National Meeting of AIChE, March 6–10, 1988, pp 1–19.
35. W. H. Douthit, Performance features of 15% MTBE/gasoline blends. SAE Paper 881667, 1998.
36. API. Alcohols and ethers: a technical assessment of their application as fuels and fuel components. API Publication 4261, July 1988, pp 23–89.
37. K. Alm, M. Ciprian. Vapor pressure, refractive index at 20°C and vapor-liquid equilibrium at 101.325 kpa with MTBE-methanol system. J Chem Eng Data, 25: 100–103, 1980.
38. R. L. Furey, K. L. Perry. Volatility characteristics of blends of gasoline with ethyl tertiary-butyl ether (ETBE). SAE Technical Paper 901114, 1990.
39. I. Čerenková, T. Boublík. Vapor pressures, refractive indexes and densities at 20°C and vapor-liquid equilibrium at 101.325 kPa, in the tertiary-amyl methyl ether–methanol system. J Chem Eng Data, 29: 425–427, 1984.
40. P. Dorn, A. M. Mourao. The properties and performance of modern automotive fuels. SAE Technical Paper 841210, 1984.
41. F. F. Pischinger. Alcohol fuels for automotive engines. 10th World Petroleum Congress, Bucharest, pp 1–10, 1979.
42. T. C. Austin, G. Rubenstein Gasohol: technical, economic or political panacea? SAE Technical Paper 800891, 1980.
43. Alternate Fuels Committee of the Engine Manufacturers Association. A technical assessment of alcohol fuels. SAE Technical Paper 820261, 1982.
44. J. C. Ingamells, R. H. Lindquist. Methanol as a motor fuel or a gasoline blending component. SAE Technical Paper 750123, 1975.
45. J. L. Keller. Alcohols as motor fuels. Hydrocarbon Proc, 58(5): 127–138, 1979.
46. W. K. Kampen. Engines run well on alcohols. Hydrocarbon Proc, Feb. 1980, pp 72–75.
47. T. O. Wagner, H. L. Muller. Experience with oxygenated fuel components. Amoco Oil Co., Naperville, IL, Preprints Order No. 820-00040, 1982.
48. E. G. Guetens Jr., J. M. Dejovine, G. J. Yogis. TBA aids methanol/fuel mix. Hydrocarbon Proc, May 1982, pp 113–117.
49. J. M. Dejovine, E. C. Guetens Jr., G. J. Yogis, B. C. Davis. Gasolines show varied responses to alcohols. Oil Gas J Technol, Feb. 14, 1983, pp 87–94.

3

Blending Properties of MTBE and Other Oxygenates in Gasoline

Mohammad Ashraf Ali and Halim Hamid
Research Institute, King Fahd University of Petroleum and Minerals, Dhahran, Saudi Arabia

I. INTRODUCTION

Oxygenated ethers and alcohols are blended with gasoline to increase octane number and to fight air pollution problems. These oxygenates have replaced alkyl lead and other metal-containing compounds in gasoline because the use of compounds such as tetraethyl lead (TEL), tetramethyl lead (TML), and methylcyclopentadienyl manganese tricarbonyl (MMT) in gasoline has created air pollution problems. The emission of their combustion products from vehicle exhausts creates atmospheric pollution causing serious health hazards. The oxygenates used are methyl *tertiary*-butyl ether (MTBE), ethyl *tertiary*-butyl ether (ETBE), *tertiary*-amyl methyl ether (TAME), *tertiary*-amyl ethyl ether (TAEE), diisopropyl ether (DIPE), methyl alcohol, ethyl alcohol, and *tertiary*-butyl alcohol (TBA). Among these oxygenates, MTBE appears to be the most effective choice because its physical, chemical, and thermal properties are compatible with those of gasoline, especially in the boiling range where gasoline typically shows lowest antiknock characteristics. In this chapter, the blending characteristics of a number of ether and alcohol oxygenates are presented.

II. BLENDING PROPERTIES OF MTBE

A. Octane Number

The output of an engine is determined by knocking. Excess knocking can damage the engine. Low-engine-speed knock is usually audible to the driver but not damaging to the engine. High-engine-speed knock, however, is often inaudible above the engine, road, and wind noise. The most severe knock, which can be very damaging, often occurs at motorway cruising speeds of 4000–5000 rpm, and modern high-compression engines increase the tendency to knock.

Many engines will fail in less than 50 hr under conditions of heavy knock, and the damaging effect of knock is cumulative [1]. The same study also concludes that the maximum engine speed associated with knock is greatly reduced with MTBE. Laboratory Research and Motor Octane rating procedures such as ASTM methods D-2699 and D-2700 are not suitable for use with neat oxygenates such as MTBE. Octane values obtained by these methods are not useful in determining knock-limited compression ratios for vehicles operating on neat oxygenates when blended with gasoline [2].

The octane value of MTBE is measured by its blending octane value (BOV) [3]. This value is calculated from the difference between the octane value of a base gasoline with a known amount of MTBE and the base gasoline without MTBE. The formula for BOV calculation is

$$\text{BOV} = \frac{\text{ON} - \text{ON}_{\text{base}}(1-x)}{x} = \text{ON}_{\text{base}} + \frac{\text{ON} - \text{ON}_{\text{base}}}{x}$$

where

ON = RON or MON of MTBE blend with base gasoline
ON_{base} = RON or MON of base gasoline
x = volume fraction of MTBE in the blend.

The range of MTBE blending octane numbers is given below [4,5]. This range is determined from the large amount of experimental data obtained in the formulation of gasolines within the specification limits.

Blending RON	115–135
Blending MON	98–110
Blending (RON + MON)/2	106.5–122.5

The blending octane numbers of MTBE are very sensitive to the composition and octane numbers of the unleaded gasoline base [6]. The MTBE blending octane number generally rises when base gasoline octane number decreases, MTBE concentration in the gasoline decreases, or the saturate content of the gasoline increases.

Addition of MTBE increases the RON and MON of a gasoline. The effect of MTBE on the antiknock properties of the three types of base gasoline have been determined. These were A, B, and C, having RONs 84.6, 90.5, and 93.7, and MONs 79.0, 83.0, and 84.0, respectively. Gasoline A was composed of 10 vol% straight-run light gasoline and 90 vol% reformate; gasoline B consisted of 50% each of C_5–C_6 isomerate and heptanes plus reformate, and gasoline C was comprised of 60 vol% reformate, 23 vol% light catalytically cracked gasoline, 6 vol% heavy catalytically cracked gasoline, and 11 vol% C_3–C_4 alkylate. MTBE in the concentration levels of 5, 10, and 15 vol% was added. An increase in RON and MON was found for all gasoline blends. The gasoline samples having higher RON and MON were found to have less increase in their octane numbers as compared to the gasolines with lower octane numbers [4]. The sensitivity (RON – MON) was higher for gasoline having higher octane numbers (Table 1).

Table 1 Octane Improvement and Effect on Sensitivity of the Lead-Free Base Gasoline Stock by MTBE Blending

Properties of gasoline			A	B	C
RVP, psi			5.26	7.96	4.55
Specific gravity at 15/4°C			0.751	0.740	0.749
% distilled at 70°C			14	36	21
% distilled at 100°C			50	55	53
Olefins, vol%			—	—	12
Aromatics, vol%			40	36	34
MTBE addition (vol%)		0	5	10	15
Gasoline A	RON	84.6	87.0	88.9	90.8
	MON	79.0	80.6	82.4	83.8
	Sensitivity (RON – MON)	5.6	6.4	6.5	7.0
Gasoline B	RON	90.5	92.2	93.7	95.2
	MON	83.0	84.0	85.1	86.4
	Sensitivity (RON – MON)	7.5	8.2	8.6	8.8
Gasoline C	RON	93.7	94.9	96.0	97.2
	MON	84.0	84.6	85.4	86.5
	Sensitivity (RON – MON)	9.7	10.3	10.6	10.7

Table 2 Effect of MTBE Addition on Octane Improvement of Reformate and Gasoline

		Octane numbers		
Fuel type and properties	MTBE (vol%)	RON	MON	(RON + MON)/2
Reformate, RON 90.0	5	92.0	87.2	89.6
41.0 vol% paraffins	10	93.5	89.5	91.5
2.0 vol% napthenes	15	96.0	90.5	93.3
57.0 vol% aromatics	20	97.5	91.5	94.5
A-380 blend (unleaded)	5	85.6	83.1	84.4
RON 83.7	10	88.0	84.0	86.0
59.7 vol% paraffins	15	89.6	84.9	87.3
2.4 vol% naphthenes	20	91.5	86.1	88.8
37.9 vol% aromatics	30	95.5	90.2	92.9
A-380 blend + 0.15 g Pb/L	5	89.5	83.5	86.5
RON 87.0	10	92.6	86.2	89.4
	15	95.0	89.5	92.3
A-380 blend + 0.28 g Pb/L	5	91.6	85.9	88.8
RON 89.5	10	95.5	89.7	92.6
A-380 blend + 0.40 g Pb/L	5	92.5	87.5	90.0
RON 90.2	10	97.0	90.5	93.8

A-380 is a gasoline produced by Saudi Aramco. The RON of A-380 lead-free gasoline was increased from 83.7 to 85.6 by adding 5 vol% MTBE and to 95.5 for 30 vol% MTBE (Table 2). The increase in RON ranged from 1.9 to 11.8 with the addition of MTBE to A-380 gasoline by 5–30 vol%, respectively [7]. For A-380 leaded gasoline having lead (Pb) concentration 0.28 g/L, only 10 vol% MTBE was needed to increase the RON to 95.5. When 0.4 g Pb/L of gasoline was present, the RON increased to 97.0.

MTBE acts as a high-octane blending stock and not like a lead antiknock agent [6]. Up to 15 vol% of MTBE was added to base gasoline with RON and MON of 93 and 83, respectively. The concentration of antiknock compounds (lead alkyls) in gasolines is much lower than in MTBE blends. It has been reported [8] that the average octane number, (RON + MON)/2, also abbreviated (R + M)/2, increases by 2.3 with the addition of 11 vol% MTBE to base gasoline having 90 (R + M)/2. Hence, the blending value, (R + M)/2, of MTBE is 110.9. The addition of 10 vol% MTBE to gasoline having RON 98.1 and MON 80.1 increases both the RON and MON by 2–3 points [9–11].

It has been observed that the octane number of 90-RON base fuel can be increased by using various concentrations of MTBE and secondary butyl alcohol [12]. The results clearly showed that the pure MTBE provided more octane to the gasoline as compared to secondary butanol. A chart has been formulated comparing the incremental gain of average octane number, (R + M)/2, in base gasoline resulting from the addition of each volume percentage of oxygenates including MTBE [13]. Based on some of the studies [12,14], 15% represented a reasonable concentration of MTBE in gasoline in terms of octane number increase, change in fuel stoichiometry (air/fuel ratio), and commercial availability of MTBE.

Fuel sensitivity is defined as the difference between RON and MON. It has been reported that the fuel sensitivity is a function of MTBE blending octane number and increases with a decrease in the blending octane number [15]. The high-octane properties of MTBE are particularly effective in blending with low-octane unleaded gasoline components [16]. Supporting this observation, the BOV of MTBE is highest in a low-octane unleaded gasoline. For example, MTBE has a blending octane number of 122 when 15 vol% is added to 82-octane unleaded gasoline.

The boiling point of MTBE is low. For this reason, MTBE provides much higher front-end octane numbers (FEON) to gasolines. FEON is the octane number of gasoline fraction that boils below 100°C. It is reported as RON at 100°C. It becomes important in cold-start conditions when the low-boiling parts of gasoline get a chance to vaporize. When there are no lower-boiling-point lead additives to increase FEON, MTBE effectively boosts the front-end octane. MTBE gives exceptionally high FEON blending numbers, generally in the range of 135 RON. The FEON of MTBE is higher than the other gasoline blending components such as butane, reformate, alkylate, and aromatics [17,18]. FEON increases engine efficiency during the low-speed acceleration stage.

When MTBE was added to an unleaded gasoline with RON = 88, MON = 81, and RON @ 100°C = 77 [19,20], its FEON was increased drastically. For example, when 15 vol% MTBE was added, the FEON reached 93 while the RON and MON increased to 93 and 86, respectively. The FEON, characterizing knocking during acceleration, shows an unparalleled octane boost. A FEON advantage of MTBE has been reported for a gasoline containing 11 vol% MTBE for which the average octane number, (R + M)/2, was increased by 8 [21]. MTBE has a very favorable effect on FEON as compared to refinery low-boiling components at IBP 100°C, which show considerably lower octane properties, e.g., gasoline having RON 98.5 has 88.5 FEON, compared to a gasoline containing 10 vol% MTBE, which has similar RON but much higher FEON (95.5).

The effects of MTBE on the antiknock properties of a large variety of gasolines and gasoline stocks have been reported. Since the improvement of octane number by MTBE addition depends on the composition of the base fuel, which contains hundreds of components, accurate values can only be determined by testing a particular gasoline. For this reason, it is important to know the hydrocarbon composition of the gasoline. It has been shown that 5, 10, and 20 vol% MTBE increase the RON of premium unleaded gasoline from 91.5 to 92.4, 94.0, and 96.2, respectively [21]. Front-end octane quality improvement has been reported for a typical premium gasoline with 98/99 RON of 50/55% distillate at 100°C containing 0.4 g Pb/L in the form of tetraethyl lead [4].

MTBE does not decrease the lead susceptibility of the alkyl lead compounds tetraethyl lead (TEL), tetramethyl lead (TML), or their blends. A study has concluded that MTBE is not affected by the lead level in gasolines. This study gives information on the production possibility of 93-RON gasoline using base gasolines with given RON, MTBE, and alkyl leads. It is possible to produce unleaded 93-RON gasoline using 88-RON base gasoline and 15 vol% MTBE. Leaded 93-RON gasoline can also be produced using 88-RON base gasoline, 10 vol% MTBE, and 0.1 g Pb/L as TML. The possibility of blending low-lead or lead-free gasoline of 93 RON using MTBE or alkyl leads can be determined. Since the octane number-improving effect of MTBE is concentrated in the low-boiling fraction due to its low boiling point, front-end octane improvement of gasolines is increased significantly. The difference between RON and FEON values drops from 6 to less than 2 when 10 vol% MTBE is added to this gasoline containing 0.6 g Pb/L [22].

The TEL response of a typical commercial gasoline containing various amounts of MTBE has been studied [23]. The RON of a gasoline of RON 92 can be increased to 99 by adding MTBE and 0.6 g Pb/L of gasoline. It has been reported that a reduction in the lead content of gasoline from 0.6 to 0.15 g/L will increase the consumption of crude oil in gasoline production by 1.73, 2.36, and 4.03% for gasolines with RONs of 94, 96, and 98, respectively [24].

The use of MTBE permits more effective utilization of petroleum raw material in gasoline production, thus increasing the gasoline output by 2.6–4% without increasing the volume of crude oil processed [25]. High-aromatic and low-olefinic gasolines reduce the blending octane value of MTBE [26].

Road Octane Number (RdON) is difficult to obtain, since it is affected by cars and test conditions. General equations for RdON and laboratory-measured antiknock properties have been published in the literature [27]. The European Fuel Oxygenates Association (EFOA) carried out a RdON performance testing of European unleaded gasoline containing MTBE [28]. Oxygenate blends containing methanol, co-solvent, and MTBE gave superior road octane performance under accelerating conditions and at low constant speeds compared to reference gasoline. At high speeds, 3500–4500 rpm, the RdON advantage of oxygenates diminished, giving similar performance compared to hydrocarbon-only gasolines. This shows that MTBE can increase gasoline FEON by approximately 4–7 numbers, in contrast to methanol/co-solvents, which have a FEON increase by 3–4 numbers.

B. Vapor Pressure

In addition to the effect of MTBE on gasoline octane numbers, there are other properties of MTBE that influence the performance of gasolines. Most notable are the Reid vapor pressure (RVP) and distillation temperatures, which are used to control both hot and cold drivability performance.

Petroleum refiners have been using increased amounts of butanes in the United States. Butane addition increases both the octane number and the RVP of the gasoline. Increased use of butane in the United States is the main reason for a 2–2.5-psi increase in the RVP over the last decade [29]. The U.S. Environmental Protection Agency (EPA) has lowered gasoline RVP by 2 in order to reduce ground-level ozone. The blending vapor pressure of MTBE is lower than that of typical commercial gasolines [15].

The RVP of MTBE is within the specifications of the gasolines produced by Saudi Aramco (specification of A-380 gasoline). The RVP of gasoline gradually increases with the addition of MTBE but remains within acceptable limits (7.11–9.24) [7]. For example, addition of 5–30 vol% MTBE increases the RVP from 9.20 to 9.24. The direction of RVP change is either up or down, depending on the original vapor pressure of the base gasoline. Most authors agree that there is only a small butane loss with MTBE and, depending on the volatility of the base gasoline, butane may be added to the blend to increase the cost effectiveness of MTBE [1].

C. Distillation Characteristics

MTBE is soluble in any ratio with gasoline, and it boils in the same temperature range as any other light refinery component. Unlike alcohols, its hydrocarbon compatibility feature does not permit it to create azeotropic effects on the distillation curve of gasoline. A comparison of distillation data shows that upon addition of MTBE to gasoline, there is generally a big decrease in the 50% boiling temperature. Addition of more MTBE produced a further decrease in the 50% boiling temperature. Ten percent of the distillation temperature of gasoline is not usually affected by addition of MTBE. For this reason, good drivability performance is maintained in hot weather. The 50% temperature of the gasoline is decreased by addition of MTBE, and it usually improves cold engine operation. It has been reported that addition of MTBE to gasoline has no effect on 10% distillation point, but provides a decrease of 8, 3, and 30°C on 50%, 90%, and end-point (EP) temperature [17,18].

Only butane and MTBE have 50% of their temperature below 93.3°C, but addition of butane increases the RVP of gasoline. Most of the high-octane components produced in a refinery are usually high-boiling-point components. Therefore, 50% temperature of the gasoline is more difficult to adjust. ASTM distillation results of base fuel and MTBE–gasoline blends (5, 10, 15, and 20) have shown that there is a decrease in percent boiling temperature for all blends, especially a sharp decrease in 50% boiling temperature [7]. The distillation characteristics of several gasolines and 7 vol% MTBE–gasoline blends were determined and the data were presented. Addition of the 7 vol% MTBE had rather small effects on the gasoline distillation temperatures [30].

The ASTM data reported also show that the addition of MTBE changes the distillation curves of the base gasoline. Indolene is a standard gasoline used in engine testing. The distortion of the distillation curve caused by MTBE is very small when compared to the distortion obtained with alcohol–gasoline mixtures [31]. The effect of adding different amounts of MTBE to a typical gasoline distillation has been reported [4]. MTBE–gasoline blend curves lie below the gasoline distillation curve. The greatest distortion in gasoline distillation occurs with methanol, and to a lesser extent with ethanol. The effect of MTBE is moderately distributed [13]. The effects of 15 vol% MTBE on the distillation characteristics of two different gasolines were also studied recently [32]. The data for the distillation characteristics of gasolines and MTBE blends were reported in this article.

D. Storage and Thermal Stability

The stability of gasolines can be evaluated by the formation of peroxides during storage. Long-term oxidation stability tests of three gasolines with 10 vol% MTBE blends were carried out at storage temperatures of up to 43.3°C [10,11]. Storage at 43.3°C for a period of 6 months can be considered to be equal to approximately 2 years of storage at ambient temperature. The gasoline containing MTBE did not produce any peroxides, whereas gasoline alone yielded substantial amount of peroxides. MTBE gasoline blends could be stored for a minimum of 2 years under proper antioxidant protection even when prepared with unstable light catalytically cracked gasoline (LCCG) gasolines. Oxidation stability of a gasoline with 10 vol% MTBE was performed according to ASTM D-525 test procedure and found no gum formation in excess of 1000 min at 100°C [17].

It has been reported that MTBE is stable during handling and storage, both as a pure compound and after addition to gasoline [5]. Storage stability was tested after 180 days [4] and no significant difference in potential gums were found between the base gasoline (having 98/99 RON and 15 vol% olefins) and base gasoline having 15 vol% MTBE. Laboratory studies confirmed that peroxides are not formed with MTBE [14]. Experiments conducted at 60 psig oxygen and 90°C temperature showed no titratable peroxide after 15 hr. An extended test performed over a period of 2 years indicated no peroxides formation in MTBE samples exposed to light and air.

E. Water Tolerance and Solubility

MTBE gasoline blends show no phase separation in distribution systems in the presence of water. Solubility of water in MTBE (1.5 wt% water) is very low compared to that of alcohols [7]. It has been indicated that water tolerance is not a problem with MTBE–gasoline blends. Clouding was observed during the course of preparing MTBE–gasoline blends for testing [31]. However, clouding was cleared entirely after 24 hr, and no other problems such as separation and residue were observed in the samples. Analysis of a reproduced sample showed that the precipitate was primarily water and MTBE, and the amount was very small.

Water tolerance of gasoline has been studied at 20 vol% MTBE and 10 vol% water. The water was settled in the mixture and there was no haziness. MTBE losses due to water contact were negligible (200–300 ppm in the water). Phase separation in MTBE–gasoline blends is not expected to cause problems as in the case of alcohol–gasoline blends [7]. The water solubility of gasolines containing MTBE and *sec*-butyl-alcohol has been studied [12]. The water solubility of these gasoline samples increased considerably with increase of *sec*-butyl-alcohol concentration in the blends. There is no hazing problem in a 15 vol% MTBE–gasoline blend up to 300 ppm water [33]. Considering that commercial MTBE contains less than 500 ppm water, a 15% MTBE–gasoline blend containing normally 75 ppm water is well below hazing condition. The water-holding capacity of MTBE is very high, e.g., a gasoline containing 15% MTBE was found to have 520 ppm of water compared to 190 ppm water for a gasoline sample having no MTBE. MTBE–gasoline blends are not as hygroscopic as alcohol–gasoline blends [34].

F. Effect on Exhaust Emissions

The effect of MTBE blending with gasoline on the exhaust emissions from internal combustion engines has been studied using different MTBE concentrations and with different composition of gasoline blends [35]. A study was conducted using a four-cylinder Opel 1.6 L engine equipped with a hydraulic brake dynamometer and fuels containing up to 11.0% MTBE in a wide range of engine operations, and a three-way catalytic converter. The addition of MTBE to gasoline resulted in a decrease in CO and HC emissions only at high engine loading. During cold start-up of the engine, MTBE, HC, and CO emissions were significant and increased with MTBE addition. At the catalytic converter outlet, MTBE was detected when its concentration in fuels was greater than 8% and only as long as the catalytic converter operated at low temperatures. Methane and ethylene emissions were comparable for all fuels tested at the engine outlet, but methane emissions remained almost at the same level while ethylene emissions were significantly decreased by the catalytic converter.

The effects of using MTBE in gasoline on exhaust emissions of a typical SI engine has been examined [36]. The MTBE was blended with a base unleaded fuel in three ratios (10, 15, and 20 vol%). The emissions of CO, HC, and NO_x were measured at a variety of engine operating conditions using an engine dynamometer setup. The results of the MTBE blends were compared to those of the base fuel and of a leaded fuel prepared by adding TEL to the base. With respect to the base fuel, the addition of MTBE decreased the CO emissions, decreased the HC emissions at most operating conditions, but generally increased the NO_x emissions. The emissions results for the leaded fuel were comparable to those of the base fuel.

The combustion by-products of pure MTBE have been evaluated in previous laboratory studies, but little attention has been paid to the combustion by-products of MTBE as a component of gasoline. MTBE is often used in reformulated gasoline (RFG), which has chemical and physical characteristics distinct from conventional gasoline. Engine-out automotive dynamometer studies have compared RFG with MTBE to nonoxygenated RFG. The findings suggest that adding MTBE to reformulated gasoline does not affect the high-temperature flame chemistry in cylinder combustion processes. A comparison of tailpipe and exhaust emission studies indicated that reactions in the catalytic converter are quite effective in destroying most hydrocarbon MTBE by-product species. Since important reaction by-products are formed in the postflame region, understanding changes in this region will contribute to the understanding of fuel-related changes in emissions [37].

An ethanol–MTBE fuel blend was used as a substitute for hydrated ethanol to improve vehicle cold start and drivability characteristics during the warm-up period. The vehicle was tested in a cold box, reaching a lower ambient temperature limit of −6°C, which represents the most severe weather condition experienced in Brazil. Different concentrations of MTBE in ethanol were investigated. The results show satisfactory drivability characteristics for the tested conditions comparable with those found in gasoline-fueled vehicles, thus overcoming the existing cold-start difficulties in production ethanol-fueled vehicles [38].

In another work, engine and tailpipe (after a three-way catalytic converter) emissions from an internal combustion engine operating on two oxygenated blend fuels (containing 2 and 11 wt% MTBE) and a non-oxygenated base fuel were characterized [39]. The engine (Opel 1.6 L) was operated under various conditions, in the range of 0–20 hp. Total unburned hydrocarbons, carbon monoxide, methane, hexane, ethylene, acetaldehyde, acetone, propanol, benzene, toluene, 1,3-butadiene, acetic acid, and MTBE were measured at each engine operating condition. Concerning the total HC emissions, the use of MTBE was beneficial from 1.90 to 3.81 hp, which were the most polluting conditions. Moreover, CO emissions in tailpipe exhaust were decreased in the whole operation range with increasing MTBE in the fuel. The greatest advantage of MTBE addition to gasoline was the decrease in ethylene, acetaldehyde, benzene, toluene, and acetic acid emissions in engine exhaust, especially when MTBE content in the fuel was increased to 11% w/w. In tailpipe exhaust, the catalyst operation diminished the observed differences. Ethylene, methane, and acetaldehyde were the main compounds present in exhaust gases. Ethylene was easily oxidized over the catalyst, while acetaldehyde and methane were quire resistant to oxidation.

III. BLENDING PROPERTIES OF ETBE

ETBE has been investigated as a potential ether on a laboratory scale to supplement MTBE in the U.S. gasoline pool [26]. The blending octane RON and MON of ETBE are 118 and 102. The volatility characteristics of ETBE are quite comparable to those of MTBE as reflected by 3–3 psi as blending RVP [26,40]. The vapor pressure and distillation characteristics of

ETBE have been studied extensively [33]. In contrast to the increase of RVP in ethanol–gasoline blends, ETBE–gasoline blends show lower RVPs than the corresponding gasoline. The RVP of MTBE is between those of ethanol and ETBE. The effects of ETBE on the distillation characteristics of gasoline are similar to those of MTBE. The ethers behaved essentially the same way as hydrocarbons of similar volatility. ETBE has favorable fate and transport properties for the environment. ETBE has the lowest tendency to evaporate into the atmosphere, and its lower solubility reduces ETBE's ability to transfer to water. ETBE can be easily air-stripped from water during remediation, and it degrades more quickly in the atmosphere. ETBE helps expand the ethanol benefits of reformulated gasoline (RFG) as well as enhances the use of ethanol in the RFG. The use of ETBE can double the ethanol demand and thus further extend gasoline supplies and energy independence. ETBE supports environmental benefits, provides more greenhouse gas benefits, protects groundwater contamination, and enhances vehicle performance such as drivability. ETBE provides the most clean diluent, octane, and front-end volatility volume for making U.S. Summer Grade RFG. ETBE displaces three times more aromatics than ethanol in RFG. A comparison of ETBE, MTBE, and ethanol showed that ETBE provides more octane benefit and relatively higher oxygenate volume.

IV. BLENDING PROPERTIES OF TAME

The blending octane numbers of TAME have not been extensively studied in the literature as compared to MTBE. Chase et al. [9–11] reported that addition of TAME to a sample of unleaded gasoline having RON 89.1 and MON 80.1 increased the research and motor octane numbers. The gasoline chemical composition (vol%) was as follows: light ends (C_5-) 10.1, paraffins 55.3, olefins 18.8, naphthenes 1.9, and aromatics 24.0. By adding 10 and 20 vol% TAME, the RON of gasoline changed to 91.2 and 93.5 and the MON was increased to 82.0 and 84.5, respectively. Torck et al. [41] worked particularly on octane determination for TAME in olefinic feeds, catalytic cracked gasoline, and alkylates. The octane blending values of TAME (Table 3) are higher when blended with alkylate (RON 122, MON 105) as compared to when blended with a steam-cracked C_5 fraction (RON 105, MON 96). In conclusion, the RON and MON octane values of TAME are within the following range: RON 105–122, MON 42–51. Likewise, it appears that the octane blending values of TAME are roughly 5% lower than MTBE values. Table 3 also presents the RON and MON of two TAME-containing unleaded

Table 3 Octane Number Data for Base Gasoline and TAME–Gasoline Blends

	RON	MON
C_5 olefinic feed (74% olefins)	95.3	88.3
+5% TAME	95.7	89.2
+10% TAME	96.4	90.0
Light cut catalytic cracked gasoline[a]	89.0	80.5
+5% TAME	90.2	81.4
Alkylate	92.0	91.0
+10% TAME	95.0	92.4
Unleaded gasoline	92.1	83.7
+10% TAME	94.2	85.3
+20% TAME	94.7	85.4
Unleaded gasoline	89.1	80.1
+10% TAME	91.2	82.0
+20% TAME	93.5	84.5

[a]Saturated hydrocarbons = 69.5%, olefins = 24%, aromatics = 6.5%.

gasoline blends [16], and the data show a substantial increase with the addition of TAME to gasoline. The Reid vapor pressure of a gasoline is reportedly not affected by the addition of 10 and 20% TAME [16,26].

V. BLENDING PROPERTIES OF METHANOL

Methanol is the simplest alcohol, containing one carbon atom. It is a colorless, tasteless liquid with a very faint odor and is commonly known as "wood alcohol." Methanol's physical and chemical characteristics result in several inherent advantages as an automotive fuel. Table 4 provides the properties of gasoline with and without methanol [52]. The blending properties of methanol change with the composition of gasoline. Table 5 gives the properties of two test fuels and methanol blends with the test fuels at 5 and 10% [53]. Fuel 1 was unleaded low-octane indolene, widely used in engine and emission control studies, and fuel 2 was a fuel of slightly lower vapor pressure that might be used commercially with methanol blends. Fuel energy content was calculated using API gravity, distillation, and aromatic content data.

Cox [54] studied the physical properties of gasoline–methanol automotive fuels. The experimental data and discussion covered four physical property areas: water tolerance, vapor pressure, distillation characteristics, and octane quality. Table 6 demonstrates the changes in blending vapor pressures (RVP), distillation, and octane numbers (RON

Table 4 Properties of Unleaded Gasoline and Gasoline–Methanol Blends

Properties	Unleaded gasoline	Unleaded gasoline having 15 vol% methanol
Gravity, °API	50.2	57.1
Sp. gravity 60/60°F	0.742	0.750
RVP, kPa	75.8	2.4
ASTM gum, mg/100 mL	0.1	0.1
Sulfur, wt%	0.05	0.04
Lead, ppm	0.5	0.4
Saturates, vol%	48	49
Olefins, vol%	11	16
Aromatics, vol%	41	35
Carbon, wt%	86.7	79.0
Research octane number	93.0	98.3
Motor octane number	85.0	86.1
ASTM distillation data (°C)		
IBP	29	32
10% evaporated	46	41
30% evaporated	72	53
50% evaporated	90	58
70% evaporated	122	114
90% evaporated	154	152
FBP	193	192

and MON) of methanol at 10 and 25 vol% blended with two different unleaded gasolines [16].

These fuels were studied to determine the effect of methanol on vapor pressure, distillation, and gravity (Table 7). A typical unleaded fuel having no methanol was used as a base gasoline. One methanol blend was prepared by adding 15 vol% methanol directly to this base fuel. The other methanol blend, also having 15 vol% methanol, was adjusted to give the same RVP as the base fuel. This required taking out all the butane and half of the pentanes from the base fuel blend before adding the methanol. It can be seen from the data that the RVP of the non-matched RVP methanol blend is much higher than that of the base fuel.

A. Octane Blending

Blending RON and MON values of methanol vary widely, but a few general statements could be made:

Blending values increase with decreasing octane numbers of the base stock.

Table 5 Properties of Methanol–Gasoline Test Fuels

	Fuel 1 (indolene fuel)		Fuel 2 (commerical fuel)	
Properties	Clear	With 10% methanol	Clear	With 10% methanol
API gravity	60.0	58.9	63.7	62.1
RVP, psi	8.3	11.1	7.2	9.7
RON	91.6	96.1	88.0	93.2
MON	83.9	85.7	82.4	85.1
Hydrocarbon contents, vol%				
Olefin	5	NA	6	NA
Aromatics	26	NA	23	NA
Energy content, 10^5 Btu/gal	1.154	1.095	1.127	1.071
Distillation data, °F				
IBP	104	96	103	109
10% evaporated	134	116	140	118
30% evaporated	176	128	176	128
50% evaporated	216	206	207	198
70% evaporated	252	247	235	229
90% evaporated	316	313	286	284
FBP	383	380	383	366

RON increases approximately linearly with alcohol concentration.

MON of blends increases with alcohol concentration much less than the RON, and the gains become progressively less with successive increments of alcohol.

Table 8 shows data for blending road octane number (RdON) values in unleaded-type base stock [55] and blending RON and blending MON values for comparison. Blending RdON, like blending RON and MON, tends to increase with decreasing base stock octane. For pure methanol, RON and MON values have been reported in the ranges 106–115 and 82–92, respectively. The variation probably comes from the difficulty in maintaining prescribed engine conditions with pure methanol.

There are many laboratory parameters to determine antiknock properties that may be related to each other. The most important one is the Road Octane Number (RdON). However, this parameter is difficult to obtain from a road test, since it is affected by the car and test conditions. The general equation between RdON and laboratory-measured antiknock properties is the most valuable one. Date et al. [27] reported the relationship between RdON, RON, and MON prediction equations.

Table 6 Effect of Methanol in Gasoline on RVP, Distillation, and Octane Numbers

Properties	Fuel A (unleaded gasoline)	10% Methanol in fuel A	25% Methanol in fuel A	Fuel B (unleaded gasoline)	10% Methanol in fuel B
Specific gravity	0.757	0.758	0.763	0.736	0.742
RVP, psia	10.3	12.0	12.3	10.4	12.9
RON	96.0	98.8	102.5	91.6	96.0
MON	85.0	86.5	87.3	82.9	85.1
Hydrocarbon contents					
Paraffins, vol%	62	—	—	66	—
Olefins, vol%	3	—	—	8	—
Aromatics, vol%	35	—	—	26	—
Distillation data (°C)					
IBP	29	31	34	29	32
10%	49	43	45	48	44
20%	63	48	51	63	49
30%	77	53	56	76	53
40%	96	67	59	92	64
50%	114	100	61	102	96
60%	132	122	63	115	111
70%	148	143	122	123	122
80%	159	157	151	138	137
90%	173	169	168	156	157
95%	187	183	182	171	172
FBP	222	208	205	199	200

The most desirable feature of methanol is its high octane quality RON response with TEL when blended with varying proportions [23]. Gething and Lestz [56] studied the knocking and performance characteristics of low-octane fuels blended with methanol. Methanol can act as an effective knock suppressant in gasoline blends with octane numbers as low as 52, and methanol–gasoline blends yield higher efficiency than pure methanol.

B. Volatility and Distillation Blending

Methanol affects the volatility of its blends with gasoline significantly. The introduction of small quantities of polar methanol into nonpolar gasoline results in a large increase in fuel vapor pressure. It has been observed that adding as little as 2% methanol gives a 3-psi increase in RVP, despite the fact that the volatility of pure methanol is much less than that

Table 7 Effect of Methanol Addition on Fuel Properties

	Base blend	Matched RVP methanol blend	Nonmatched RVP methanol blend
Methanol, vol%	0.0	15.0	15.0
RVP, psia	11.9	11.7	16.0
Percent distilled			
at 70°C	30	50	55
at 100°C	48	53	59
at 150°C	87	87	89
Gravity (g/cm³)	0.76	0.78	0.76

of typical gasoline [57]. Typical changes in RVP with addition of methanol were also reported [55]. Methanol as well as ethanol form low-boiling azetropes with C_5 and higher hydrocarbons. In a blend, the azeotrope increases the vapor pressure and depresses the front-end distillation temperatures below that of either the gasoline hydrocarbons or the alcohol. Methanol also changes the shape of the gasoline distillation curve drastically. It has been observed that the presence of methanol leads to a much greater fraction distilled up to 70°C. It causes the blend to distill much faster at temperatures below methanol's boiling point, that is, 65°C [13,55,57–60].

Table 8 Blending Octane Numbers of Base Gasoline and with Methanol

	Base gasoline			Blending octane of base gasoline with 10 vol% methanol		
Fuel no.	RON	MON	RdON	RON	MON	RdON
1	90.9	82.0	88.9	132.9	100.0	116.9
2	91.8	83.5	89.6	134.8	99.5	112.8
3	90.7	82.1	89.2	136.7	107.1	120.5
4	91.1	84.6	88.5	—	—	113.5
5	87.1	82.1	86.4	—	—	117.0
6	83.8	79.7	82.7	—	—	131.7
7	90.9	82.5	87.3	133.9	96.0	100.6
8	90.6	83.5	—	134.8	93.5	—
9	84.0	—	138.0	120.0	—	—
10	83.4	—	136.2	113.1	—	—
11	82.4	—	137.0	106.4	—	—

VI. BLENDING PROPERTIES OF ETHANOL

Ethanol is an alcohol typically fermented from grain and has about 35 wt% oxygen. It is an octane enhancer added to motor fuel at up to 10%. It increases octane 2.5–3 points at a 10% concentration. Ethanol is a fuel oxygenate. Ethanol suspends and removes moisture as it is used in the fuel system, and eases cold-weather starting. Ethanol reduces deposits in the induction system and in the combustion chamber. It increases the octane (antiknock) rating when blended with gasoline. Ethanol improves engine combustion. It reduces carbon monoxide (CO) emissions and, by a modest amount, reduces hydrocarbon (HC) emissions as well. Table 9 indicates the changes in blending vapor pressures, and octane numbers (RON and MON) of ethanol at 10 vol% blended with unleaded base gasoline. The change in distillation properties is shown in Figure 1 [16].

A. Octane Blending

Table 10 lists blending Road Octane Numbers (RdON) of ethanol in unleaded regular-type base stocks. Table 10 also shows for comparison of blending RON and MON values. Blending RdOn tends, like blending RON and MON, to increase with decreasing base stock octane [55].

The effect of ethanol on blending octane values for different base gasolines has been studied. The addition of ethanol has a greater effect on RON than on MON, and on the octane numbers of gasoline having lower octane numbers than on gasoline having higher octane numbers [16]. In experiments conducted with ethanol blends, the study of Monti et al. [61] included two experimental base hydrocarbon fuels—fuel 1, with olefins, having 90 RON, and fuel 2, without olefins, having 90.5 RON, to which lead was added at 0.0, 0.15, and 0.4 g/L levels. The compositions

Table 9 Effect of Ethanol on the Characteristics of the Gasoline

Properties	Unleaded gasoline	Unleaded gasoline with 10% ethanol
RVP, psia	10.3	10.4
Paraffins (vol%)	62	—
Olefins (vol%)	3	—
Aromatics (vol%)	35	—
RON	96.0	99.0
MON	85.0	86.6
Specific gravity	0.757	0.760

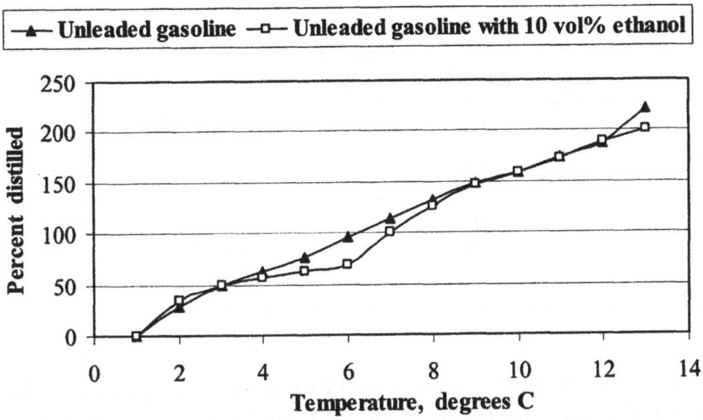

Figure 1 Effect of ethanol on the distillation characteristics of the gasoline.

of these two base fuels are reported in Table 11. The results indicated that ethanol octane blending value decreases with increasing lead and olefin content.

B. Volatility Blending

The typical changes of RVP in ethanol–gasoline blends [56] have been studied and show that a significant increase in RVP results even with a small addition of alcohol, due to positive deviations from Raoult's law, i.e., alcohol vapors exist at concentrations disproportionate to the alcohol

Table 10 Blending Octane Values of Base Gasoline and Ethanol–Gasoline Blends

	Base gasoline octane			Blending octane of base gasoline with 10 vol% ethanol		
Base fuel	RON	MON	RdON	RON	MON	RdON
1	90.8	83.6	84.8	—	—	116.8
2	86.3	79.3	80.5	—	—	115.5
3	81.7	74.9	76.2	—	—	122.2
4	90.9	82.5	87.3	132.4	102.0	—
5	90.6	83.5	—	129.9	95.6	—
6	92.0	84.0	—	136.0	112.0	—
7	92.7	83.4	—	128.3	110.5	—
8	92.0	82.4	—	135.0	108.4	—

Table 11 Composition of Two Fuels in Volume Percent

Fuel components	Fuel 1	Fuel 2
Reformate	41	65
Catalytic naphtha	34	—
Isomerate	—	30
Alkylate	10	—
Butane	4	5
Virgin naphtha	11	—

concentration in the blend [23]. These observations are contrary to what would be expected, because the RVPs of pure alcohols are lower than those of gasoline, and the RVP of a mixture should therefore be less than that of gasoline. This can be explained by the fact that because alcohol molecules are more polar than gasoline molecules, the alcohol content of vapor above the gasoline–alcohol mixture exceeds the concentration of alcohol in the mixture [62]. It has been illustrated that the addition of ethanol to gasoline increased front-end volatility [59].

VII. BLENDING PROPERTIES OF ARCONOL

Arconol, a commercial TBA for gasoline blending, is a mixture of *t*-butanol and mixed butanes. Table 12 presents the composition and some of the properties of Arconol [16]. Table 13 illustrates the blending RON and MON of Arconol on leaded and unleaded gasolines [5,16]. The data show that adding TBA increases the Reid vapor pressure and the volatility of the gasoline blends. Table 14 shows the change in Reid vapor pressure and volatility by adding 5% Arconol in a base gasoline [16].

VIII. BLENDING PROPERTIES OF OXINOL

Oxinol is a mixture of methanol and TBA that has been utilized as an octane enhancer. The typical composition and properties of Oxinol are given in Table 15 [16]. An unusual characteristic of Oxinol and other alcohols is the change of blending RVP with their concentration in the gasoline blend. Higher concentration of Oxinol gives lower blending RVP. For example, 9.5% Oxinol gives an RVP of 34, and 5.5% Oxinol gives an RVP of 50 [63]. The effect of addition of 10% Oxinol to a base gasoline (having RVP 11.5) increased the RVP to 13.6. The effect on distillation characteristics is shown in Figure 2 [16].

Table 12 Properties of Arconol

Composition (wt%)	
t-Butanol	91–94
Butanes	5–7
Water	0.8–1.2
Acetone	0.5–1.0
Other oxygenated hydrocarbons	0.5–0.7
Peroxides	Less than 50 ppm
Properties	
Color	White
API gravity	48–49
Reid vapor pressure (psia)	6.5–7.5
Freezing point (°F)	40–45
Boiling–range distribution, ASTM D-1078 (°F)	
IBP	160–170
10%	174–178
90%	179–180
EP	180–183
Gum, ASTM D-381	Less than 1
Solubility (77°F)	Miscible in all proportions with gasoline and water
Recommended level of addition	Up to 5% to avoid possible haze formation and loss of performance[a]

[a]An inhibitor package is commercially available to reduce haze formation and corrosion [16].

Gasoline grade *tertiary*-butyl alcohol (GTBA) improves the water tolerance of methanol in the same way as it improves the RVP blending of methanol gasoline blends. Methanol has very low water tolerance in the absence of GTBA. Methanol will separate from gasoline when the water content is over 0.04 vol%. The GTBA increases the water tolerance to over 0.2 vol%. Higher percentages of GTBA exhibit the best water tolerance [63].

Table 13 Blending RON and MON of Arconol-Added Gasolines

	Base gasoline		TBA	Blending	
	RON	MON	(%)	RON	MON
Leaded premium	98–100	90–92	5	109–114	101–107
Leaded regular	92–95	83–86	4.5	106–117	94–95
Unleaded premium	92–109	82–97	5	98–108	89–98
Unleaded regular	87–93	79–84	4.5	109	89–92

Table 14 RVP and Distillation Characteristics of Base Gasoline and Arconol-Added Gasoline

	Base gasoline	With 5% Arconol
RVP (psia)	12.8	13.2
Distillation temperature (°F)		
IBP	77	78
10%	103	99
20%	127	121
30%	155	147
50%	214	197
90%	322	318
EP	404	396

ARCO has found that methanol levels of up to 3%, with up to 7% GTBA as co-solvent, are quite acceptable in European cars [64].

IX. A COMPARISON OF BLENDING PROPERTIES OF OXYGENATED FUELS

Table 16 gives blending clear octane numbers for MTBE, methanol, ethanol, TBA, ETBE, and TAME [4]. These results were obtained after adding 10 vol% of the oxygenate to a gasoline having the following octane properties: RON clear = 94.3, MON clear = 84.3.

Methanol and ethanol have higher RON but lower MON than MTBE. RONs of MTBE and ETBE are comparable. However, ETBE

Table 15 Composition and Properties of Oxinol

Typical composition (wt%)	
Methanol	45–55
TBA	45–55
Butanes	1.5–2.0
Water	0.5–0.8
Other oxygenated hydrocarbons	0.2–0.5
Typical properties	
Density at 15°C, lb/gal	6.62
RVP (psi)	7.0
Flash point (°C)	1.0
Heat of combustion (Btu/lb)	8020
Solubility	Miscible in water and gasoline

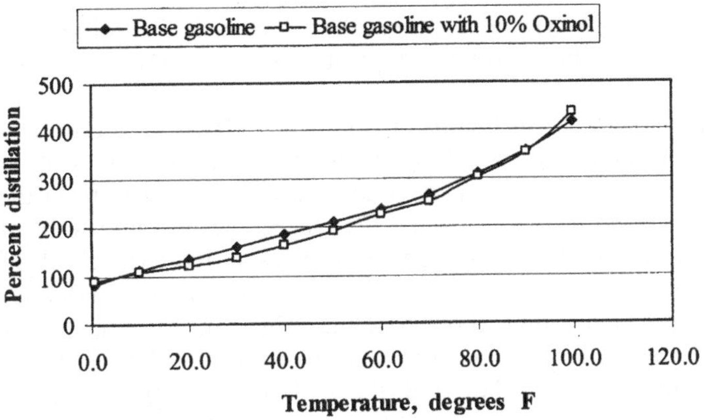

Figure 2 Effect of 10% Oxinol addition on distillation characteristics of a gasoline.

MON is higher than MTBE MON. A comparison of distillation characteristics has been made between alcohols (methanol, ethanol, TBA) and MTBE with gasoline [62,65]. It was observed that MTBE has less effect on distillation properties (ASTM D 86) than methanol and ethanol.

Table 16 Blending Properties of Oxygenate Ethers and alcohols

Oxygenate	Blending RVP (psi)	Average octane (R+M)/2	Blending limit vol% (wt%)[a]	Blending RON	Blending MON
Ethers					
MTBE	8.0	110	15 (2.7)	116	98
ETBE	4.0	111	13 (2.0)	118	105
TAME	2.0	105	13 (2.0)	111	94
TAEE	2.0	100	13 (2.0)	105	95
DIPE	0.7	107	13 (2.0)	105	95
Alcohols					
Methanol	60	120	[d]	123	91
Ethanol	18	115	10 (3.7)	123	96
Methanol/GTBA blend[b]	31	108	9.5 (3.7)	—	—
GTBA[c]	12	100	16 (3.7)	106	89

[a]Volume percent oxygenate in the fuel blend (weight percent oxygen).
[b]Methanol must be used with an appropriate co-solvent.
[c]Gasoline-grade *t*-butyl alcohol.
[d]Varies with the type of co-solvent.

The concentration of oxygenates in gasoline blends is 7.0 vol%. The lower-boiling-point alcohols produce a greater bulge on the distillation curve.

The effect of alcohols (methanol, ethanol, and TBA) and MTBE on a low-RVP gasoline (7.9) has been studied [62,65]. In addition, heat of vaporization, the other major property that separates the ethers from the alcohols, is their blending RVP. Unlike the ethers, the blending RVP of alcohols is much higher than their true vapor pressure [66]. The RVPs of the pure alcohols are lower than those of gasoline, but the RVPs of gasoline–alcohol blends are higher than those gasoline. This can be explained by the polarity differences between alcohols and hydrocarbons. The hydroxyl group of alcohols is very polar, and the hydrocarbons (gasoline) are nonpolar. For this reason, the alcohol content of the vapor mixture exceeds the concentration of alcohol in the mixture. The higher the oxygen content of alcohol (more polar), the higher will be the increase of blending RVP [62,65].

To produce low RVP in gasolines with an alcohol, a significant amount of nonaromatic butane which has high RVP and octane should be removed from the gasoline pool. For this reason, the net volume contribution of an alcohol is significantly reduced after correcting for RVP [66].

It has been shown that methanol and ethanol, blends at the maximum allowable oxygen levels contribute only around 8 vol% nonaromatic volume to gasoline. However, even at lower oxygen levels, ethers contribute twice as much nonaromatic volume. In addition to the high volume contribution of nonaromatics, ethers do not impart unfavorable characteristics such as high oxygen, high heat of vaporization, or sensitivity to water to gasoline. Reduction of the aromatic content in gasoline contributes to lower carbon monoxide and hydrocarbon emissions [66].

X. CONCLUSIONS

The improvement of blending properties by the addition of MTBE depends on the composition of the base gasolines, which contain hundreds of components in different concentrations. MTBE blending octane number generally rises with a decrease in the octane number of base gasoline, a decrease in MTBE concentration of gasoline blends, and an increase in saturated hydrocarbons of base gasoline. MTBE provides much higher front-end octane numbers (FEON) to gasolines. FEON increases engine efficiency during the low-speed acceleration stage. MTBE does not decrease the lead susceptibility of alkyl lead compounds, tetraethyl lead (TEL), tetramethyl lead (TML), or their blends.

MTBE has little effect on the distillation and vapor pressure characteristics of gasoline. There is no evidence of significant azeotrope formation, as is the case when alcohols (methanol or ethanol) are blended with gasolines. Pure MTBE forms azeotropes with water, but no MTBE–gasoline azeotropes have been reported. MTBE–gasoline blends, even in the presence of water, show no separation problems in the distribution system. Phase-separation problems cause driveability problems in addition to corrosion. MTBE is stable during handling and storage. No difference in potential gums was found between base gasoline and MTBE blends after extended storage.

ACKNOWLEDGMENT

The authors wish to acknowledge the support of the Research Institute of the King Fahd University of Petroleum and Minerals, Dhahran, Saudi Arabia, for this work.

REFERENCES

1. N. R. Gribble. Alcohols and other oxygenates in automotive fuels. Ph.D. thesis, The University of Aston in Birmingham, England, 1987.
2. API. Alcohols and ethers: a technical assessent of their application as fuels and fuel components. API Publication 4261, pp 23–89, 1988.
3. J. M. Dejovine, E. C. Guetens, Jr., G. J. Yogis, B. C. Davis. Gasolines show varied responses to alcohols. Oil Gas J Technol, Feb. 14, 1983, pp 87–94.
4. G. Marceglia, G. Oriani. MTBE as alternative motor fuel. Chem Econ Eng Rev, 14(4) (No. 157): 39–45, 1982.
5. G. Pecci, T. Floris. Ether ups antiknock of gasoline. Hydrocarbon Proc, Vol. 56, No. 12, pp 98–102, 1977.
6. T. J. Russel. Petrol and diesel additives. Petroleum Rev, pp 35–42, 1988.
7. M. F. Ali, A. Bukhari, A. Amer. Evaluation of MTBE as a high octane blending component for gasoline. Arabian Journal of Science and Engineering, Vol. 12, No. 4, pp 451–458, 1987.
8. W. E. Morris. Gasoline compositions in no-lead era. Oil Gas J, pp 99–106, 1985.
9. J. D. Chase, B. B. Galvez. Maximize blend ethers with MTBE and TAME. Hydrocrabon Proc, pp 89–94, 1981.
10. J. D. Chase, H. J. Woods. Process for high octane oxygenated gasoline components. Symposium on Octane in the 1980's, ACS Miami Beach Meeting, Sept. 10–15, 1978, pp 1072–1082.
11. J. D. Chase. MTBE and TAME—a good octane boosting combination. Oil Gas J, April 9, 1979, pp 149–152.

12. R. Csikos. Low-lead fuel with MTBE and C_4 alcohols. Hydrocarbon Proc., Vol. 55, pp 121–125, 1976.
13. P. Dorn, A. M. Mourao. The properties and performance of modern automotive fuels. SAE Technical Paper 841210, 1984.
14. R. W. Reynolds, J. S. Smith, I. Steinmetz. Methyl ether (MTBE) scores well as high-octane gasoline component. Oil Gas J, June 16, 1975, pp 50–52.
15. ARCO. MTBE octane enhancer. Tech Bull, ARCO Chemical Company, 1985.
16. SRI. SRI Report 158, SRI International, Menlo Park, CA, 1983.
17. W. J. Piel. The role of MTBE in future gasoline production. Paper 47d, Spring National Meeting of AIChE, March 6–10, 1988, pp 1–19.
18. W. J. Piel. The role of MTBE in future gasoline production. Energy Prog, Vol. 8, No. 4: 201–204, 1988.
19. F. Obenaus, W. Droste. Huls process: methyl tertiary butyl ether. Erdol und Kohle- Erdgas, 33(6): 271–275, 1980.
20. Huls. The Huls MTBE process. Tech Brochure, Chemische Werke Huls AG, 1980.
21. D. L. Trimm. Catalyst in petroleum refining 1989. Proceedings of the Conference on Catalysts in Petroleum Refining, Kuwait University, Kuwait, p 46, 1990.
22. R. Csikos, I. Pallay, J. Laky. Practical use of MTBE produced from C_4 fraction. 10th World Petroleum Congress, Bucharest, pp 167–175, 1979.
23. P. L. Dartnell, K. Campbell. Other aspects of MTBE/methanol use. Oil Gas J, Nov. 13, 1978, pp 205–212.
24. V. S. Azev, B. P. Kitski, S. R. Leebedev. Conservation of fuels and lubricants in equipment operation. Chem Technol Fuels Oils, pp 709–711, 1981.
25. L. M. Norieko. Economic efficiency of utilization of MTBE as component of high octane automotive gasoline. Chem Technol Fuels Oils, pp 338–340, 1980.
26. G. H. Unzelman. Ethers have good gasoline-blending attributes. Oil Gas J, April 10, 1989, pp 33–44.
27. K. Date. Road octane number trends of Japanese passenger cars. SAE Paper 770811, 1977.
28. M. Mays. Road octane performance of unleaded gasoline containing oxygenates. European Fuel Oxygenates Association, Third Conference, Madrid, Spain, March 8–9, 1989.
29. R. G. Clark, R. B. Morris, D. C. Spence, E. Lee Tucci. Unleaded octanes from RVP changes: MTBE from butane. Energy Prog, Vol. 7, No. 3, pp 164–169, 1987.
30. R. L. Furey. Volatility characteristics of gasoline-alcohol and gasoline-ether fuels blends. SAE Technical Paper 852116, 1985.
31. R. T. Johnson, B. Y. Taniguchi. Methyl tertiary-butyl ether, evaluation as a high octane blending component for unleaded gasoline. Symposium on Octane in the 1980's, ACS Miami Beach Meeting, Sept 10–15, 1978.
32. R. L. Furey, K. L. Perry. Volatility characteristics of blends of gasoline with ethyl tertiary-butyl ether (ETBE). SAE Technical Paper 901114, 1990.

33. W. H. Douthit. Performance features of 15% MTBE/gasoline blends. SAE Paper 881667, pp 3.981–3.997, 1988.
34. F. F. Pischinger. Alcohol fuels for automotive engines. 10th World Petroleum Congress, Bucharest, pp 1–10, 1979.
35. S. Poulopoulos, C. Philippopoulos. Influence of MTBE addition into gasoline on automotive exhaust emission, Atmos Environ, 34 (28): 4781–4786, 2000.
36. A. A. Al-Farayedhi, A. M. Al-Dawood, P. Gandhidasan. Effects of blending MTBE with unloaded gasoline on exhaust emissions of SI engine, J Energy Resources Technol, Trans ASME, 122(4): 239–247, 2000.
37. P. M. Franklin, C. P. Koshland, D. Lucas, R. F. Sawyer. Evaluation of combustion by-products of MTBE as a component of reformulated gasoline. Chemosphere, 42 (5–7): 861–872, 2001.
38. N. R. Silva, J. R. Sodre. Cold start and drivability characteristics of an ethanol-methyl-*t*-butyl ether blend fuelled vehicle. Proc Inst Mech Eng D, J Automobile Eng, 215 (D5): 645–649, 2001.
39. S. G. Poulopoulos, C. J. Philippopoulos, Speciated hydrocarbon and carbon monoxide emissions from an internal combustion engine operating on methyl tertiary butyl ether-containing fuels. J Air Waste Mgmt Assoc, 51 (7): 992–1000, 2001.
40. G. H. Unzelman. Reformulated gasolines will challenge product-quality maintenance. Oil Gas J, April 9, 1990, pp 43–48.
41. B. Trock, A. Convers, A. Chauvel. Methanol for motor fuel via the ethers route. Chem Eng Prog, pp 36–45, 1982.
42. J. M. DeJovine. Performance testing of gasolines blended with methanol and gasoline grade TBA, chemistry of oxygenates in fuels. Division of Petroleum Chemistry, American Chemical Society, Kansas City, MO, 1982.
43. F. H. Palmer, G. H. Lang. Fundamental volatility/driveability characteristics of oxygenated gasolines at high underbonnet temperatures. SAE Technical Paper 831705, 1983.
44. Widener University. Automobilie canister performance evaluation. Widener University School of Engineering, April 20, 1987.
45. Widner Universtiy. Effects of gasoline mixtures on automobile canister performance. Widener University School of Engineering, April 25, 1988.
46. L. Olsson. Motor vehicle pollution control and regulations of Sweden. SAE Technical Paper 871081, pp 85–91, 1987.
47. S. R. Reddy. Evaporative emission from gasolines and alcohol containing gasolines with closely matched volatilities. SAE Technical Paper 861556, 1986.
48. T. J. Wallington. Gas-phase reactions of hydroxyl radicals with the fuel additives. Environ Sci Technol, 22 (7): 842–844, 1988.
49. W. P. L. Carter. Investigation of the atmospheric chemistry of MTBE. Final report to Auto/Oil Air Quality Improvement research program, University of California, Riverside, CA, 1991.
50. Suntech. Technical Information on use of oxygenated hydrocarbons in gasolines., Sept. 29, 1978.

51. M. W. Jackson. Effects of some engine variables and control system on composition and reactivity of exhaust hydrocarbons. SAE Vehicle Emission, Part II, PT-12., 1968.
52. G. Publow, L. Grinberg, Performance of late model cars with gasoline-methanol fuel. SAE Technical Paper 780948, 1978.
53. J. R. Allsup. Methanol/gasoline blends as automotive fuel. SAE Paper 750763, 1975.
54. F. W. Cox. The physical properties of gasoline/alcohol automotive fuels. III International Symposium on Alcohol Fuels Technology, San Antonio, TX, Paper II-22, May 29–30, 1979.
55. J. L. Keller. Alcohols as motor fuels. Hydrocarbon Proc, Vol. 58, No. 5, pp 127–138, 1979.
56. J. A. Gething, S. S. Lestz. Knocking and performance characteristics of low octane primary reference fuels blended with methanol. SAE Paper 780079, 1978.
57. E. E. Wigg, R. S. Lunt. Methanol as a gasoline extender—fuel economy, emission and high temperature driveability. SAE Technical Paper 741008, 1974.
58. P. W. Mccallum, T. J. Timbario, R. L. Bechtold. Alcohol fuels for highway vehicles. Chem Eng Prog, pp 52–59, 1982.
59. J. C. Ingamells, R. H. Lindquist. Methanol as a motor fuel or a gasoline blending component. SAE Technical Paper 750123, 1975.
60. R. L. Furey, J. B. King. Emissions, fuel economy and driveability effects of methanol/butanol/gasoline fuel blends. SAE Technical Paper 821188, 1982.
61. F. Monti, G. F. Marchesi, S. Leoncini. Oxygenated compounds in gasoline: octane blending value and vehicle requirements response. Twelfth World Petroleum Congress, Houston, TX, April 26–May 1, 1987, pp 151–157.
62. M. F. Ali, M. M. Hassan, A. Amer. The effect of oxygenates addition on gasoline quality. Arabian Journal of Science and Engineering, Vol. 9, No. 3, pp 221–226, 1984.
63. E. G. Guetens, Jr., J. M. Dejovine, G. J. Yogis. TBA aids methanol/fuel mix. Hydrocaron Proc., pp 113–117, 1982.
64. P. Butler. Gasoline additives—methanol's only hope. Mftg Chem, pp 22–23, 1984.
65. M. F. Ali, A. Bukhari. MTBE as a high octane blending component for gasoline, Proceedings of the Second National Meeting of Chemists, Riyadh, Saudi Arabia, pp 225–239, 1987.
66. W. J. Piel, R. X. Thomas. Oxygenates for reformulated gasoline. Hydrocarbon Proc, pp 68–72, 1990.

4
Zeolites as Catalysts for Ether Synthesis

François Collignon
Katholieke Universiteit Leuven, Leuven, Belgium

Georges Poncelet
Université Catholique de Louvain, Louvain-la-Neuve, Belgium

I. INTRODUCTION

A. Zeolites as Crystalline Microporous Alumino-Silicates

Zeolites are crystalline alumino-silicates, natural or most often synthesized under autogenous water pressure. The zeolite structure consists of an uninterrupted tridimensional network of tetrahedra centered on Si or Al atoms and characterized by a [Al + Si]/O molar ratio of 1:2. These silicates have the following general formula:

$$M_V^{n+}(AlO_2)_x(SiO_2)_y \cdot zH_2O$$

where M^{n+} is a cation of valence n and $v = x/n$.

The negative charges resulting from the tetrahedral Al-for-Si isomorphic substitutions are neutralized by cations of variable nature located in the pores or cavities. These compensation cations are exchangeable and can be replaced by protons, which confer the Brønsted acid properties to the zeolites. Protonation of zeolites may be achieved by direct exchange with a solution of a mineral acid, or via exchange with an

Table 1 Crystallographic Data for Zeolites Studied

Zeolite	Pore opening (Å)	Number of tetrahedra per opening[a]	Unit cell
ZSM-5	5.3 × 5.6	[010]; **10**	Orthorhombic
	5.1 × 5.5	[100]; **10**	
Mordenite	6.5 × 7.0	[001]; **12**	Orthorhombic
	2.6 × 5.7	[010]; **8**	
Beta-zeolite	7.6 × 6.4	[001]; **12**	One tetragonal and two monoclinics
	5.5 × 5.5	[100]; **12**	
US-Y	7.4	[111]; **12**	Cubic
Mazzite	7.4	[001]; **12**	Hexagonal
	3.4 × 5.6	[001]; **8**	
SAPO-5	7.3	[001]; **12**	Cubic

[a] [001] means channel in the crystallographic direction 001; **10** means 10 tetrahedra per opening.

ammonium salt followed by a thermal decomposition at temperatures between 400 and 500°C, during which ammonia is eliminated.

The crystal structures of zeolites generate porous networks made up of channels and cavities smaller than 0.8 nm. For this reason zeolites are also called molecular sieves. The porous network is monodimensional when the pores or channels are independent of each other, two-dimensional when the pores are interconnected along two directions, and three-dimensional when the interconnections occur in three directions.

This review is centered on the catalytic efficiency for ether synthesis of some zeolites available on the market and commonly used in the petrochemical industry, namely, Y-faujasite, mordenite, ZSM-5, mazzite (omega zeolite), and beta-zeolite. Table 1 provides an outline of the crystallographic characteristics of these zeolites [1]. A representation of the structure of Y-type zeolite (Y-faujasite) is shown in Figure 1.

B. Aluminum Content and Acid Properties of Zeolites for Catalysis

The Brønsted acid properties of a zeolite are related to the content of framework aluminum: the higher the tetrahedral aluminum content, the higher is the number of the Brønsted acid sites. The strength of the Brønsted acid sites is an important parameter for proton-catalyzed reactions and depends on the type of zeolite and its Si/Al ratio. When the framework Al content decreases, the strength of the acid sites increases; it reaches a maximum when the Al tetrahedra are far enough from each other. This is the so-called next-nearest-neighbor effect [2,3]. The Al content of a zeolite can be

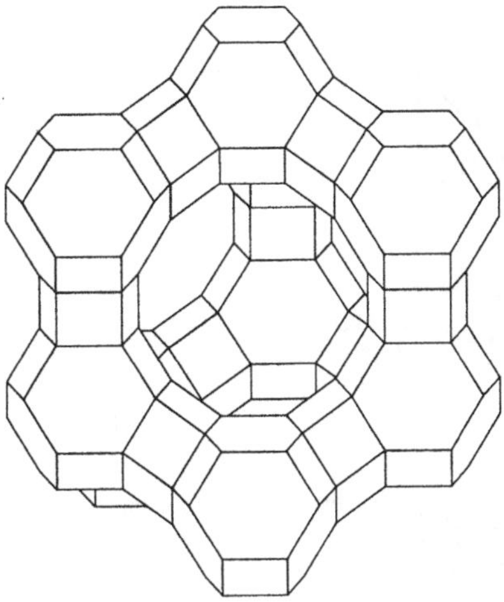

Figure 1 Representation of the structure of Y-type faujasite.

modulated by *direct* synthesis or by *post*synthesis dealumination. Aluminum-deficient zeolites are obtained by chemical treatment (acid leaching, reaction with chelating agent, etc.) or by hydrothermal treatment (steaming, self-steaming) [4].

Dealumination, especially by steam treatment, generates cationic (e.g., Al^{3+}, AlO^+, $Al_2O_2^{2+}$, etc.) or neutral [Al_2O_3, $Al(OH)_3$, etc.], extra-framework aluminum (EFAL) species, which poison the Brønsted acid sites and/or restrict the accessibility of the active sites to the reactant(s) [4]. These species can also generate acid sites with an affinity for paired electrons of a molecule. Such sites constitute the Lewis acidity.

Dealumination of a zeolite (e.g., by acid leaching or steaming) increases the content of silanol groups. This brings about changes of the hydrophobicity of the solid. Another effect of dealumination is destruction of a part of the zeolite framework and the formation of a secondary (meso-)porous volume, which improves the diffusion of reactant(s) and product(s).

In summary, a postsynthesis modification of the framework Si/Al ratio of a zeolite results in changes of a number of physicochemical parameters which are important to catalysis. A schematic representation of the dealumination of a mordenite is shown in Figure 2.

C. Zeolites as Catalysts for Ether Synthesis

Sulfonic acid resins are well-performing catalysts used in commercial production units. However, release of thermally fragile sulfonic groups

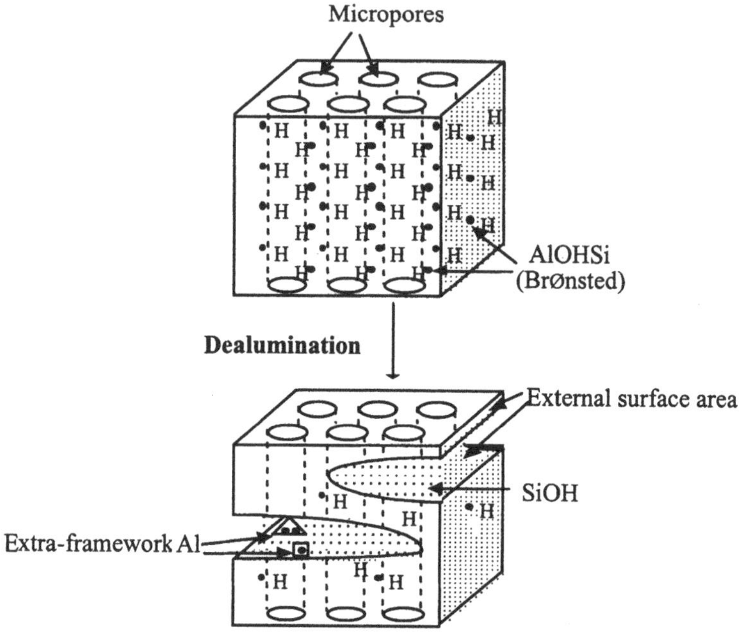

Figure 2 Representation of dealumination effect on texture of a mordenite crystal.

causing loss of performance and corrosion problems [5], and other minor limitations such as swelling and use of methanol/isobutene molar ratios in excess of stoichiometry, requiring recycle operations, are the main drawbacks. Alternative acid catalysts exempt from the drawbacks of sulfonic resins have been investigated, principally zeolites and, to a lesser extent, sulfate-modified zirconias [6–8], amorphous silico-aluminas [7,9], montmorillonites [10–12], layered Zr- and Ti-phosphates [13], titanium silicate [14], cloverites [15,16], and heteropoly acids [17–23]. In particular, silica-supported tantalum oxide [24] and heteropolyacid in a membrane reactor have been found to be efficient catalysts for the reverse reaction [25,26].

In most of the reported studies, methyl *t*-butyl ether (MTBE) was synthesized from methanol and isobutene, both reactants used in commercial processes. Methanol–isobutanol mixtures also have been studied [27–32]. Ethyl *t*-butyl ether (ETBE) from ethanol and isobutene over acid zeolites has been investigated by several authors [15,16,33,34].

Chu and Kühl [35] compared various acid zeolites (mordenite, beta-zeolite, Y-zeolite, ZSM-5, and ZSM-11) in vapor-phase synthesis of MTBE and found that ZSM-5 and ZSM-11 were more selective but less active than Amberlyst-15. The higher selectivity of those zeolites is due to their smaller pores, which prevent isobutene dimerization. These zeolites have a high thermal stability, are less sensitive to the MeOH/IB ratio than the resin, and do not deactivate. Beta-zeolite and mordenite were the least selective catalysts.

Pien et al. [36] studied the synthesis of MTBE in liquid-phase conditions over ZSM-5 with Si/Al ratio = 50 and confirmed the high selectivity of this zeolite and the minor influence of the MeOH/IB ratio on isobutene conversion. Ti-silicalite was found by Chang et al. [14] to be even more selective than ZSM-5.

Ahmed et al. [37] showed that the catalytic activity of ZSM-5 zeolites (synthesized with Si/Al ratios in the 25–90 range) was proportional to the number of acid sites. Ali et al. [38] studied both mass transfer effects and kinetic parameters in liquid phase (batch autoclave) in presence of a MFI (with a Si/Al molar ratio of 10). An activation energy of 140.8 kJ/mol, comparable with work of Al-Jarallah et al. [39], was found.

Nikolopoulos et al. [9] observed that, at temperatures higher than 100°C, US-Y and ZSM-5 zeolites were more selective than Amberlyst-15. As most acid zeolites are thermally stable, they should constitute suitable catalysts, particularly for integrated processes from syngas to methanol to MTBE at temperatures where the reaction is limited by thermodynamics [40,41]. This operation could be conducted at high temperature, with an inorganic acid catalyst (a zeolite) showing high selectivity and weak deactivation. The cited authors observed that the formation of secondary products (mainly di-isobutenes and their cracking products) increased with temperature. These products have a significant impact on deactivation. Of all the zeolites tested, ZSM-5 appears to be most appropriate with respect to selectivity and lifetime.

Dealuminated US-Y zeolites were tested by Nikolopoulos and co-workers in vapor-phase synthesis of MTBE [9,40–43]. The catalytic activity increased and then decreased with increasing degree of dealumination, with an optimum activity for US-Y with Si/Al ratio of 14.1. According to the authors, the aluminum content of their samples was too low as to explain the increase of the catalytic activity by the next-nearest-neighbor effect. This increase was attributed to interactions between Brønsted acid sites and extra-framework aluminum (EFAL). Extra-framework Al-to-framework Al ratios (EFAL/FAL) lower than 0.5 increased in a proportional way the turnover frequency (TOF, activity per catalytic site).

Kogelbauer et al. [43,44] studied the gas-phase synthesis of MTBE over US-Y zeolites partially exchanged with alkaline ions in order to isolate the effect of acid strength from other parameters which can play a role in the reaction (structure, framework composition, and site density). A cation exchange was insufficient to increase the catalytic activity [43,44].

Le Van Mao et al. [45] observed, in the presence of ZSM-5 and Y zeolites impregnated with triflic acid (trifluoromethanesulphonic acid, TFA), that the production of di-isobutylenes is much smaller compared to Amberlyst-15, as a consequence of a reduction of the pore opening after impregnation with TFA. However, the MTBE yield remains lower than for Amberlyst-15. Optimal TFA loading is around 2–3 wt% [44–46]. Higher TFA contents decrease the accessibility to the micropores and to the acid sites, and thus the activity. The improvement of the activity seems to be related to the formation of extra-framework aluminum after TFA impregnation. Similar conclusions were reported when using ammonium fluoride as impregnation agent [47]. However, in the recent paper of Le Van Mao et al. [48], the enhanced activity by F^- is believed to be due to formation of new Brønsted acid sites and strengthening of some acid sites of the parent zeolite.

Collignon et al. [49] showed that beta-zeolites are more active than US-Y zeolites, mordenite, ZSM-5, omega-, and SAPO-5, and exhibit similar performances to Amberlyst-15, in both vapor-phase and liquid-phase conditions. The efficiency of zeolites is favored by a high external surface area, partial dealumination, and low content of extra-framework Al. The superior catalytic behavior of beta-zeolites compared with other zeolites have been confirmed by Hunger and co-workers [50–54] in vapor-phase conditions. Beta-zeolites are also very active catalysts in liquid-phase conditions [55,56].

The adsorption behavior of methanol and isobutene over zeolites and acid Amberlyst-15 has been studied in order to evaluate its impact on MTBE synthesis [57]. In the case of zeolites, 2.5 methanol molecules are adsorbed per acid site, while isobutene forms a 1:1 complex. The two reactants are adsorbed in equimolar amounts over Amberlyst-15 but at a rate of one molecule for every three catalytic sites. The authors account for this result by the slow permeation of the reactants in Amberlyst-15. The greater quantity of methanol adsorbed on zeolites results in a higher selectively to MTBE. Preadsorption of methanol prior to isobutene introduction increases the formation rate of the ether [57] because it avoids isobutene oligomerization [58].

Methanol sorption has been studied by ^1H MAS-NMR over beta-, Y-, and ZSM-5 zeolites. It has been observed that: (1) beta-zeolite adsorbs more methanol than ZSM-5 and Y (with 8 methanol molecules per OH group for beta-zeolite, 1 for US-Y, and 3 for ZSM-5); (2) methanol molecules are less tightly bonded on silanols than on bridging AlOHSi; and (3) on beta-zeolite, 70% of the bonded methanol is adsorbed on silanol groups [50–52]. These methanol clusters act as a reactant reservoirs for MTBE synthesis. The intervention of silanol groups was recently confirmed by in situ MAS-NMR spectroscopy of MTBE synthesis under continuous flow conditions [53,54].

The synthesis of ETBE has been investigated by Larsen et al. [33] over mordenite. The protective effect of ethanol clusters on the acid sites of mordenite are responsible for the inhibiting effect of ethanol; they also prevent isobutene oligomerization and ensure zeolite stability.

Recently, the synthesis of a novel ether, 3-methoxy-3-methylpentane, obtained from 2-ethyl-1-hexene and methanol, has been reported [59,60].

In the following, the influence of various textural (external surface area, porosity) and physicochemical parameters (acid sites, silanol groups, EFAL, etc.) on ether synthesis (MTBE, ETBE) carried out over various types of dealuminated zeolites is examined.

II. SCREENING OF ACID ZEOLITES IN VAPOR-PHASE SYNTHESIS OF ETHERS

Catalytic screening in vapor-phase conditions is a useful and easy way to obtain a first evaluation of the efficiency of various catalysts, especially when one of the reactants is a vapor at atmospheric pressure and room temperature. Several zeolites are considered in this screening study: large- and small-port mordenites, ZSM-5, US-Y, omega-zeolites, and beta-zeolties. The synthesis of the ethers was performed over protonated forms. The reference catalyst was Amberlyst-15, a sulfonic acid resin (from Aldrich), with specific surface area of $50\,m^2/g$ and cation-exchange capacity of 4.9 mEq/g.

MTBE synthesis was carried out at atmospheric pressure in a fixed-bed microreactor operated at atmospheric pressure, with methanol/isobutene molar ratio of 1, and WHSV (weight hourly space velocity) of $3.2\,h^{-1}$ [49].

A typical curve showing the variation of the yield of isobutene (IB) to MTBE (Y_{MTBE}) and dimerization products (2,4,4-trimethyl-2-pentene and 2,4,4-trimethyl-1-pentene) (YC8) as a function of reaction temperature is provided in Figure 3. It was obtained over a beta-zeolite washed with 0.1 M nitric acid, followed by calcination.

The yield of MTBE increases with temperature, reaches a maximum, and decreases beyond this maximum. The decreasing part of the curve is imposed by the thermodynamic equilibrium, which is important in vapor-phase reaction when temperature is higher than 70°C. A representation of the thermodynamic curve established from the data of Tejero et al. [61] is

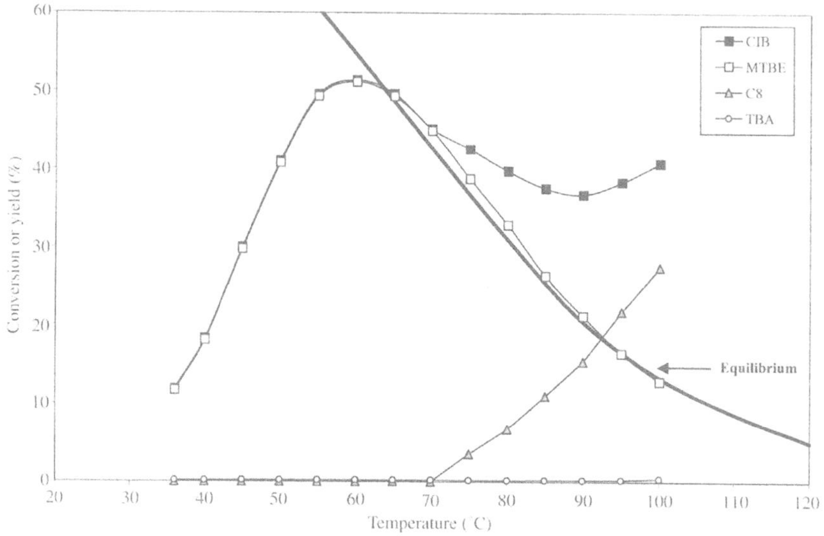

Figure 3 Yield of isobutene into MTBE, di-isobutenes (C8), and *t*-butyl alcohol (TBA) versus reaction temperature over a beta-zeolite [sample pqBt(12.6) 0.1c]; thermodynamic equilibrium curve.

shown in Figure 3. Beyond the maximum of MTBE formation, where thermodynamics imposes the reverse reaction, total conversion of isobutene (and yield of MTBE) decreases, and dimerization of isobutene takes over, becoming, with further increase of temperature, the main reaction. As shown in Figure 3, the formation of *tert*-butyl alcohol is negligible. Dimethyl ether from methanol dehydration on acid sites is another reaction by-product, representing less than 1% of the methanol transformed [49].

Figure 4 shows the IB-to-MTBE conversion versus temperature obtained over a series of commercial US-Y zeolites with various Si/Al ratios (samples from PQ-zeolite). All the curves exhibit the typical "volcano" profiles, with the experimental values of the decreasing parts of the curves aligning on the same curve (thermodynamic curve). The most active catalysts are, of course, those which operate the reaction at the lower temperatures.

The less dealuminated Y-zeolite (CVB500), with Si/Al ratio of 2.6, is the least active among the series. The catalytic activity increases for more dealuminated samples (CBV-712, -720, and -740, with Si/Al ratios of 5.8, 13, and 21, respectively). Sample CBV-760, with Si/Al ratio of 30, is the most active one. The more dealuminated sample with Si/Al ratio of 37 (CBV-780) exhibits intermediate activity.

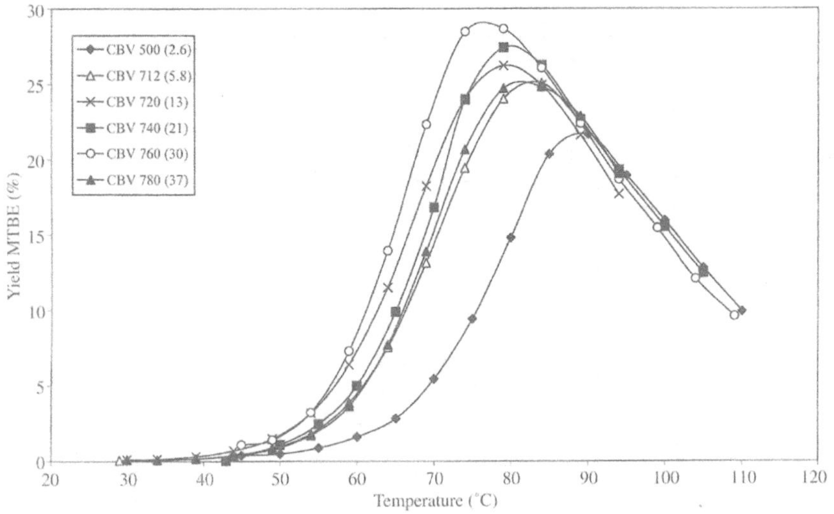

Figure 4 Zeolites US-Y: yield of MTBE versus reaction temperature.

Similar behavior with respect to Si/Al ratio was also observed for omega-zeolites, ZSM-5, and "small-" and "large-" port mordenites [49]. Table 2 gives the lowest and highest yields of MTBE achieved over the different types of zeolites. The most active sample among the series of ZSM-5 zeolites has a Si/Al ratio of 25, close to that of the most active US-Y zeolite (Si/Al = 30), with a yield of ether at maximum IB-to-MTBE conversion of 27% (versus 29% for US-Y).

The starting mordenites were from (former) Société Grande Paroisse, France (small-port variety, SP), and Tosho Corp., Japan (large-port variety, LP). These two mordenites were dealuminated either by a steaming treatment and subsequent acid leaching with nitric acid to remove extra-framework Al species, or by calcination followed by acid leaching and a second calcination at the same temperature. Dealumination by acid attack alone or followed by a steaming treatment was also performed [62]. The results obtained over "small-" and "large-" port mordenites are indicated in Table 2. Among the different samples of large-port mordenites, with Si/Al ratios between 8.1 (starting zeolite) and 195, the sample with Si/Al ratio of 35 was the most active. For the dealuminated "small-port" mordenites, with Si/Al ratios between 5.8 (starting sample) and 121, the most active sample has a Si/Al of 13, but with a maximum yield of MTBE markedly lower.

Table 2 Zeolites Used for MTBE Screening Test, Origin or Preparation Method, Crystallite Diameter (μm), Range of Si/Al Ratio Studied, Dealumination Method, Yield MTBE Obtained over Nondealuminated Zeolite, Maximum Yield MTBE (mol%), and Optimum Si/Al Ratio Found for Each Kind of Zeolite

Zeolite	Origin prep. method	Crystallite diameter	Si/Al studied	Yield MTBE without dealum.[b]	Dealumination[a]	Yield MTBE max[b]	Si/Al optimum
ZSM-5	PQ	0.1–1.5	16–137	24	?	27	25
Mordenite "small port"	G[rde]Paroisse	0.25–7	5.8–121	8	S, H, C	17	13
Mordenite "large port"	Tosho	0.08–0.2	8.1–195	12	S or H, C	21	35
US-Y	PQ	0.4–0.6	2.6–37	22	S & H	29	30
Mazzite	UOP	?	4.2–52	10	S & H	26	26
Beta [pqB1(13.2)]	PQ (batch 1)	0.02	13.2–194	47	H	51	26
Beta [pqB1(13.4)]	PQ (batch 2)	0.02	13.4–28	34	H	41	21
Beta [pqBt]	PQ template form	0.02	11–13	48	C & H	51	11
Beta [pqB(36)]	PQ	0.02	36	—	No	43	—
Beta [scB(24.6)]	Süd-Chemie	?	24.6	—	No	43	—
Beta [1Bt]	Borade et al. (68)	?	7–18	37	C & (H or E)	48	9
Beta [1BF (6.7)]	Caullet et al. (71)	1–3	6.7	—	C	30	—

[a]S or H means dealumination by steaming or acid leaching. C means that the sample was calcined. E means NH_4^+ exchange.
[b]Reaction conditions: $P = 1$ bar; [MeOH:IB:He] = [1:1:4.4]; WHSV = 3.2 hr^{-1}; $36 < T°C < 100°C$.

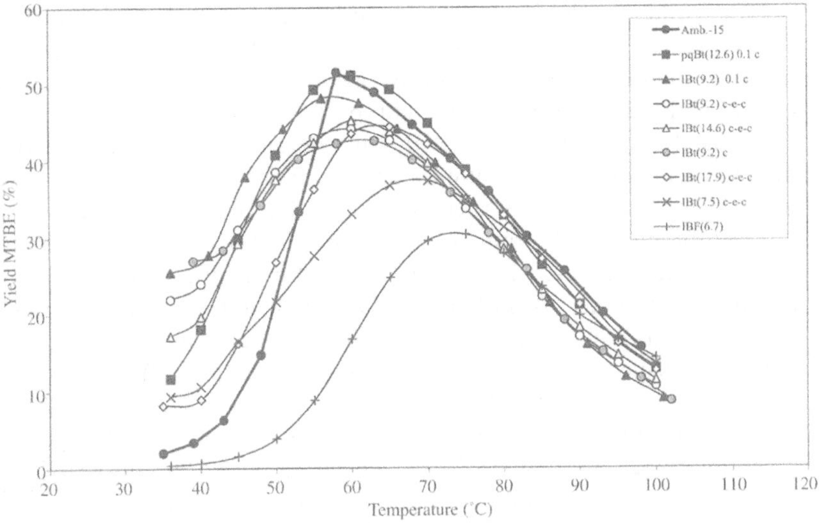

Figure 5 Amberlyst-15 and lab-synthesized beta-zeolites: yield of MTBE versus reaction temperature.

Nondealuminated omega-zeolite (LZ6 from UOP) exhibits poor performance. Dealumination of this zeolite by steaming followed by acid leaching substantially improves its performance, with the most active form having a Si/Al ratio of 26, but poor crystallinity.

As seen in Figure 5 and Table 2, the beta-zeolites show the best catalytic performance. Indeed, under identical reaction conditions, the yields of the ether produced at maximum IB-to-MTBE conversion are comparable to those obtained over the reference Amberlyst-15 resin, with even the least active beta-zeolite being more efficient than the best of the US-Y zeolites. The catalytic performance of dealuminated beta-zeolite is sensitive both to sample origin or batch and to concentration of the acid solution used for the dealumination treatment.

Mild dealumination of samples pqB1 (13.2) and pqB2 (13.4) (two different batches of the same beta-zeolite from PQ) by acid leaching using the procedure described in Ref. 63, has a beneficial effect on MTBE production. This is due to the removal of the EFAl species, which poison the Brønsted acid sites or block their access, with, as a result, an increase of both the content of the acid sites [64,65] and catalytic activity. A more severe acid treatment (higher acid concentration) leads to more highly dealuminated samples [66,67], thus with less acid sites and lower activity.

Lab-synthesized beta-zeolites with Si/Al ratios of 9.2, 14.6, and 17.9 (samples 1Bt in Table 2 and Figure 5) were obtained with tetraethyl ammonium hydroxide as structuring agent (template), following the procedure of Borade and Clearfield [68]. pqBt(11.8) (Si/Al of 11.8) was an as-synthesized beta-zeolite (from PQ), still containing Na^+ and tetraethyl ammonium template (t). The samples denoted 0.1c were previously washed with 0.1 M nitric acid before calcination (c) to remove the template.

The optimum Si/Al ratio of 9.2 in this series is lower than those reported for dealuminated beta-zeolites [49]. No significant effect on the formation of the ether (MTBE and ETBE) was noted whether the lab-synthesized beta-zeolites were ammonium-exchanged [sample 1Bt(9.2)cec] or simply calcined to remove the template [sample 1Bt(9.2)c]. This Si/Al ratio of 9.2 is close to that reported by Vaudry et al. [69] (10.8), for which all the template molecules are associated with framework Al, with the Na^+ cations being located at structural defects (SiO^- Na^+). Bourgeat-Lami et al. [70] have shown that the ammonium ions resulting from the decomposition of the template (tetraethyl ammonium) occupy negative $SiOAlO_3Si_3$ centers, with the protons left after deammoniation neutralizing the negatively charged framework sites. These observations may explain the catalytic results. For these samples, the similar amounts of acid sites determined by NH_3-TPD, on the one hand, and the corresponding ether (MTBE and ETBE) yields, on the other hand, are therefore consistent. The activity is improved when template-containing beta-zeolites are treated with 0.1 M nitric acid prior to the oxidation of the template [pqBt(12.6)0.1 c and 1Bt(9.2)0.1 c].

Other factors than the Si/Al ratio (amount of Brønsted acid sites) of beta-zeolites also have an influence on the ether synthesis. Indeed, sample 1BF(6.7), with larger crystallite size (1–3 μm instead of 0.02 μm, Table 2 and Fig. 5), synthesized following the procedure of Caullet et al. [71], is the least active among the beta-zeolites. Thus, the crystallite size or, more precisely, the extent of the outer surface area is also an important factor in ether synthesis.

Similar behavior is observed in vapor-phase synthesis of ETBE (ethanol/isobutene molar ratio of 1, and WHSV of $2\,h^{-1}$) over various zeolites [64]. The conversion-temperature curves of the most active samples among the different series of zeolites are compared in Figure 6. In this figure, beta-zeolites pqB1(13.2), and pqB2(13.4) were commercial samples (from PQ-zeolite), with pqB1(13.2) corresponding to ZB25 in previous studies [49,56].

As for the synthesis of MTBE, all the IB-to-ETBE conversion curves exhibit a similar volcano shape, with, again, the superimposition of the decreasing part of the curves being imposed by the thermodynamic equilibrium of the reaction at each temperature. The higher activity of the

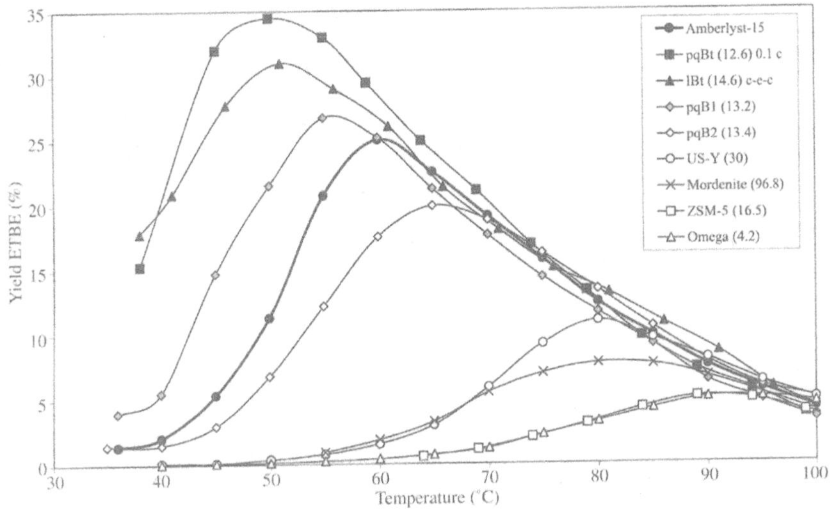

Figure 6 Amberlyst-15 and various zeolites: yield of ETBE versus reaction temperature.

beta-zeolites compared with the US-Y, ZSM-5 zeolites, and acid resin is obvious. For example, the yield of ETBE over pqBt(12.6)0.1c reached 32% at 45°C, which is much higher compared with the other catalysts (commercial samples and acid resin). For all the tested zeolites, oligomerization products (di-isobutenes) appeared at temperatures above that of maximum IB-to-ETBE conversion.

US-Y with Si/Al ratio of 30 (CBV 760) was the most active one of the Y series, with a maximum ETBE yield of 10–11% at about 80°C. This zeolite also was the most active in MTBE synthesis. The ZSM-5 zeolites were all less efficient than the US-Y, with a yield at maximum conversion barely exceeding 5% (at around 90°C), namely, half the value obtained over the most efficient US-Y. Finally, the yields of ETBE at maximum IB conversion obtained over a dealuminated large-port mordenite (with Si/Al of 96.8) and omega-zeolite (LZ6, with Si/Al of 4.2) were 8% and 5%, thus in the range of the values found for US-Y and ZSM-5.

III. CONTRIBUTION OF THE EXTERNAL SURFACE AREA TO THE REACTION

The beta-zeolites have substantially larger external surface areas relative to the other series of dealuminated zeolites [49,64]. Partial dealumination

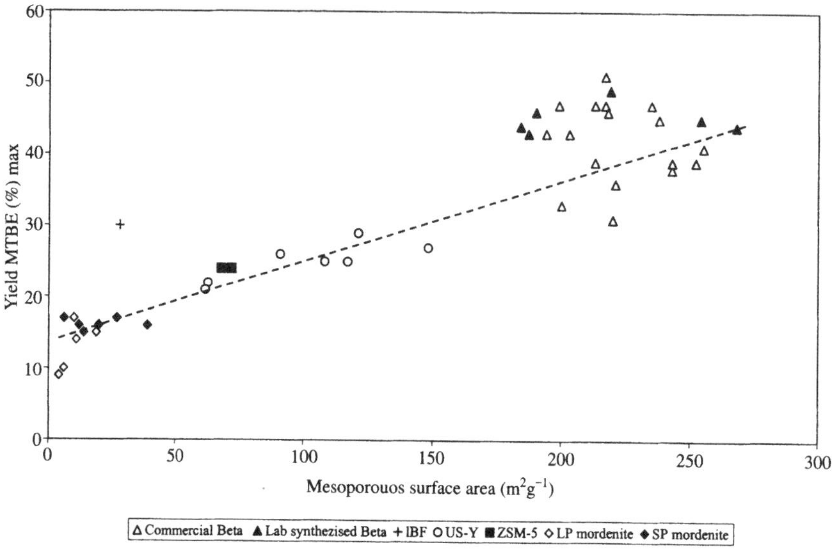

Figure 7 Yield of MTBE at maximum isobutene conversion for various zeolites versus mesoporous surface area (m^2/g).

(increasing Si/Al ratio) enhances the ether (MTBE and ETBE) formation by increasing the accessibility of the acid sites (e.g., US-Y zeolites) and/or removing some EFAL species (e.g., the acid-leached forms of beta-zeolite). 1BF, a beta-zeolite with large crystallite size (small external surface area), shows the lowest ether yield. In Figure 7, the yields of MTBE (at maximum conversion) obtained for the series of beta-zeolite, US-Y, ZSM-5, and mordenites are plotted against the surface areas developed by mesopores [72] (pore diameter between 2 and 50 nm). These mesopores are produced by the destruction of a part of the crystals during dealumination (see Fig. 2) and the intercrystalline void spaces [73]. The smaller the crystal diameter, the higher is the outer surface area per unit weight ($S_{m2g-1} = f\ [6/d]$). The macroporous surface area developed by pores with diameter greater than 50 nm is negligible.

There is an obvious link between these two parameters. The beta-zeolites with the highest mesoporous surface area (200–250 m^2/g) are more active than the other zeolites. The fact that the relationship does not pass through the origin is an indication that the reaction occurs partly in the microporous volume of the zeolite. A similar relationship is found with the total external surface area (meso- and macroporous surface area, calculated according to the *t*-plot [74] or Remy [75] method). A similar trend is also observed for the vapor-phase synthesis of ETBE [64] and liquid-phase synthesis of MTBE [56]. The external surface area is also an important parameter in the alkylation of isobutane with 1-butane [76], and phenol with *t*-butyl alcohol [77]. Dispersion of heteropolyacids on silica (257 m^2/g) [22] or active carbon (958 m^2/g) [17] enhanced the formation of MTBE. The sole external surface area, however, cannot account for the differences of activity observed among the beta-zeolites, except for the sample prepared by the fluoride method (1BF), which, as mentioned earlier, gives large crystal sizes (1–3 μm). Indeed, the series of beta-zeolites has similar mesoporous surface area and yet some zeolites derived from template-containing beta-zeolite (1Bt or pqBt) are substantially more efficient than the other commercial beta-zeolites and their acid-leached forms, which points to the intervention of other physicochemical factors than the textural characteristics alone.

IV. CONTRIBUTION OF THE BRIDGING AlOHSi AND SiOH GROUPS, AND EXTRA-FRAMEWORK ALUMINUM, TO CATALYSIS

The acid content of the zeolites can by quantified by several methods, such as adsorption and/or desorption of a base molecule (e.g., temperature-programmed desorption of ammonia, NH_3-TPD [78]) or, more directly, by spectroscopic studies (XPS, FTIR, etc.) of adsorbed bases such as ammonia, pyridine, lutidine, etc. FTIR spectroscopy is also a useful tool to examine and quantify the hydroxyl groups in the OH stretching region. A total of five bands has been observed in the OH stretching region of beta-zeolites, at 3780, 3745, 3738, 3660, and 3610 cm^{-1}. The typical Brønsted acid sites (bridging AlOHSi groups) are characterized by an OH stretching band at 3610 cm^{-1}. Figure 8 shows the relationship between the yield of MTBE at 40 and 60°C and the integrated area of the band at 3610 cm^{-1} normalized to the sample weight (proportional to the total amount of AlOHSi groups).

The most efficient samples [1Bt(9.2 to 17.9) and pqBt(11.8)] exhibit higher content of bridging AlOHSi than the less active commercial beta-zeolites. A similar relationship is obtained when the yields of MTBE (or ETBE) are plotted against the acid contents determined by NH_3-TPD [64]. Beta-zeolites differentiate from other zeolites by a high content of silanol groups (OH bands at 3745 and 3738 cm^{-1}) contributed by the small crystallite size and the coexistence of 3-unit cells inducing structural defects [79–81]. Hunger and co-workers [50–54] proposed that the SiOH groups act as reservoirs of reactant, forming clusters of alcohol molecules near the active sites. Compared with the commercial beta-zeolite samples, the

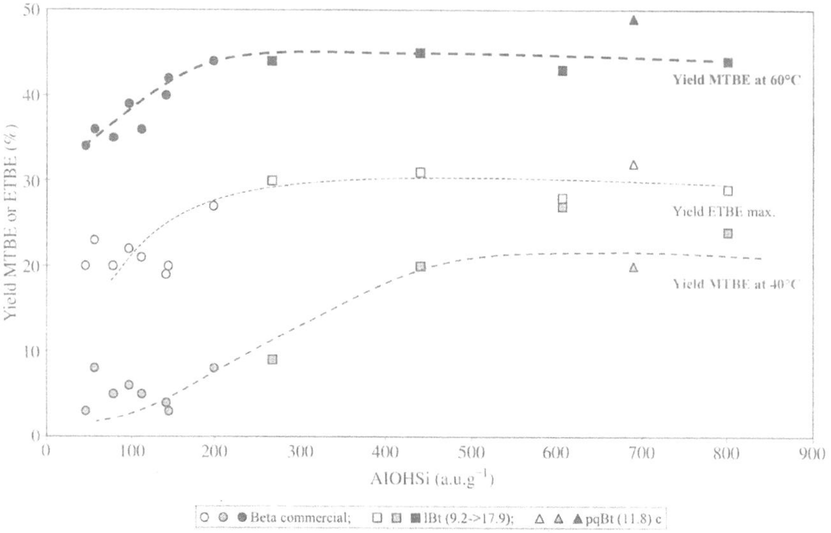

Figure 8 Yield of MTBE (at 40 and 60°C) and ETBE at maximum conversion versus AlOHSi normalized band area of various beta-zeolites.

lab-synthesized zeolites (samples 1Bt) have fewer SiOH groups, with a higher proportion at external ($3745\,cm^{-1}$) than internal ($3738\,cm^{-1}$) position [67,82–87]. Plotting the yields of ETBE and MTBE versus the SiOH/AlOHSi normalized band area ratio of various beta-zeolites (Fig. 9) shows that those with large ratios are less active and, conversely, the more active ones have smaller ratios. This suggests that ether synthesis over beta-zeolites is favored by a high content of bridging AlOHSi groups (active sites) and a moderate amount of SiOH groups at the external surface, preventing the formation of too large and/or numerous alcohol clusters which perturb the access of isobutene to the AlOHSi active sites.

A mild acid leaching of pqB2(13.4) (with 0.02 to 0.06 M HNO_3) favors the elimination of extra-framework Al species which poison or block the acid sites. Removal of these species is reflected by an increase of the band intensity of the bridging AlOHSi groups [64,65]. These samples also have high amounts of OH groups related to the bands at 3780 and $3660\,cm^{-1}$. According to Jia et al. [85], these bands are more developed when hot spots or self-steaming conditions are achieved during calcination (deep-bed conditions and/or temperature higher than 500°C). The band at $3780\,cm^{-1}$ has recently been attributed to a hydroxyl group on tricoordinated Al partially connected to the framework [88]. These OH groups have a weaker acid character than those absorbing at 3610 and $3660\,cm^{-1}$ [83,89].

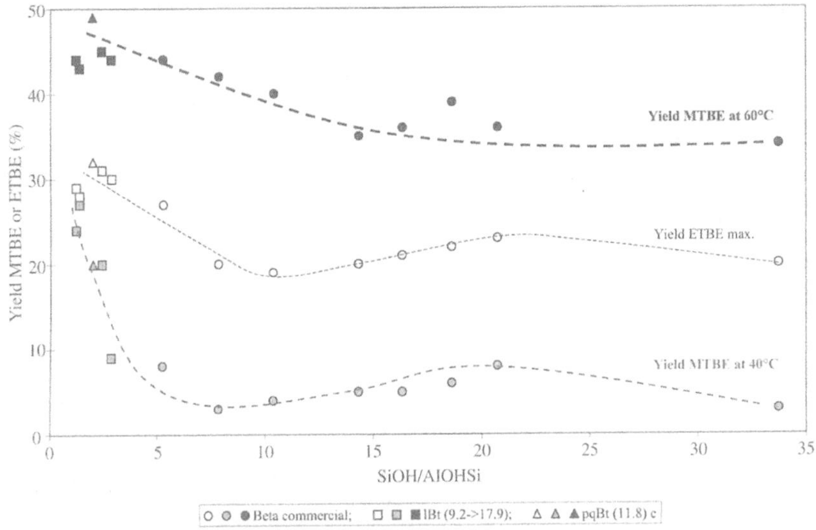

Figure 9 Yield of MTBE (at 40 and 60°C) and ETBE at maximum conversion versus SiOH/AlOHSi normalized band area ratio of various beta-zeolites.

The band at 3660 cm^{-1} has been assigned to OH of extra-framework AlOH species [83,85] or OH groups bonded to tetrahedral Al partially disconnected from the framework [90]. These species are likely those which block the acid sites. These also could promote, as the SiOH groups, the formation of alcohol clusters.

Figure 10 illustrates, for the US-Y zeolites (and mordenite), the effect of the EFAL species on the formation of MTBE, by plotting the yields at 60°C versus the ratio of pentacoordinated Al or distorted tetrahedral Al (Alx) to framework tetrahedral Al (Al^{IV}). The EFAL species are characterized by a ^{27}AlMAS NMR signal at 30 ppm, whereas the framework Al has a signal at 54 ppm.

The unfavorable influence of Alx on the reaction is obvious. OH groups connected to distorted framework Al^{IV} could have a similar effect to nonbridging OH groups in beta-zeolites as a result of the formation of too large and/or too stable alcohol clusters, without excluding the poisoning effect of extra-framework Al on the active sites. Indeed, XPS (X-ray photoelectron spectroscopy) [91] and, recently, three-dimensional transmission electron microscopy (TEM) studies [73] have shown that the outer surface of steamed US-Y samples (with low Si/Al ratio) is Al-enriched. The majority of this extra-framework Al at the outer surface is nonacidic [91]. In the acid-leached samples (the most active one

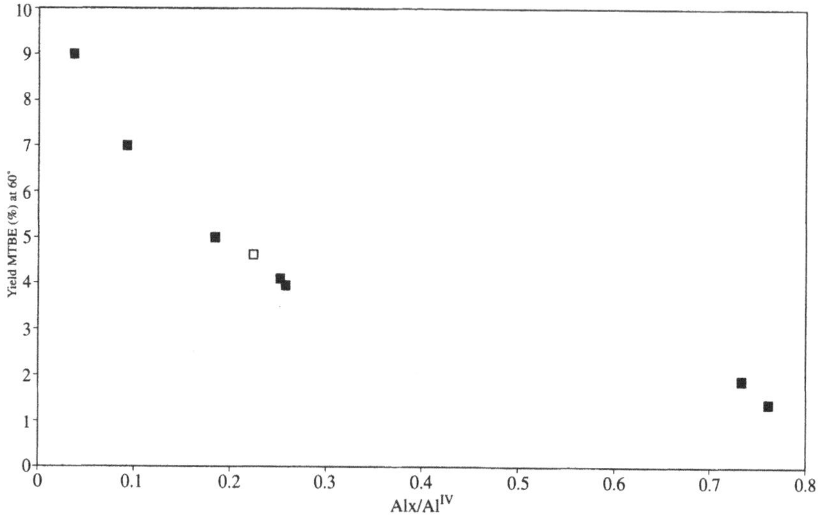

Figure 10 Zeolites US-Y (solid squares) and dealuminated mordenite (open square): yield of MTBE at 60°C versus Alx/Al^{IV} ratio (framework Al^{IV} with ^{27}Al NMR signal at 60 ppm; Alx: Al^{V} or distorted Al^{IV} with NMR signal at 30 ppm).

with Si/Al ratio around 30), the outer surface is Al-depleted [73,91] and thus relatively more acidic, because the sites are free from extra-framework species [91].

V. SYNTHESIS OF MTBE IN LIQUID-PHASE CONDITIONS

Only a few studies reported in the literature have been devoted to the liquid-phase synthesis of ethers in the presence of acid zeolites. Chu and Kuhl [35] investigated the liquid-phase synthesis of MTBE from isobutene and methanol over ZSM-11 and ZSM-5. In their study, ZSM-5 had a higher selectivity to MTBE than Amberlyst-15 and was less sensitive to the methanol/isobutene molar ratio of the feed. The high selectivity of ZSM-5 zeolite was confirmed by Pien et al. [36]. However, both ZSM-5 and ZSM-11 need higher reaction temperatures ($T > 100$°C) to reach the same isobutene conversion as the reference Amberlyst-15. In other words, these zeolites produce less MTBE than Amberlyst-15 at temperatures below 100°C. A similar observation was reported by Tau and Davis [92] for the liquid-phase synthesis of ETBE over ZSM-5. Briscoe et al. [55] studied MTBE synthesis in batch conditions over Nu-2 zeolites. A comparison of various commercial beta-zeolites, US-Y (with Si/Al of 30), and Amberlyst-15 was extended to liquid-phase conditions in the temperature range 30–120°C [27]. Results are reported in Table 3.

Between 40 and 100°C, higher isobutene conversions are achieved over the commercial beta-zeolites with small crystal size compared with Amberlyst-15, 1BF, and US-Y (CBV-760). Sample 1BF, with large crystal size (Tables 2 and 3) and low external surface area, is the least active beta-zeolite at 50°C. At this temperature, the yield of MTBE over the beta-zeolites is related to their external surface area (Table 3 and Ref. 56), confirming vapor-phase results. Efficiency differences of samples pqB1(13.2) and pqB(36), with similar external surface area, are accounted for by a higher silanol content of the latter. Sample pqB(36), with Si/Al ratio of 36, and optimum silanol and bridging hydroxyl content, leads to stoichiometric methanol and isobutene adsorption, and gives the highest yields of MTBE.

At temperatures higher than 65°C, the selectivity to MTBE (S_{MTBE}) diminishes and the yield of the ether (Y_{MTBE}) is lower than over the resin. Therefore, the influence of reactant ratio (MeOH/IB), pressure, and space time on the conversion and MTBE selectivity was investigated [56]. Side reactions of isobutene are more important over the beta-zeolites than over Amberlyst-15. However, isobutene dimerization can be cut down by using MeOH/IB ratios higher than 1:1 or by decreasing the contact time. The WHSV (on methanol and isobutene basis) should be equal or superior to 14 hr^{-1} in order to maintain a good yield and selectivity to MTBE [56]. In the pressure range investigaged (6–20 bar) and under optimized reaction conditions (60°C; WHSV = 14 hr^{-1}; MeOH/IB = 1.2), a yield of 87% MTBE ($S_{\mathrm{MTBE}} = 97\%$) was obtained with beta-pqB(36). This is substantially higher than the result obtained over Amberlyst-15 ($Y_{\mathrm{MTBE}} = 67\%$ and $S_{\mathrm{MTBE}} = 96\%$) in similar conditions.

On beta-zeolite, no deactivation was observed during a period of more than 50 hr on stream at 65°C, 14 bar pressure, WHSV of 14 hr^{-1}, and MeOH/IB ratio of 1. With these operating conditions, a total weight of 638 g MTBE per gram of catalyst was produced. Under industrial conditions, considering a feed containing 12% isobutene, a catalyst lifetime of 2 years and a WHSV close to 1 hr^{-1}, the MTBE production should approximate 4000 $\mathrm{g/g_{cat}}$. The present data suggest that MTBE productivity with a beta-zeolite might reach a similar value to an acid resin.

Another set of screening tests was performed over a beta-zeolite which showed good performance in vapor-phase synthesis. In those liquid-phase experiments, pressure, WHSV, and MeOH/IB ratio were 14 bar, 14 hr^{-1}, and 1.2, respectively. The samples and the corresponding yields of MTBE

Table 3 Isobutene Conversion, Yield of Isobutene to MTBE, and Selectivity to MTBE at 50, 65, and 90°C for Various Catalysts; Outer Surface Area (S_{ext}, m^2/g), SiOH/AlOHSi Ratio, Relative Quantity of SiOH, and AlOHSi Groups (a.u. g^{-1})

					X IB (%)			Ymtbe (%)			Sel. (%)		
Catalyst	S_{ext}[a]	SiOH/AlOHSi	SiOH	AlOHSi	50°C	65°C	90°C	50°C	65°C	90°C	50°C	65°C	90°C
Amb.-15	—	—	—	—	26	72	90	25	68	**87**	**98**	**94**	**97**
pqB1(13.2)	219	5.3	1045	199	52	85	95	**49**	**76**	65	93	89	68
pqB(36)	218	10.4	1466	141	68	94	96	**62**	**82**	65	92	87	68
scB(24.6)	124	7.9	1132	144	33	85	95	**32**	**81**	83	97	**95**	87
1BF	30	n.d.	n.d.	n.d.	7	30	82	7	28	74	97	93	90
US-Y(3)	143	n.d.	n.d.	n.d.	3	7	71	3	7	63	100	100	89

Reaction conditions: $P = 20$ bar; MeOH/IB = 1; WHSV = 14 hr^{-1}.
[a]Calculated according to Ref. 75.

Table 4 Si/Al Molar Ratio, Total Acidity, SiOH/AlOHSi Ratio, Concentration of Silanol Groups, and Yield of MTBE (mol%) Obtained in Liquid Phase at 40, 50, and 90°C

Catalyst[a]	Acidity[b]	SiOH/AlOHSi	[SiOH][c]	Y40	Y50	Y90
Amb.- 15	—	—	—	54	**84**	**90**
1Bt(9.2) c	1.12	1.3	813[d]	**75**	**95**	86
1Bt(9.2) c-e-c	1.43	1.2	953[d]	**93**	**94**	83
pqBt(11.8) c	0.76	—	—	**66**	**89**	74
pqBt(11.8) c-e-c	1.02	—	—	**95**	**98**	79
pqBt(12.6) 0.1 c	1.24	—	—	**69**	**94**	75
1Bt(14.6) c	0.95	—	—	**93**	**98**	**90**
scB(24.6)[e]	0.45	7.9	1132	18	53	81
pqB(36)	0.41	10.4	1466	**66**	**89**	80

[a](Si/Al) determined from ICPS.
[b]Total acidity measured by NH_3-TPD (mmol/g).
[c]Concentration of SiOH groups measured by FTIR (a.u./g).
[d]SiOH groups in external position ($\nu \geq 3743\,cm^{-1}$).
[e]External surface area $< 200\,m^2/g$.
Reaction conditions: $P = 14$ bar; MeOH/IB = 1.2; WHSV = $14\,hr^{-1}$.

obtained at different temperatures are compiled in Table 4. Amberlyst-15 and two commercial zeolites were also tested for comparison purpose.

The beta-zeolite with low external surface area (sample scB from Süd-Chemie, see Table 2) is the least active catalyst. The most active beta-zeolites have high external surface area and Si/Al ratio between 9.2 [sample 1Bt(9.2)] and 14.6 [sample 1Bt(14.6)]. High yields were produced over PqBt(11.8) and 1Bt (samples provided with template and activated according to Ref. 64). Figures 11 and 12 show, for the most active zeolites, the variation of isobutene conversion and yield of MTBE versus temperature, respectively. The isobutene conversion reached a maximum at 65°C. As shown in Figure 11, all the conversion curves are superimposed at reaction temperatures above 60°C. Below 60°C, all the tested zeolites are more active than reference Amberlyst-15, with similar results for 1Bt(9.2) c-e-c and 1Bt(14.6) c.

The yields of MTBE (Fig. 12) at temperatures above 65°C are lower on the zeolite than on the acid resin. At those temperatures, the total IB conversions are constant and the resin is more selective than the zeolites. Below 65°C, the curves of conversion and yields follow a similar variation because the formation of secondary products (C8, C12) remains very low: conversions and yields are nearly identical. Up to about 70°C, higher yields of MTBE are obtained over the zeolites compared with the resin.

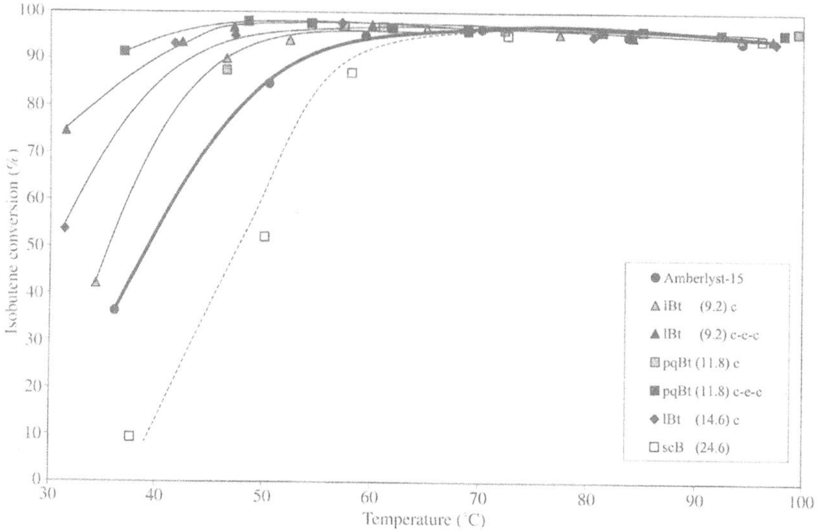

Figure 11 Total isobutene conversion versus reaction temperature on Amberlyst-15 and optimized beta-zeolites (high external surface area, low SiOH content, $9 < Si/Al < 14$). Reaction conditions: $WHSV = 14\,hr^{-1}$; 14 bar; MeOH/IB = 1.2.

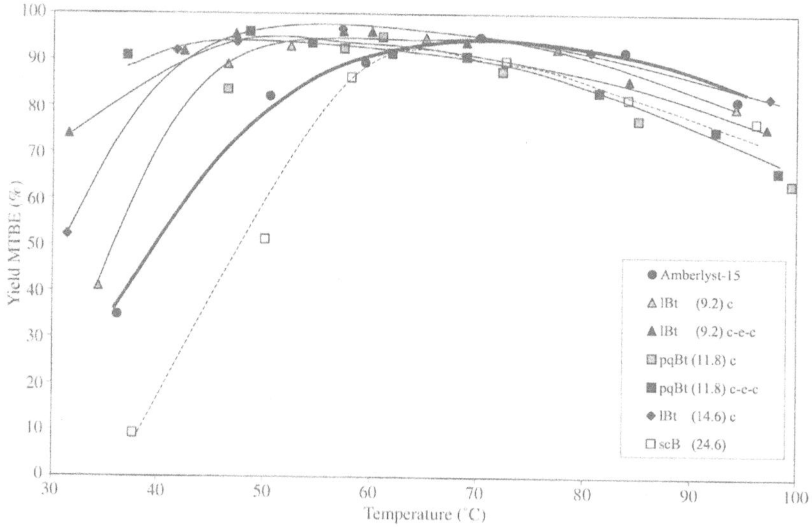

Figure 12 Yield of isobutene into MTBE versus reaction temperature on Amberlyst-15 and optimized beta-zeolites (high external surface area, low SiOH content, $9 < Si/Al < 14$). Reaction conditions: $WHSV = 14\,hr^{-1}$; 14 bar; MeOH/IB = 1.2.

Above 70°C, the resin produces slightly more ether. The nonexchanged beta-zeolite with Si/Al ratio of 14.6 [1Bt(14.6)c] and the resin have similar activity at high temperature (Table 4 and Fig. 12). This shows that in liquid-phase conditions, beta-zeolite can combine both high conversion and high selectivity for ether (MTBE) synthesis.

Samples exchanged with ammonium acetate and recalcined, 1Bt(9.2) c-e-c and pqBt(11.8) c-e-c (with c-e-c standing for calcination, ammonium exchange, and calcination to remove ammonia) are more active at low temperature than their nonexchanged forms [1Bt(9.2)c and pqBt(11.8)c] (Figs. 11 and 12, and Table 4). An opposite result was observed in vapor-phase synthesis of MTBE and ETBE. This different behavior can have several explanations. 1) During the reaction, the Na^+ ions associated with SiO^- groups [69] in samples 1Bt(9.2)c and pqBt(11.8)c could be make a complex with polar methanol molecules and poison the catalytic sites, according to the possible following sequence:

$$SiO^-Na^+ + nCH_3OH \rightarrow SiO^- + \{Na^+(CH_3OH)_n\}$$
$$\{Na^+(CH_3OH)_n\} + AlO^-H^+Si \rightarrow AlO^-Na^+Si + \{H^+(CH_3OH)_n\}$$
$$\{H^+(CH_3OH)_n\} + SiO^- \rightarrow SiOH + nCH_3OH$$

2) Ether synthesis would be more sensitive, in liquid phase, to the amount of silanol groups responsible for the preadsorption of methanol and isobutene [50–54]. An exchange with ammonium would increase the quantity of silanols by "washing" the SiO^-Na^+ groups. This second assumption is preferred because there is no evidence in the literature making it possible to corroborate explanation 1. In addition, the SiOH groups are, indeed, more abundant after ammonium exchange and calcination [64,65].

These results confirm the activities obtained for the MTBE synthesis in vapor-phase conditions. In the light of those results, an acid zeolite for optimal synthesis of MTBE should meet the following requirements:

The preferred zeolite is by far a beta-zeolite.
The size of the elementary crystals should be as small as possible.
The framework Si/Al molar ratio should be preferably between 10 and 15.
An exchange with ammonium acetate is advisable but not indispensable.
The silanol content should be as low as possible, with SiOH groups preferably situated at external position (outer surface).

VI. CONCLUSIONS AND SUGGESTIONS FOR FUTURE WORK

Synthesis of ethers has been performed over acid zeolites (beta-zeolite, US-Y, omega-zeolite, ZSM-5, "large-" and "small-" pore mordenites) from isobutene and an alcohol. For the zeolites, the higher the mesoporous (or external) surface area, the higher the ether production. Dealumination has a beneficial effect on the reaction, owing, on the one hand, to the removal of extra-framework aluminum species which act as poison of the Brønsted acid sites, and, on the other hand, to the increase of the external surface area of the zeolite which allows easier access of the reactants to the active sites. Beta-zeolites with small crystal size (large external surface area) are the most active catalysts. They can produce more ether than the acid resin Amberlyst-15, the industrial catalyst.

The reaction occurs on bridging AlOHSi acid sites, with higher yields of ether being obtained for beta-zeolites with low SiOH/AlOHSi ratios. The silanols groups enhance alcohol adsorption. Silanol groups should be present at the outer surface of the crystallites, and not in too large quantity in order to avoid the formation of large alcohol clusters which perturb the access of isobutene to the acid active sites (AlOHSi).

Future work based on the synthesis of zeolites with very small crystal size could lead to more active catalysts [93]. The major problem could be the stability of tetrahedral aluminum in the zeolite framework when the crystal diameter is small, especially for beta-zeolites [94]. This drawback could be overcome, however, by a selective extraction of the template before calcination [95], as, for example, the one reported in this study.

In the coming years, ethers with higher molecular weight may be needed, since they are less water-soluble than MTBE. Future catalyst design should extend to materials with high mesoporous volume. "Dispersion" of small crystals of beta-zeolite in a mesoporous material (MCM-41 and the like) could also give access to promising catalysts, especially for the synthesis of heavier ethers. Such catalysts could be prepared by using, e.g., MCM-41 as silica source instead of amorphous silica for the zeolite synthesis [96].

ACKNOWLEDGMENTS

F. Collignon gratefully acknowledges FRIA (Belgium) for a doctoral grant. The authors are indebted to P. A. Jacobs and J. A. Martens (K. U. Leuven, Belgium) for helpful discussions and for providing access to the pressure reactor for the liquid-phase experiments.

REFERENCES

1. M. M. Meier, D. H. Olson. In: Atlas of Zeolite Structure, International Zeolite Association, 2nd ed. Butterworth, London, 1987.
2. S. J. De Canio, J. R. Sohn, P. O. Fritz, J. H. Lunsford. J Catal, 101: 132–141, 1986.
3. P. V. Shertukde, G. Marcelin, G. A. Sill, K. H. Hall. J Catal, 136: 446–462, 1992.
4. J. Scherzer. J Am Chem Soc, 248: 157–200, 1984.
5. G. T. Vicencio, A. O. Ramos, J. R. Villanueva, G. P. Lopez. Rev Inst Mexicano Petrol, 24: 63–67, 1992.
6. M. E. Quiroga, N. S. Figoli, U. A. Sedran. Chem Eng J, 67:199–203, 1997.
7. K. Klier, R. Herman, M. A. Johansson, O. C. Feeley. Am Chem Soc Div Fuel Chem, 37(1): 236–246, 1992.
8. O. C. Feeley, M. A. Johansson, R. Herman, K. Klier. Am Chem Soc Div Fuel Chem, 37(4): 1817–1824, 1992.
9. A. Nikolopoulos, T. P. Palucka, P. V. Shertukde, R. Oukaci, J. G. Goodwin Jr, G. Marcelin. In: L. Guczi, F. Solymosi, P. Tétényi, eds. New Frontiers in Catalysis. Elsevier, Amsterdam, 75: 2601–2604, 1992.
10. J. M. Adams, K. Martin, R. W. McCabe, S. Murray. Clays Clay Miner, 34: 597–603, 1986.
11. A. Bylina, J. M. Adams, S. H. Graham, J. M. Thomas. J Chem Soc Chem Commun, 1003–1004 (1980).
12. J. Wang, J. Merino, P. Aranda, J. C. Galvan, E. Ruiz-Hitzky. J Mater Chem, 9: 161–168, 1999.
13. S. Cheng, J.-T. Wang, C.-L. Lin. J Chin Chem Soc, 3: 529–534, 1991.
14. K.-H. Chang, Geon-Joong Kim, Wha-Seung Ahn. Ind Eng Chem, Res 31: 125–130, 1992.
15. R. Fricke, M. Richter, H. L. Zubowa, E. Schreier. In: Studies in Surface Science and Catalysis. Elsevier, Amsterdam, 105: 655–662, 1997.
16. M. Richter, H. L. Zubowa, R. Heckelt, R. Fricke. Microporous Mater, 7: 119–123, 1997.
17. W. L. Chu, X. G. Yang, X. K. Ye, Y. Wu. React Kinet Catal Lett, 62: 333–337, 1997.
18. A. Malecka, J. Pozniek. C R Acad Sci Paris, t 1, sér II c: 361–368, 1998.
19. A. Bielanski, R. Dziembaj, A. Malecka-Lubanska, J. Pozniek, M. Hasik, M. Drozdek. J Catal, 185: 363–370, 1999.
20. G. Baronetti, L. Briand, U. Sedran, H. Thomas. Appl Catal, A: Gen 172: 265–272, 1998.
21. S. Shikata, T. Okuhara, M. Misono. J Mol Catal, 100: 49–59, 1995.
22. S. Shikata, S. Nakata, T. Okuhara, M. Misono. J Catal, 166: 263–271, 1997.
23. A. Malecka, J. Pozniczek, A. Micek Ilnicka, A. Bielanski. J Mol Catal A: Chem, 138: 67–81, 1999.
24. T. Ushikubo, K. Wada. Appl Catal A: Gen, 124: 19–31, 1995.
25. J. K. Lee, I. K. Song, W. Y. Lee. Catal Lett, 29: 241–248, 1994.
26. I. K. Song, W. Y. Lee. Appl Catal A: Gen, 96: 53–63, 1993.

27. C. P. Nicolaides, C. J. Stotijn, E. R. A. van der Veen, M. S. Visser. Appl Catal A: Gen, 103: 223–232, 1993.
28. K. Klier, Q. Sun, O. C. Feeley, M. Johanson, R. G. Herman. In: Studies in Surface Science and Catalysis. Elsevier, Amsterdam, 101: 601–610, 1996.
29. A. Molnar, C. Keresszegi, B. Török. Appl Catal A: Gen, 189: 217–224, 1999.
30. J.-Y. Chen, A. W. Ko. Reaction Kinet Catal Lett, 71: 77–83, 2000.
31. Z. Ziyang, K. Hidajat, A. K. Ray. J Catal, 2: 209–211, 2001.
32. P. Brandao, A. Philippou, J. Rcha, M. W. Anderson. Catal Lett, 73: 59–62, 2001.
33. G. Larsen, E. Lotero, M. Marquez, H. Silva. J Catal, 157: 645–655, 1995.
34. M. Richter, H. Fischer, M. Barztoszek, H. L. Zubowa, R. Fricke. Microporous Mater, 8: 69–78, 1997.
35. P. Chu, G. H. Kühl. Ind Chem Res, 26: 365–369, 1987.
36. S. I. Pien, W. J. Hatcher. Chem Eng Commun, 93: 257–265, 1990.
37. S. Ahmed, M. Z. El Faer, M. M. Abdillahi, J. Shirokoff, M. A. B. Siddiqui, S. A. I. Barri. Appl Catal A: Gen, 161: 47–58, 1997.
38. M. A. Ali, B. J. Brisdon, W. J. Thomas. Appl Catal A: Gen, 197: 303–309, 2000.
39. A. M. Al-Jarallah, M. A. B. Siddiqui, A. K. Lee. Can J Chem Eng, 66: 802, 1989.
40. A. A. Nikolopoulos, R. Oukaci, J. G. Goodwin Jr., G. Marcelin. Catal Lett 27: 149–157, 1994.
41. A. A. Nikolopoulos, R. Oukaci, J. G. Goodwin Jr., G. Marcelin. In Symposium on Octane and Cetane Enhancement Processes for Reduced Emissions Motor Fuels. Div Petrol Chem, Am Chem Soc, San Francisco, 37(3): 787–792, 1992.
42. A. A. Nikolopoulos, A. Kogelbauer, J. G. Goodwin Jr., G. Marcelin. Appl Catal A: Gen, 119: 69–81, 1994.
43. A. Kogelbauer, M. Ocal, A. A. Nikolopoulos, J. G. Goodwin Jr., G. Marcelin. J Catal, 148: 157–163, 1994.
44. A. Kogelbauer, A. A. Nikolopoulos, J. G. Goodwin Jr., G. Marcelin. In: Studies in Surface Science and Catalysis. Elsevier, Amsterdam, 84: 1685–1692, 1994.
45. R. Le Van Mao, R. Carli, H. Ahlafi, V. Ragaini. Catal Lett 6: 321–330, 1990.
46. A. A. Nikolopoulos, A. Kogelbauer, J. G. Goodwin Jr., G. Marcelin. J Catal 158: 76–82, 1996.
47. A. A. Nikolopoulos, A. Kogelbauer, J. G. Goodwin Jr., G. Marcelin. Catal Lett 39: 173–178, 1996.
48. R. Le Van Mao, T. S. Le, M. Fairbairn, A. Muntasar, S. Xiao, G. Denes. Appl Catal A: Gen 185: 41–52, 1999.
49. F. Collignon, M. Mariani, S. Moreno, M. Remy, G. Poncelet. J Catal 166: 53–66, 1997.
50. M. Hunger, T. Horvath, J. Weitkamp. In: Proceedings DGMK Conference C4 Chemistry—Manufacture and Uses of C4 Hydrocarbons. German Soc Petrol Coal Sci Technol, 65–72, 1997.
51. M. Hunger, T. Horvath. Catal Lett, 49: 95–100, 1997.
52. M. Hunger, T. Horvath, J. Weitkamp. Microporous Mesoporous Mater, 22: 357–367, 1998.
53. M. Hunger, M. Seiler, T. Horvath. Catal Lett, 57: 199–204, 1999.

54. T. Horvath, M. Seiler, M. Hunger. Appl Catal A: Gen, 193: 227–236, 2000.
55. N. A. Briscoe, J. L. Gasci, J. A. Daniels, D. W. Johnson, M. D. Shannon, A. Stewart. In: Studies in Surface Science and Catalysis. Elsevier, Amsterdam, 49: 151–160, 1989.
56. F. Collignon, R. Loenders, J. A. Martens, P. A. Jacobs, G. Poncelet. J Catal, 182: 302–312, 1999.
57. A. Kogelbauer, A. A. Nikolopoulos, J. G. Goodwin Jr., G. Marcelin. J Catal, 152: 122–129, 1995.
58. A. Kogelbauer, J. G. Goodwin Jr., J. A. Lercher. J Phys Chem, 99: 8781–8777, 1995.
59. R. S. Karinen, A. O. I. Krause, E. Y. O. Tikkanen, T. T. Pakkanen. J Mol Catal A: Chem, 152: 253–255, 2000.
60. R. S. Karinen, A. O. I. Krause. Catal Lett, 67: 73–79, 2000.
61. J. Tejero, F. Cunill, J. F. Izquierdo. Ind Eng Chem Res, 27: 338–343, 1988.
62. S. Moreno, G. Poncelet. Microporous Mater, 12: 197–222, 1997.
63. F. Fajula, E. Bourgeat-Lami, C. Zivkov., T. Des Courières, D. Anglerot. French Patent 2,069,618 (1992).
64. F. Collignon, G. Poncelet. J Catal, 202: 68–77, 2001.
65. F. Collignon, G. Poncelet. In: Studies in Surface Science and Catalysis. Elsevier, Amsterdam, 135: 308, 2001.
66. C. Coutanceau, J. M. Da Silva, M. F. Alvarez, F. R. Ribero, M. Guisnet. J Chim Phys, 94: 765–781, 1997.
67. M. Guisnet, P. Ayrault, C. Coutanceau, M. Fernanda Alvarez, J. Dataka. J Chem Soc Faraday Trans, 93: 1661–1665, 1997.
68. R. B. Borade, A. Clearfield. Microporous Mater, 5: 289–297, 1996.
69. F. Vaudry, F. Di Renzo, F. Fajula, P. Schulz. In: Studies in Surface Science and Catalysis. Elsevier, Amsterdam, 84: 163–170, 1994.
70. E. Bougeat-Lami, F. Di Renzo, F. Fajula. J Phys Chem, 96: 3087–3091, 1992.
71. P. Caullet, J. Hazm, J. L. Guth, J. F. Joly, F. Raatz. Zeolites, 12: 240–250, 1992.
72. E. P. Barret, L. G. Joyner, P. H. Hallenda. J Am Chem Soc, 71: 373–380, 1951.
73. A. H. Janssen, A. J. Koster, and K. P. de Jong. Angew Chem, 113(6): 1136–1138, 2001.
74. B. C. Lippens, J. H. de Boer. J Catal, 4: 319–323, 1964.
75. M. J. Remy, G. Poncelet. J Phys Chem, 99: 773–779, 1995.
76. R. Loenders, J. A. Martens, P. A. Jacobs. J Catal, 176: 545–551, 1998.
77. K. Zhang, C. Huang, H. Zhang, S. Xiang, S, Liu, D. Xu, H. Li. Appl Catal A: Gen, 166: 89–95, 1998.
78. M. Niwa, N. Katawa, M. Sawa, Y. Murikami. J Phys Chem, 99: 8812–8816, 1995.
79. J. M. Newsam, M. M. Treacy, W. T. Koetsier, C. P. de Gruyter. Proc Roy Soc Lond A 420: 375–405, 1988.
80. J. B. Higgins, R. B. LaPierre, J. L. Schenkler, A. C. Rohrman, J. D. Wood, G. T. Kerr, W. J. Rohrbaugh. Zeolites 8: 446–452, 1988.
81. M. M. Treacy, J. M. Newsam. Nature 332: 249–251, 1988.

82. E. Bourgeat-Lami, P. Massiani, F. Di Renzo, F. Fajula, T. Des Courières. Catal Lett 5: 265–272, 1990.
83. I. Kiricsi, C. Flego, G. Pazzuconi, W.O. Parker Jr., R. Millini, C. Perego, G. Bellussi. J Phys Chem 98: 4627–4634, 1994.
84. L. W. Beck, J. F. Haw. J Phys Chem, 99: 1076–1079, 1996.
85. C. Jia, P. Massiani, D. Barthomeuf. J Chem Soc Faraday Trans, 89: 3659–3665, 1993.
86. M. Maache, A. Janin, J. C. Lavalley, J.-F. Joly, E. Benazzi. Zeolites, 13: 419–426, 1993.
87. C. Pazé, S. Bordiga, C. Lamberti, M. Salvalaggio, A. Zecchina, G. Bellussi. J Phys Chem, 101: 4740–4751, 1997.
88. A. Vimont, F. Thibault-Starzyk, J.-C. Lavalley. J Phys Chem B, 104: 286–291, 2000.
89. B.-L. Su, V. Norberg. Zeolites, 19: 65–74, 1997.
90. E. Loeffler, U. Lohse, Ch. Peuker, G. Oelhmann, L. M. Kustov, V. L. Zholobenko, V. B. Kazansky. Zeolites, 10: 266–271, 1990.
91. M. J. Remy, D. Stanica, G. Poncelet, E. J. P. Feijen, P. J. Grobet, J. A. Martens, P. A. Jacobs. J Phys Chem, 100: 12440–12447, 1996.
92. L.-M. Tau, B. H. Davis. Appl Catal, 53: 263–271, 1989.
93. I. Schimdt, C. Madsen, C. J. H. Jacobsen. Inorg Chem, 39: 2279–2283, 2000.
94. F. Vaudry, F. Di Renzo, F. Fajula, P. Schulz. J Chem Soc Faraday Trans, 94: 617–621, 1998.
95. S. Hitz, R. Prins. J Catal, 168: 194–206, 1997.
96. T. Takewaki, S.-J. Hwang, H. Yamashita, M. E. Davis. Microporous Mesoporous Mater 32: 265–278, 1999.

5
Fluorinated Zeolite-Based Catalysts

Raymond Le Van Mao and Tuan Si Le
Concordia University, Montreal, Quebec, Canada

I. INTRODUCTION

The commercial catalyst Amberlyst-15 (A-15) resin and its sister catalyst, Amberlyst-35, presently used in the synthesis of methyl *tertiary*-butyl ether (MTBE), although giving high performance [1] in the liquid-phase synthesis, still suffer from several drawbacks, such as thermal instability, acid leaching from the resin surface [2], and the production of isobutene oligomers as by-products. The latter aspect constitutes a non-negligible economic disadvantage in terms of loss of isobutene reactant, which cannot be recycled as is methanol in the industrial process [3]. Because of these restrictive factors, considerable efforts have been focused on the search for alternative catalysts. From this perspective, inorganic materials such as acidified clays [4] or zeolites [2] were thus proposed to carry out the etherification. In particular, zeolite catalysts, which have high chemical and thermal stability, do not present the problem of acid leaching at the surface as encountered with the commercial A-15 resin. In addition, the pore systems of these microporous materials, the most representative being the (large-pore-sized) Y and the (medium-pore-sized) ZSM-5 zeolites, result in advanced product shape selectivity, thus reducing sharply the production of bulky isobutene oligomers [2]. However, the relatively lower surface acidity and the microporous nature of the zeolite catalysts require a much higher reaction temperature to compensate for the lower reactivity of the zeolite acidic surface and a more demanding diffusion of reactant and product molecules. Higher reaction temperature also means that the reaction is now controlled by thermodynamics, since decomposition of the product MTBE, occurring at high temperature, increasingly limits the MTBE yield. Attempts were thus made to enhance the surface acidity of the zeolitic materials in order to permit the reaction to be carried out at relatively low temperature. The two main procedures for zeolite modification which have been investigated with some success are the following: loading of the well-known superacidic triflic acid onto solid silica-based materials; and fluorination of the zeolite surface.

II. INCORPORATION OF TRIFLIC ACID INTO SOLID SILICA-BASED MATERIALS

Many attempts have been made in the past to incorporate acid species into solid silica-based materials [5], including silica impregnated with triflic acid (trifluoromethane sulfonic acid) or FSO_3H, developed by Topsoe for the aliphatic alkylation [6], and Y-zeolite treated with FSO_3H, developed by Texaco for the MTBE synthesis [7]. In our laboratory, the quite stable superacidic triflic acid, CF_3SO_3H, has been incorporated into various zeolite materials. These include the H-ZSM-5 and H-Y zeolites (both in acidic form).

Thus, in order to compare the catalytic performance of the Amberlyst-15 resin with that of the H-Y and the H-ZSM-5, three series of tests of MTBE synthesis were carried out using these catalysts [8]. Over the 60–95°C temperature range and at 1 atm total pressure, the maximum of MTBE yield depended on the catalyst tested (Fig. 1), suggesting that there was some cross-effect of the acid properties (density and strength) and the thermodynamic limitations due to the reaction temperature used. All this resulted in the following outcome in terms of MTBE production performance:

A-15 > H-ZSM-5 > H-Y

Interestingly, the yield of isobutene isomers depends strongly on the openness of the porous catalyst (Fig. 2), thus confirming the great influence of the catalyst pore size on the product selectivity.

When triflic acid (TfA) was incorporated into these two zeolites, there was a very significant increase in the MTBE yield [8,9]. However, as shown in Figure 3, the TfA loading should not exceed 3 wt% because otherwise the MTBE yield could decrease markedly due to the pore size narrowing upon incorporation of the rather bulky TfA molecules (see also Fig. 2) and also probably to some partial structural collapse. The latter phenomenon is a

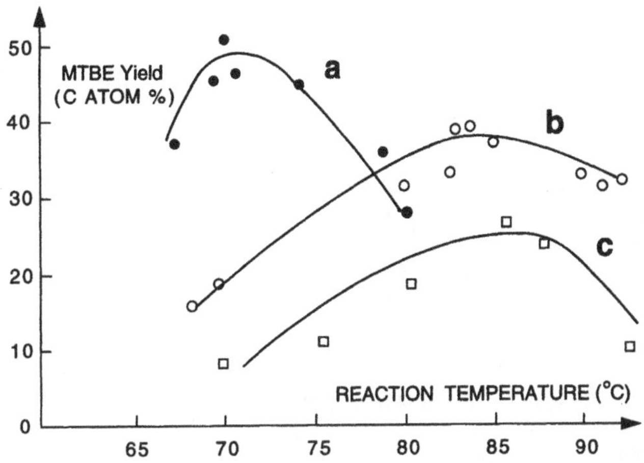

Figure 1 MTBE yields versus reaction temperature, obtained with Amberlyst-15 (a), H-ZSM-5 (b), and H-Y (c) catalysts. (From Ref. 8, with permission from the editor of *Catalysis Letters*.)

consequence of the dealumination effect of the strong acidic TfA on the zeolite structure. It is worth noting that a 3 wt% loading of the HY zeolite was far from providing a complete layer of TfA covering the H-Y zeolite surface. Recently, Kogelbauer et al. [10,11] attributed the activity enhancement to the formation of extra-lattice Al rather than to the direct presence

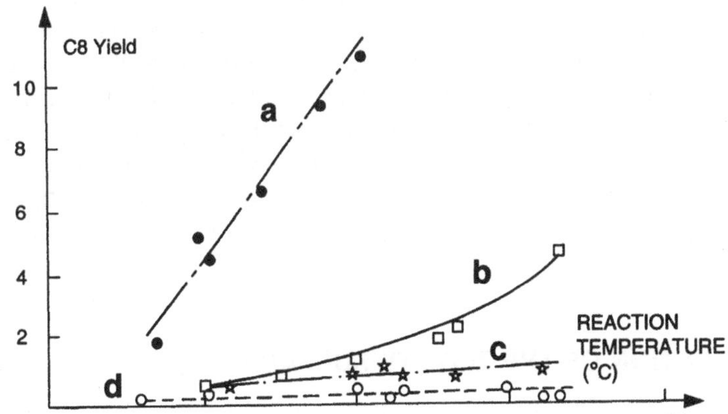

Figure 2 C8 by-products yields versus reaction temperature, obtained with Amberlyst-15 (a), H-Y (b), H-Y/TFA (3) (c), and H-ZSM-5/TFA (3) (d). (From Ref. 8, with permission from the editor of *Catalysis Letters*.)

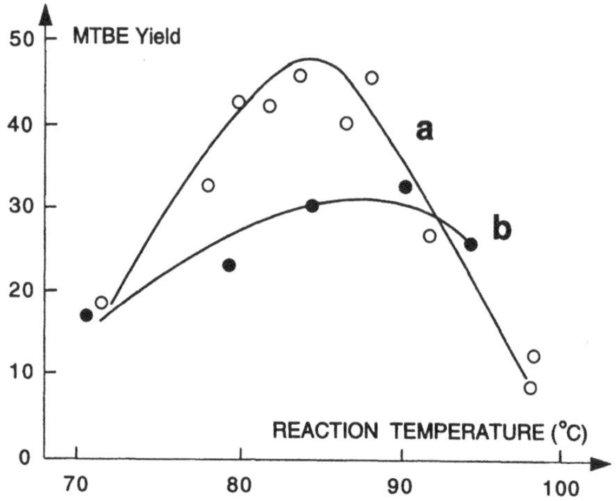

Figure 3 MTBE yields versus reaction temperature, obtained with H-Y/TFA (3) (a) and H-Y/TFA (4) (b). (From Ref. 8, with permission from the editor of *Catalysis Letters*.)

of TfA. However , current consensus is not in favor of the assumption which ascribes the higher acid properties to the increasing number of these extra-lattice Al species [12,13]. Instead, for other researchers, these Al species were found to act as powerful sorption sites [14] due to their electron-accepting properties [15], which were capable of positively affecting the reaction kinetics [14]. These extra-framework Al sites could also have synergistic interactions with the zeolite Brønsted acid sites [13].

Nevertheless, the H-Y zeolite loaded with 3 wt% TfA, when compared to its ZSM-5 counterpart, showed a higher production of MTBE, quite comparable to that of the commercial A-15 tested at 1 atm. However, the zeolite catalyst still needed a higher reaction temperature, 85–90°C (Fig. 3), versus 65–70°C for the A-15 resin (Fig. 1). This was definitely not an advantage for the zeolite-based catalyst because under such a condition, thermodynamic limitation started having a serious consequence for the MTBE yield.

III. FLUORINATION OF THE ZEOLITE SURFACE

Fluorine treatment of a zeolite material, particularly with HF, carried out in order to enhance the surface acidity or selectivity, is a well-known technique [16]. However, even with diluted HF aqueous solutions (0.1–0.2 M), appreciable aluminum removal from the framework was observed for zeolites such as mordenite. By using ammonium fluoride solution under quite mild conditions, strong Lewis acid sites corresponding to presumably low-coordinate Si ions were obtained [15,17–19]. The mordenites so-treated exhibited enhanced catalytic activity (mostly product selectivity) in the cumene cracking and toluene disproportionation [15], and also in other acid-catalyzed reactions [17–19].

In our laboratory, fluorinated ZSM-5 zeolites were used for the catalytic synthesis of MTBE [20,21]. Thus, the ZSM-5 zeolite surface was modified by the F-containing species incorporated initially by impregnation with an aqueous solution of ammonium fluoride. The concentration of this aqueous solution was relatively low. The resulting zeolite was subsequently activated in air following a well-defined temperature programming procedure which ended up at 450°C [20]. It was observed that there was a "proton attack" of the zeolite surface by the chemisorbed (H^+F^-) ion pair at that temperature, resulting in the formation of new acid sites and the strengthening of some of the (original) acid sites of the parent zeolite.

The structure of the parent zeolite was preserved because of the quite small amount of F species used. A mechanism showing the formation of new active sites is proposed in Figure 4 [20]. It is worth noting that such a mechanism is not very different from that proposed by Sanchez et al. [22].

With such new acidic materials, a very significant increase in production of MTBE was reported [20,21], while the product shape selectivity of the parent ZSM-5 zeolite was maintained.

The combination of two modification techniques developed for the ZSM-5 zeolite—i.e., a mild desilication of the zeolite surface (selective removal of Si atoms, which results mainly in some increase in the density of the acid sites [23,24]), and the incorporation of fluorine in the NH_4F form followed by a stepwise thermal treatment—produced a highly efficient catalyst for the synthesis of MTBE (Table 1) [21]. Both the density and the strength of the acid sites are significantly increased. In fact, the new material, referred to as H-DZSM-5/F sample, showed the following changes of acidic properties with respect to the parent zeolite: a density of acid sites higher by almost 80%, and a greater number of strong surface acid sites [21].

IV. CONCLUSION

The acidic properties of the shape-selective zeolites can be enhanced in two different ways:

1. Loading the zeolite surface with an organic superacid such as triflic acid.

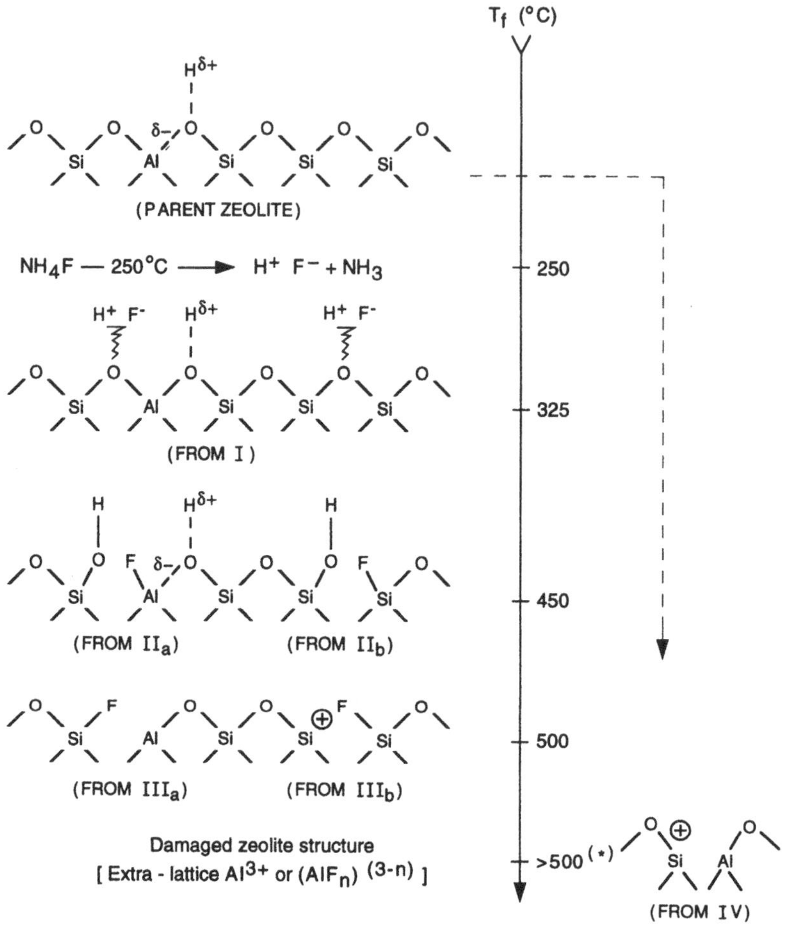

Figure 4 Mechanism of formation of active sites as a function of activation temperature. (From Ref. 20, with permission from Elsevier Science.)

2. Impregnating the surface of a desilicated zeolite with ammonium fluoride and then activating in air stepwise to 450°C.

Both procedures result in catalysts which have a surface acidity sufficient to provide the same level of MTBE production as the commercial Amberlyst 15. However, the microporous structure of these zeolite materials considerably reduces the amount of unwanted oligomers of isobutene produced. In addition, both catalysts, particularly that containing the

Table 1 MTBE Yield Obtained with Commercial Amberlyst-15, Parent H-Y and H-ZSM-5 Zeolites, H-ZSM-5/F, TfA/HY (TfA loading = 3 wt%), and H-DZSM-5/F Catalysts, at Reaction Temperature Giving the Maximum Activity (total pressure = 1 atm)

Catalyst	Reaction temperature (°C)	MTBE yield (wt%)	Yield of isobutene oligomers (wt%)
A-15	70	47.5	13.9
H-Y	85–90	22.5	2.5
H-ZSM-5	80	34.2	0.0
H-ZSM-5/F	80	44.9	0.0
TfA/H-Y	85–90	47.8	1.2
H-DZSM-5/F	80	49.9	0.1
	75	47.7	0.0

F-treated desilicated ZSM-5 zeolite, are chemically and thermally more stable than the commercial counterpart.

ACKNOWLEDGMENTS

The authors thank the Natural Sciences and Engineering Research Council of Canada for financial support, and the editors of *Catalysis Letters* (Kluwer Academic/Plenum) and *Applied Catalysis* (Elsevier Science) for permission to reproduce some copyright-reserved materials from Refs. 8 and 20, respectively.

REFERENCES

1. R. Chavez, R. Olsen, M. Ladisch. Oil Gas J, 10: 66, 1994.
2. P. Chu, G. H. Kuhl. Ind Eng Chem Res, 26: 365–369, 1987.
3. E. M. Elkanzi. Chem Eng Process, 35: 131–139, 1996.
4. G. D. Yadav, N. Kirthivasan. J Chem Soc Chem Commun, 203–204, 1995.
5. J. F. Joly. In: P. Leprince, ed. Conversion Processes. Technips, Paris, 2001, pp 257–289.
6. S. I. Hommeltoft, H. Topsoe. US Patent 5,245,100 (Sept. 14, 1993) assigned to Haldor Topsoe S.A.
7. J. F. Knifton, US Patent 5,081,318 (Jan. 14, 1992), assigned to Texaco Chem Comp.
8. R. Le Van Mao, R. Carli, H. Ahlafi, V. Ragaini. Catal Lett, 6: 321–330, 1990.

9. R. Le Van Mao, H. Ahlafi, T. S. Le. In: M. E. Davis, S. L. Suib, eds. Selectivity in Catalysis, ACS Symp Ser 517, American Chemical Soceity, Washington, DC, 1993, pp. 233–243.
10. A. Kogelbauer, A. A. Nikolopoulos, J. G. Goodwin Jr., G. Marcelin. In: J. Weitkamp, H. G. Karge, H. Pfeifer, W. Holderich, eds. Zeolites and Related Microporous Materials: State of the Art 1994. Elsevier, Amsterdam, 1994, pp. 1685–1692.
11. A. A. Nikopoulos, A. Kogelbauer, J. G. Goodwin Jr., G. Marcelin. J Catal, 158: 76–82, 1996.
12. F. Collignon, M. Mariani, S. Moreno, M. Remy, G. Poncelet. J Catal, 166: 53–66, 1997.
13. R. A. Beyerlein, G. B. McViker. Preprints 17th North American Catalysis Society Meeting, Toronto, Canada, June 2001, p 30.
14. R. Le Van Mao, M. A. Saberi. Appl Catal A: Gen, 199: 99–107, 2000.
15. S. Kowalak, A. Yu Khodakov, L. M. Kustov, V. B. Kazansky. J Chem Soc Faraday Trans, 91(2): 385–388, 1995.
16. A. K. Ghosh, R. A. Kydd. J Catal, 103: 399–406, 1987.
17. H. N. Sun. US Patent 5,932,512 (Aug. 3, 1999), assigned to Exxon Chemical Patents Inc.
18. F. Avendano, J. Tejada. Am Chem Soc Div Petrol Chem, 41(4): 1996, pp 723–725. Preprints.
19. A. G. Panov, V. Gruver, J. J. Fripiat. J Catal, 168(2): 321–327, 1997.
20. R. Le Van Mao, S. T. Le, M. Fairbairn, A. Muntasar, S. Xiao, G. Denes. Appl Catal A: Gen, 185: 41–52, 1999.
21. T. S. Le, R. Le Van Mao. Micropor Mesopor Mater, 34: 93–97 2000.
22. N. A. Sanchez, J. M. Saniger, J. B. d'Espinose de la Caillerie, A. L. Blumenfeld, J. J. Fripiat. Micropor Mesopor Mater, 50(1): 41–52, 2001.
23. R. Le Van Mao, S. T. Le, D. Ohayon, F. Caillibot, L. Gelebart, G. Denes, Zeolites, 19: 270–278, 1997, and refs therein.
24. C. Doremieux-Morin, A. Ramsaran, R. Le Van Mao, P. Batamack, L. Heeribout, V. Semmer, G. Denes, J. Fraissard. Catal Lett, 34: 139–149, 1995.

6

Heteropolyacids as Catalysts for MTBE Synthesis

Adam Bielański
Jagiellonian University and Institute of Catalysis and Surface Chemistry, Polish Academy of Sciences, Krakow, Poland

Anna Małecka-Lubańska, Joanna Poźniczek, and Anna Micek-Ilnicka
Institute of Catalysis and Surface Chemistry, Polish Academy of Sciences, Krakow, Poland

I. INTRODUCTION

Synthesis of methyl-*tertiary*-butyl ether (MTBE) by the electrophilic addition of methanol to isobutene as well as by etherification of methanol–*t*-butanol mixture needs catalysts exhibiting strong protonic acidity. In industry, MTBE is produced in the liquid phase using ion-exchange resins such as Amberlyst-15 at temperatures above 373 K. At these conditions sulfonated resins are not stable enough and gradually lose sulfuric acid, thus decreasing activity and creating the possibility of environmental damage. For this reason the search for more convenient catalysts has been undertaken in many laboratories. Various mineral substances have been used: zeolites, silica-alumina, acid-treated clays, and also heteropolyacids. The latter, among the strongest mineral acids, are the subject of the present chapter.

Heteropoly compounds are a group of acids and their salts containing oxoanions of the general formula $[E_aZ_bO_c]^{n-}$, in which Z denotes the "polyatom," most frequently an atom of Mo, W, V, or Nb, Ta in electronic configuration d^0 or d^1, and E is a "heteroatom," a metal or nonmetal in the

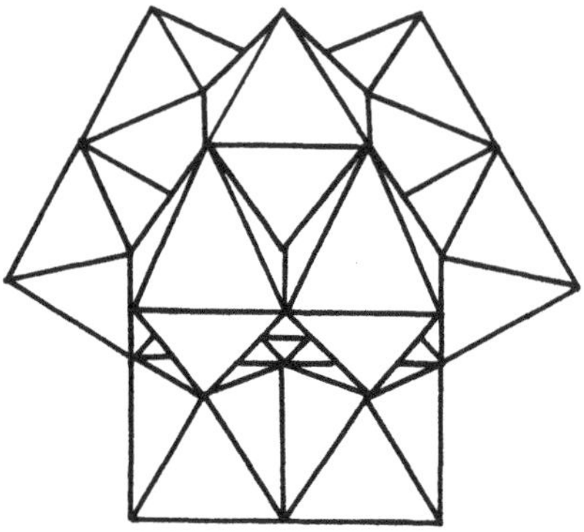

Figure 1 Structure of dodecatungstophosphoric acid $H_3PW_{12}O_{40}$ (Keggin structure).

positive oxidation state. Polyatoms are always hexacoordinated, forming $[ZO_6]$ octahedra bonded by the common corners, edges, or faces. On the other hand, heteroatoms exhibit various coordinations by O atoms: tetrahedral, octahedral, antiprismatic, and icosahedral [1]. Most catalytic experiments have been carried out with heteropolyacids (HPAs) and salts containing anion $[E^{m+}Z_{12}^{6+}O_{40}]^{(8-m)-}$. Typical examples of these compounds are $H_3PMo_{12}O_{40}$, dodecamolybdophosphoric (or dodecaphosphoromolybdic according to IUPAC nomenclature), $H_3PW_{12}O_{40}$, dodecatungstophosphoric, and $H_4SiW_{12}O_{40}$, dodecatungstosilicic acids. The structure of the dodecaheteropolyanion, called Keggin structure (or Keggin unit), is shown in Figure 1. In this structure, 12 $[ZO_6]$ octahedra are present, grouped in four subunits each of three octahedra sharing common edges. The subunits sharing common corners form a cluster anion of T_d symmetry. Heteroatom E is situated in its geometric center, around which four O atoms, each belonging to a different triple subunit, are tetrahedrally coordinated.

In the anhydrous state, Keggin-type anions in heteropolyacids are interbonded by the intermediation of hydrogen bonds. In the presence of polar molecules such as H_2O, CH_3OH, *n*-butylamine, etc., a supramolecular secondary structure is spontaneously formed by the penetration of polar molecules into the bulk of HPA crystallites. In the case of hydrates of such acids as $H_3PMo_{12}O_{40}$ or $H_4SiW_{12}O_{40}$, up to about 30 molecules of water per one Keggin unit (KU) may be taken up. This water of crystallization may be totally or partly removed, depending on the pressure as in the case of zeolites. However, lower hydrates of fixed composition, such as $H_3PW_{12}O_{40} \cdot 6H_2O$ can be obtained. Physicochemical and catalytic properties of heteropolyacids were recently reviewed by Okuhara et al. [2].

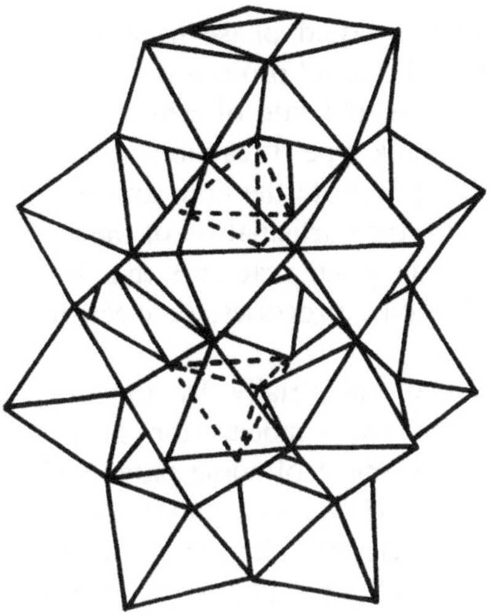

Figure 2 Structure of octadecatungstodiphosphoric acid $H_6P_2W_{18}O_{62}$ (Dawson structure).

Dodecaheteropolyacids are among the strongest mineral acids, reaching acid strengths of some solid superacids due to the fact that the negative charge of anion is smeared over 36 external oxygen atoms and attraction of protons is much easier than, e.g., in the case of sulfuric acid anion. Brønsted acidity of heteropolyacids is the reason for their high catalytic activity in acid–base-type catalytic reactions. It should be observed here that some HPAs and their salts, particularly those containing molybdenum, are strong oxidizing agents and are active in catalytic oxidation reactions.

In recent years catalytic properties of the heteropolyacids $H_6P_2W_{18}O_{62}$ and $H_6P_2Mo_{18}O_{62}$ have been investigated. Their structure, called Dawson structure, contains two heteroatoms and can be represented as obtained by fusion of two Keggin fragments generated by loss of three adjacent corner-shared $[ZO_6]$ octahedra from each of two $H_3PW_{12}O_{40}$ molecules [1]. This new structure, exhibiting D_{3h} symmetry, is represented in Figure 2.

Okuhara et al. [2] distinguished three different classes of catalytic reactions occurring with heteropoly compounds as catalysts: surface catalysis, bulk type I catalysis, and bulk type II catalysis. In the first case the catalytic reaction occurs entirely at the external surface of heteropoly compound crystallites, without any participation of the species contained in their bulk. In this case only surface-active centers participate in catalysis. In the bulk type I catalysis, polar reactants migrate into the bulk of HPA crystallites, thus forming a so-called pseudo-liquid phase in which they undergo catalytic transformation. In an ideal case all protons present in the bulk may take part in the catalytic reaction, the rate of which becomes proportional to the catalyst volume. In the bulk type II reaction at least one of the reactants, being nonpolar, does not penetrate the bulk and remains adsorbed at the surface, where it reacts with the participation of the species supplied from the bulk. These may be other polar reactants that form a pseudo-liquid phase, protons in acid–base-type processes, or electrons in redox ones.

As already said, MTBE is obtained industrially by the electrophilic addition of methanol to isobutene. The condensation of *t*-butyl alcohol with methanol recently became the object of some patents [3–5]. Fundamental studies on the application of HPAs as catalysts for both processes are reviewed in the present chapter, in which gas- and liquid-phase catalyses are described on unsupported and supported heteropoly compounds.

II. GAS-PHASE SYNTHESIS OF MTBE FROM ISOBUTENE AND METHANOL ON HPA-BASED CATALYSTS

Following a U.S. patent [6] and a paper by Igarashi et al. [7] in which the possibility of MTBE synthesis in both gas and liquid phases on heteropolyacids was announced, Ono and Baba [8] applied this process as a test reaction in their study of catalytic properties of heteropolyacids: $H_3PMo_{12}O_{40}$, $H_3PW_{12}O_{40}$, and $H_4SiW_{12}O_{40}$ supported on active carbon and their silver, copper, and aluminium salts. All catalysts have shown activity for MTBE formation in the gas phase at 353 K. In the case of $H_3PW_{12}O_{40}$, $H_4SiW_{12}O_{40}$, and copper and silver salts of $H_4SiW_{12}O_{40}$, small amounts of isooctenes were co-produced. The rest of the catalysts showed 100% selectivity of MTBE. The formation of dimethyl ether was not observed. Generally, heteropolyacids had higher activity than their salts. However, the activity of Ag and Cu phosphotungstates could be distinctly enhanced by pretreatment in hydrogen in 523 K. The authors concluded that this effect was due to the reduction of Ag^+ or Cu^{2+} ions with the simultaneous formation of Brønsted acid centers according to the reaction

$$Ag^+ + \frac{1}{2}H_2 \rightarrow Ag^0 + H^+ \tag{1}$$

No such activation could be obtained in the case of aluminum salt.

It should be observed that heteropolyacids and their salts may also be active in the synthesis of other ethers used as octane enhancers, as was

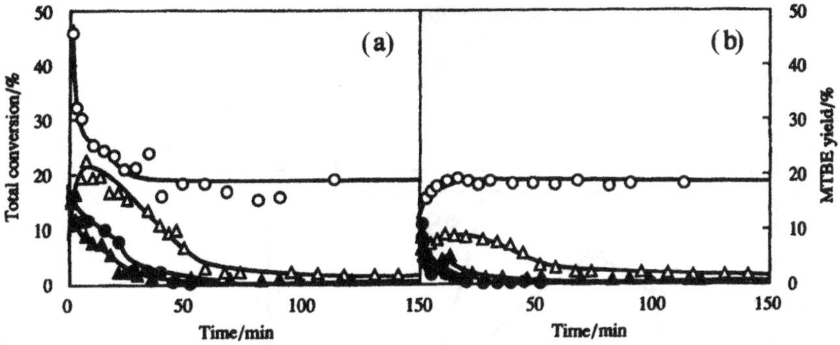

Figure 3 Time courses of MTBE synthesis from methanol and isobutene over heteropolyacids at 323 K: (a) total conversion of methanol; (b) yield of MTBE. (○) $H_6P_2W_{18}O_{62}$, (●) $H_3PW_{12}O_{40}$, (Δ) $H_4SiW_{12}O_{40}$, (▲) $H_4GeW_{12}O_{40}$. Catalyst weight 0.5 g, methanol:isobutylene:N_2 = 1:1:3 (total flow rate 90 cm^3/min, W/F = 2.3 g-h (mol of total feed)$^{-1}$. (From Ref. 13.)

pointed in [9–12], where, besides MTBE, also synthesis of TAME and ETBE was studied.

A systematic investigation of MTBE gas-phase synthesis over heteropolyacids was undertaken by Shikata et al. [13]. In their paper a study of catalytic properties of unsupported Keggin-type $H_nXW_{12}O_{40}$ (X = P, Si, Ge, B, and Co) and also Dawson-type $H_6P_2W_{18}O_{62}$ heteropolyacids were described and their behavior compared with that of other mineral catalysts such as SO_4^{2-}/ZrO_2, SiO_2-Al_2O_3, and HZSM-5. Figure 3, from Ref. 13, shows the time course of the total conversion of methanol and the yield of MTBE [(ratio of the flow rate of MTBE at the outlet to the flow rate of methanol at the inlet) × 100%] at 323 K. It is seen that in the case of Dawson-type HPA the high initial conversion of methanol decreases and after some 40 min reaches a constant level of about 20%. On the other hand, MTBE yield is initially low but subsequently increases up to nearly the same value as CH_3OH conversion, which indicates selectivity close to 100%. The initial high conversion of methanol has been ascribed to its sorption by the catalyst. Parallel to the sorption of methanol increased the yield of MTBE. At the stationary state nearly all methanol adsorbed was transformed into MTBE. The behavior of Keggin-type HPAs was more complicated. Generally their activity was definitely lower than that of $H_6P_2W_{18}O_{62}$, and in the stationary state both conversion of methanol and MTBE yield were less than 1.5%.

Figure 4 shows the temperature dependence of MTBE yield in the stationary state [13]. The dotted line introduced by the authors of the

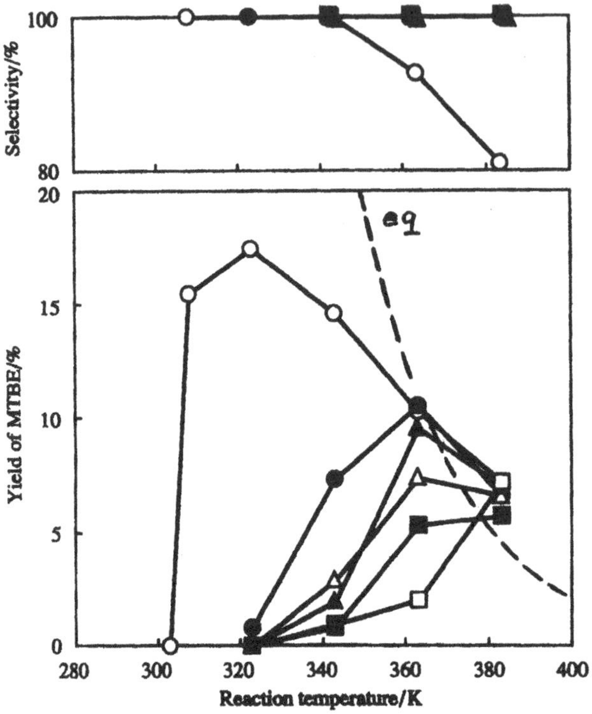

Figure 4 Dependence of MTBE yield and selectivity on reaction temperature of MTBE synthesis: (○) $H_6P_2W_{18}O_{62}$, (●) $H_3PW_{12}O_{40}$, (Δ) $H_4SiW_{12}O_{40}$, (▲) $H_4GeW_{12}O_{40}$, (□) $H_5BW_{12}O_{40}$, (■) $H_6CoW_{12}O_{40}$. Catalyst weight 0.5 g, methanol:isobutylene:N_2 = 1:1:3 (total flow rate 90 cm^3/min). Dotted line (equilibrium yield) introduced by the authors of the present review. (Adapted from Ref. 13.)

present review indicates the equilibrium yield calculated on the basis of an equation given in [14]. It is seen that the Keggin-type heteropolyacids exhibited appreciable activities only above 343 K. On the other hand, $H_6P_2W_{18}O_{62}$ was very active even at 308 K, and at 323 K it reached the highest MTBE yield, equal to 17%. At 383 K all HPA catalysts gave the equilibrium yield. At the same experimental conditions, other mineral catalysts, SiO_2-Al_2O_3, HZSM-5, and SO_4^{2-}/ZrO_2, were also weakly active at the initial stage of the reaction and gave conversion less than 0.1%.

As the authors of Ref. 13 observed, the remarkably good activity of $H_6P_2W_{18}O_{62}$ is not due to its high acid strength, which was higher ($H_0 = -3.6$) than in other investigated heteropolyacid ($H_0 = -3.4$ to -0.6), nor to the high surface area (2 m^2/g), which was lower than that of

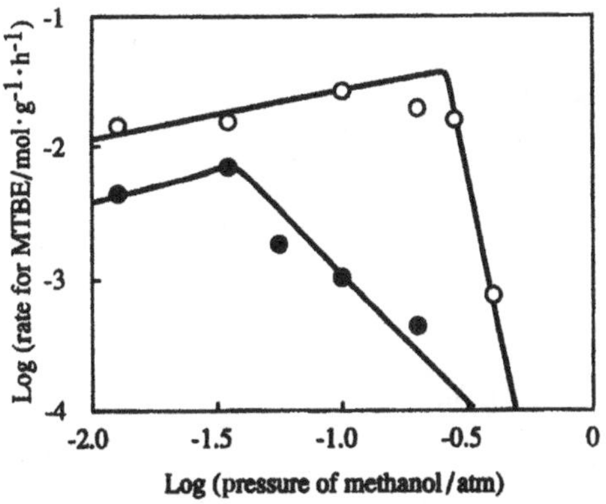

Figure 5 Dependence of rate for MTBE synthesis on methanol pressure at 323 K. (From Ref. 13.)

other HPAs ($9\,m^2/g$ in $H_3PW_{12}O_{40}$ and $H_4SiW_{12}O_{40}$). This fact became one of the arguments indicating that methanol–isobutene catalytic reaction occurs in the bulk of Dawson-type heteropolyacids in the so-called pseudo-liquid phase.

The pseudo-liquid behavior of the catalytic system became the explanation of the unique dependence of the reaction rate on the methanol partial pressure which is shown in Figure 5 [13]. It is seen that in the case of both catalysts $H_6P_2W_{18}O_{62}$ and $H_3PW_{12}O_{40}$, in the region of lower methanol pressures reaction rate does increase with p_{CH_3OH}, but above a certain limit ($\log p = -1.5$ for $H_3PW_{12}O_{40}$ and $\log p = -0.5$ for $H_6P_2W_{18}O_{62}$) it decreases; reaction order with respect to methanol switches from positive to negative values. Such behavior is very similar to that observed in the case of dehydration of ethanol on $H_3PW_{12}O_{40}$ [15], which was proved to occur in the pseudo-liquid phase in the bulk of HPA crystallites. The unique dependence on ethanol pressure was interpreted as the result of changes in the concentration of the different protonated ethanol clusters. Monomers $C_2H_5OH_2^+$ and dimers $(C_2H_5OH)_2H^+$ appearing at low ethanol vapor pressures were assumed to be chemically active species, while higher clusters prevailing at high pressure were assumed to be inert. This strongly suggests that a similar mechanism involving the formation of protonated methanol clusters $[(CH_3OH)_nH^+]$ of different size and activity is also operating in the case of MTBE formation on heteropolyacids. Volume sorption of methanol has been confirmed by the authors in a series of separate experiments. It is interesting to observe that during MTBE synthesis at 323 K, $H_3PW_{12}O_{40}$ weakly active absorbs in stationary state 4 methanol molecules per one proton, while highly active $H_6P_2W_{18}O_{62}$ absorbs only 2 CH_3OH molecules.

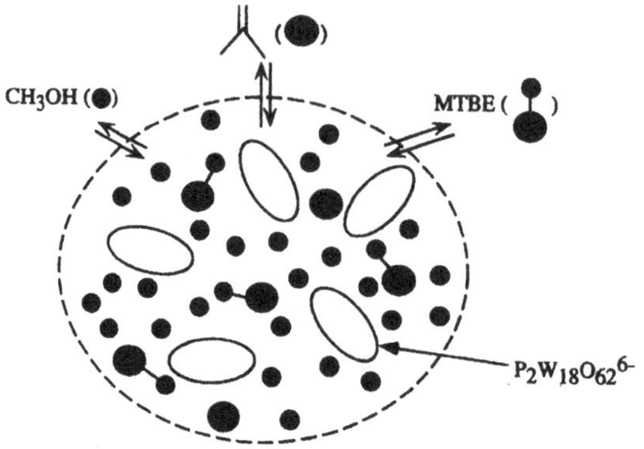

Figure 6 Model of pseudo-liquid phase for MTBE synthesis with $H_6PW_{12}O_{62}$. (From Ref. 13.)

Hence, the authors postulate the existence of two states of the catalyst: a high-activity state reached at lower methanol pressure and a low-activity state at higher pressure.

While the sorption of methanol by both catalysts is well documented, "the question may arise," as the authors write, "that isobutene is a non polar molecule, and ordinarily such a molecule is not absorbed into the bulk [13]. If one considers that isobutene is absorbed into the aqueous solutions of heteropolyacid it may be possible that isobutene is absorbed into the pseudo-liquid phase. The pseudo-liquid phase containing a significant amount of methanol may enhance the absorption due to the organophilic interaction." The model of the reaction system involving both the presence of methanol and isobutene in the pseudo-liquid phase proposed by the authors [13] is shown in Figure 6.

The behavior of $H_3PW_{12}O_{40}$ and $H_6P_2W_{18}O_{62}$ catalysts in the course of MTBE synthesis was also compared in an interesting study by Shikata and Misono [16], in which the X-ray diffraction (XRD) examination of both catalysts pretreated at 423 or 523 K was carried out before and after the catalytic reaction carried out at 323 K.

As Figure 7 shows, rapid increase of MTBE yield was observed for 0–10 min for both catalysts, which has been ascribed to the rapid absorption of reactants and the formation of MTBE in the solid bulk, as discussed previously. The stationary yield in the case of Dawson-type catalyst was high and not influenced by the pretreatment temperature. On the other hand, in the case of Keggin-type catalyst, MTBE yield after reaching a maximum dropped, the effect being dependent on the pretreatment. Higher values were obtained after the treatment at 523 K. Both catalysts, before and after catalytic reaction, independent of pretreatment, exhibited unchanged

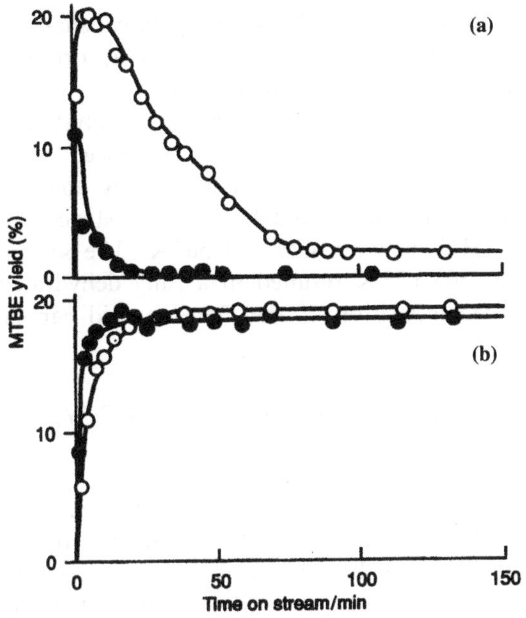

Figure 7 Effects of pretreatment temperature on MTBE synthesis catalyzed by $H_3PW_{12}O_{40}$ (a) and $H_6P_2W_{18}O_{62}$ (b) pretreated at 423 K (•), 523 K (○) in an N_2 flow. Catalyst mass 0.5 g, reaction temperature 232 K, methanol:isobutene:N_2 = 1:1:3, flow rate 90 cm^3/min. (From Ref. 16.)

IR spectra, which indicated that the primary structure, i.e., the structure of polyanion, remained unchanged during all operations. Simultaneously, XRD investigation indicated that Dawson-type catalyst exhibiting high activity was amorphous, thus showing the lack of ordered secondary structure. On the contrary, the Keggin-type catalyst $H_3PW_{12}O_{40}$, exhibiting generally much lower activity when pretreated at 423 K, was crystalline cubic, before and after catalytic reaction. The same catalyst pretreated at 523 K was amorphous, and in the catalytic reaction initially gained high activity which then slowly decreased, and after reaching quite low stationary state has been shown by XRD to be crystalline. Hence it is justified to interpret the observed decrease in activity as being due to the crystallization of the catalyst.

All these experiments led the authors to the conclusion that the essential difference between the behavior of Keggin- and Dawson-type heteropolyacids as the MTBE formation catalysts is connected with their secondary structure. "Elliptical shape of Dawson anion does not seem to favor the formation of stable crystalline structure and leads to an amorphous and flexible structure, the adsorption-desorption being easier and catalytic activity high. On the other hand the nearly spherical shape of the Keggin anion favours a crystalline cubic structure, where adsorption-desorption processes are slow and catalytic activity low" [16].

A study of physicochemical properties of $H_6P_2W_{18}O_{62}$ catalysts has been published by Baronetti et al. [17]. The acid crystallized from water contained 24 H_2O molecules per anion. TG and DTA analyses have shown gradual loss of water between 373 and 423 K and existence of a stable product, $H_6P_2W_{18}O_{62} \cdot 2H_2O$ between 423 and 563 K. The second step of sample dehydration at 563–673 K resulted in a fully dehydrated product. Hence the thermal decomposition of the Dawson acid can be represented by

$$H_6P_2W_{18}O_{62} \cdot 24H_2O \xrightarrow{373-423\,K} H_6P_2W_{18}O_{62} \cdot 2H_2O \xrightarrow{>573\,K} H_6P_2W_{18}O_{62} \quad (2)$$

The XRD spectra of the samples calcined at different temperatures show gradual changes of the X-ray pattern from that characteristic for hydrated sample preheated at 373 K to that characteristic for anhydrous acid stable up to 873 K. They reflect the changes of the secondary structure of Dawson acid, while the primary structure, the structure of anion, as a FTIR investigation has shown, remains essentially preserved at least up to 623 K.

Baronetti et al. [17] also studied the effect of calcination temperature on the catalytic properties of $H_6P_2W_{18}O_{62}$, which were tested at 383 K using two reactions in gas phase: dehydration of methanol to dimethyl ether, and MTBE synthesis from isobutene and methanol. In both cases the samples calcined at temperatures between 373 and 473 K exhibited high and constant activity. The samples calcinated at 573 K and presumably containing two H_2O molecules per anion, were distinctly less active, and those calcinated at 673 K and completely dehydrated exhibited only very weak catalytic activity. The exposure of the acid sample to an air stream saturated with water vapor for 4 hr at room temperature resulted in complete regeneration of its catalytic activity for dimethyl ether formation and confirmed the role of water molecules as the factor controlling the catalytic activity.

The properties of $H_6P_2W_{18}O_{62}$ as the catalyst for gas-phase MTBE synthesis were also described in [18].

The catalytic properties of SiO_2 supported $H_6P_2W_{18}O_{62}$ were investigated by Shikata et al. [19]. The supported Dawson-type heteropolyacid exhibited at 323 K a yield of MTBE four times and at 383 K twice higher than that in the unsupported state. Its activity was comparable to that of sulfonated resin Amberlyst-15.

Despite dispersion over the surface of the support, $H_6P_2W_{18}O_{62}$ preserved its unique dependence of methanol pressure, which has already been described. Such an effect being connected with the pseudo-liquid behavior indicates that heteropolyacid existed on the surface of SiO_2 in the form of microcrystalline aggregates.

Very interesting was the comparison of silica-supported $H_6P_2W_{18}O_{62}$ and $H_3PW_{12}O_{40}$. At 323 K, both exhibited similar activity, despite the fact that in the unsupported state the Dawson-type acid was distinctly more active than the Keggin-type one. This fact has been explained by the estimation of the catalyst's acid strength. As already said, unsupported $H_3PW_{12}O_{40}$ with Hammett constant $H_0 = -3.6$ had higher acid strength than $H_6P_2W_{18}O_{62}$ with $H_0 = -2.9$. In Ref. 19 the acid strength was characterized by calorimetry of NH_3 absorption. These measurements showed that supporting $H_3PW_{12}O_{40}$ rendered it a distinctly weaker acid, while the strength of $H_6P_2W_{18}O_{62}$ was almost unchanged when it was supported on silica. Due to this different behavior, the acid strengths of $H_6P_2W_{18}O_{62}/SiO_2$ and $H_3PW_{12}O_{40}/SiO_2$ became closer. The authors concluded that this may be responsible, at least in part, for the comparable activity of both catalysts at 323 K, since the difference in the amount of acid sites between the supported catalysts was small.

In their papers [13,16,19], Shikata et al. have shown that gas-phase synthesis of MTBE from methanol and isobutene on heteropolyacid catalysts is governed by some processes occurring in the bulk of catalysts in the pseudo-liquid phase. The participation of pseudo-liquid phase has been also assumed to be the cause of unique methanol pressure dependence of the reaction rate. In all these cases we deal with catalysts which are Brønsted acids of high acid strength comparable to that of solid superacids and the protons are expected to participate in the catalytic reaction, forming protonated intermediates which may be present both as chemisorbed at the external surface and absorbed in the bulk, thus forming pseudo-liquid state.

An alternative explanation of the unique behavior of heteropolyacids in MTBE synthesis and the changes of reaction order with respect to methanol was proposed for the case of dodecatungstosilicic acid, $H_4SiW_{12}O_{40}$, in a series of papers [20–24] aiming to elucidate the role of protons in MTBE catalytic formation. This research comprised kinetic study of MTBE formation preceded by an investigation of water, methanol, and isobutene sorption in which the formation and disintegration of hydrogen bonds was followed using IR technique.

Heteropolyacids are usually obtained in a hydrated form in which the number of water molecules may be as high as 30 per formula unit.

As already said, the hydrated dodecaheteropolyacids are supramolecular structures in which the heteropolyacid plays the role of the host and water molecules in the positions between HPA anions the role of the guests. The interaction between host and guests results in the formation of an array of hydrogen bonds between the hydrated protons, water molecules, and HPA anions. In a few cases, X-ray and neutron diffraction studies enabled determination of the positions of water molecules in the crystal lattice and, based on interatomic distances, proposed localization of hydrogen bonds. In particular, this has been done for $H_3PW_{12}O_{40} \cdot 6H_2O$ [25] and $H_3PW_{12}O_{40} \cdot 21H_2O$ [26]. In the former case it was stated that a dioxonium ion, $H_5O_2^+$, links four neighboring Keggin units by forming hydrogen bonds with terminal $W=O_d$ oxygen atoms (Fig. 1). Hexahydrates isomorphous with $H_3PW_{12}O_{40} \cdot 6H_2O$ also exist in other dodecaheteropoly-acids, such as tungstophosphoric or tungstoboric ones. Only three of four protons in the first case and three of five in the second one form dioxonium ions and, hence, their proper formulas should be given as $(H_5O_2^+)_3(HSiW_{12}O_{40}^{3-})$ and $(H_5O_2^+)_3(H_2BW_{12}O_{40}^{3-})$. Brown et al. [25] suggested that such nonhydrated protons would occupy positions between the terminal O_d atom of one Keggin unit and the bridging O_c oxygen of the other neighboring one, thus forming O_d–H^+–O_c interanionic hydrogen bonds.

It may be supposed that upon dehydration, protons bonded in dioxonium ions migrate to the positions of such "extra" nonhydrated protons, thus forming hydrogen bonds between Keggin anions. In fact, a sample of $H_4SiW_{12}O_{40} \cdot 15.6H_2O$ obtained by the evaporation of a few drops of water solution of HPA on a silicon plate exhibited a broad band at $3550\,cm^{-1}$ within the spectral region characteristic for hydrogen bonds, which has been ascribed to hydrogen bonding between O_d oxygen atoms of the Keggin unit and dioxonium ion, $H_5O_2^+$. The presence of the latter was also manifested by the 1700 and $1100\,cm^{-1}$ bands [22]. All these bands vanish in the case of anhydrous $H_4SiW_{12}O_{40}$. Simultaneously, however, a new bond appears at $3106\,cm^{-1}$, ascribed to hydrogen bonding between neighboring HPA anions, O_d–H^+–O_c. During the dehydration the IR spectrum typical of $(SiW_{12}O_{40})^{4-}$ anions ($\nu_{W=O_d} = 980\,cm^{-1}$, $\nu_{Si-O_a} = 925\,cm^{-1}$, $\nu_{W-O_b-W} = 980\,cm^{-1}$, and $\nu_{W-O_c-W} = 770\,cm^{-1}$) remains essentially unchanged with the exception of the $\nu_{W=O}$ band, which becomes split into a doublet at 987 and $1010\,cm^{-1}$, reflecting the change of the hydrogen bonds in which O_d oxygen atoms are involved. Similar effects were observed in Ref. 23.

Study of the kinetics of methanol–isobutene synthesis of MTBE on $H_4SiW_{12}O_{40}$ in gas phase was preceded by study of $H_4SiW_{12}O_4$–methanol and $H_4SiW_{12}O_4$–isobutene systems, i.e., study of the interaction

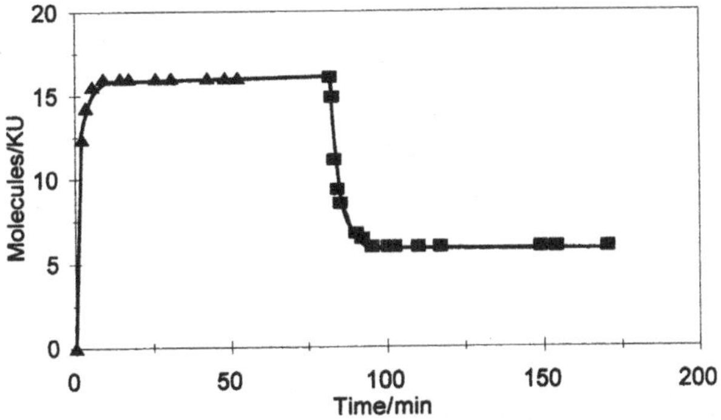

Figure 8 Typical time course of sorption and desorption of methanol from the gas phase. (From Ref. 24.)

of reaction substrate with dehydrated catalyst. Dehydrated catalyst was also used in the subsequent catalytic experiments [20].

Figure 8 shows a kinetic curve representing sorption of methanol on $H_4SiW_{12}O_4$ observed in a sorption balance at 292 K [24]. Already after 2 min about 12 CH_3OH molecules per HPA anion were taken and the final composition, close to $H_4SiW_{12}O_{40} \cdot 16CH_3OH$, was reached after 9 min. At room temperature this sorption was only partly reversible, and after evacuation, hexamethanolate, $H_4SiW_{12}O_{40} \cdot 6CH_3OH$, was obtained. The analysis of sorption and desorption isobars obtained in [24] indicated the tendency to formation of stable secondary structures in which the number of alcohol molecules per anion is an integral multiple of the number of protons, such as $H_4SiW_{12}O_{40} \cdot 4CH_3OH$, $H_4SiW_{12}O_{40} \cdot 12CH_3OH$, and $H_4SiW_{12}O_{40} \cdot 16CH_3OH$. Such phases correspond to the formation of methanol protonated clusters in the HPA crystallites bulk: $CH_3OH_2^+$, $(CH_3OH)_2H^+$, $(CH_3OH)_3H^+$, and $(CH_3OH)_4H^+$. The tendency of absorbed polar molecules in HPAs to be integral multiples of the number of protons was described, e.g., in Ref. 27.

FTIR spectra of methanol adsorbed in $H_4SiW_{12}O_{40}$ exhibit many analogies to the spectra of water described in Ref. 22. The broad band at $\sim 3100\,cm^{-1}$ of the hydrogen bond linking Keggin units in anhydrous HPA diminishes upon methanol sorption while simultaneously a new band at $3420\,cm^{-1}$ is growing, the frequency of which is similar to that of the $3445\,cm^{-1}$ band ascribed to the hydrogen bond between the Keggin unit and protonated water clusters as described above. The presence of protonated methanol species $CH_3OH_2^+$ is also confirmed by the $1550\,cm^{-1}$ band observed when methanol was sorbed on HZSM-5 zeolite [28] and on $H_3PW_{12}O_{40}$ [29]. In both cases this

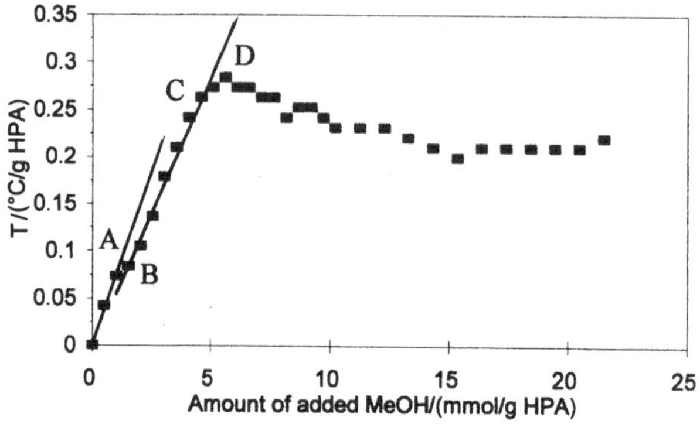

Figure 9 Thermometric titration curve of $H_4SiW_{12}O_{40}$ with methanol. (From Ref. 24.)

band was assigned to the asymmetric C—O—H deformation vibration in $CH_3OH_2^+$.

All experiments with sorption of water and methanol on dehydrated heteropolyacids indicate that the penetration of these polar molecules into the bulk of HPAs is connected with the transfer of loosely bonded protons from the positions in hydrogen bonds between Keggin units to the protonated clusters of H_2O or CH_3OH molecules. In the case of methanol sorption it was also possible to gain some information concerning the heat of such processes using thermometric titration of the dehydrated HPA suspension in toluene [30,24] with the solution of methanol in the same solvent. At sufficiently fast mixing, the sorption of methanol from solution was fast and temperature after a rapid increase reached a constant level within 2 min. The titration results were represented by curves showing the sum of temperature increments observed after addition of successive portions of methanol plotted against the volume of added solution. An example of such curve is shown in Figure 9 [24]. This figure shows two approximately linear sections, 0A and BC. Point A corresponds to the neutralization of one proton by one methanol molecule, i.e., to the formation of protonated monomer (0.99 ± 0.05 CH_3OH per H^+). Knowing the heat capacity of the calorimeter, one can calculate from the slopes of linear sections of the enthalpies of formation of the protonated monomer and also that of the formation of protonated dimer from protonated monomer. The values thus obtained were $\Delta H_{1(l)} = -20.0$ kJ/mol H^+ and $\Delta H_{2(l)} = -15.6$ kJ/mol H^+. These data can be compared with the results of the $H_4SiW_{12}O_{40}$ titration with toluene solution of *n*-butylamine: $\Delta H_{1(l)} = -102$ kJ/mol H^+ and $\Delta H_{2(l)} = -71.9$ kJ/mol H^+ [31].

Using a thermochemical cycle, the enthalpies of the formation of protonated methanol species from gaseous CH_3OH and anhydrous HPA were calculated, i.e., the enthalpies of the reactions

$$CH_3OH_{(g)} + H^+_{(s)} = CH_3OH^+_{2(s)} \qquad \Delta H_{1(g)} = -58.9\,\text{kJ/mol} \quad (3)$$

$$CH_3OH^+_{2(s)} + CH_3OH_{(g)} = (CH_3OH)_2H^+_{(s)} \qquad \Delta H_{2(g)} = -53.5\,\text{kJ/mol} \quad (4)$$

Such relatively high values indicate that proton in protonated methanol clusters is much more strongly bonded than in the anhydrous acid, where it is involved in the interanionic hydrogen bond.

The second substrate for MTBE formation was isobutene. Using sorption balance, it has been stated [21] that at room temperature and 8.53 kPa the amounts of isobutene adsorbed corresponded to 0.15 of the monolayer, and upon evacuation the coverage decreased to 0.07 of monolayer, thus indicating that no volume sorption occurred. At 353 K the coverage was still smaller and at the same partial pressure reached only 0.12 of monolayer. However, it has been recently suggested [13] that sorption of isobutene on $H_6P_2W_{18}O_{62}$ may be enhanced by presorption of methanol. In our case an experiment has been carried out in which anhydrous $H_4SiW_{12}O_{40}$ at room temperature was saturated with methanol up to the composition $H_4SiW_{12}O_{40} \cdot 12.3CH_3OH$. Subsequent evacuation resulted only in partial desorption and the solid of the composition $H_4SiW_{12}O_{40} \cdot 6.3CH_3OH$ remained on which isobutene was adsorbed. The uptake of isobutene corresponded to 0.17 monolayer and was within the limits of experimental error, the same as in experiments without methanol presorption.

Figure 10 shows the course of a typical catalytic run on dehydrated $H_4SiW_{12}O_{40}$ carried out at 313 K and isobutene/methanol molar ratio $R = 0.73$ in the gas phase. The initial increase of the reaction rate was not observed in the case of $R = 0.09$, i.e., at a high excess of methanol over isobutene. It was also not observed at 353 K.

Mass balance calculated from particular chromatographic analysis indicated a certain deficit of both methanol and isobutene. As Figure 10c shows, the deficit of methanol calculated for one portion of gas taken for chromatographic analysis (0.25 cm^3) was highest at the beginning of the run and then decreased. It was always higher in the case of methanol than in that of isobutene, which is evidently due to the fact that polar molecules of methanol penetrate the bulk of HPA crystallites while the nonpolar isobutene may only be adsorbed on the external surface. Oligomerization of isobutene results in the formation of coke, the presence of which is

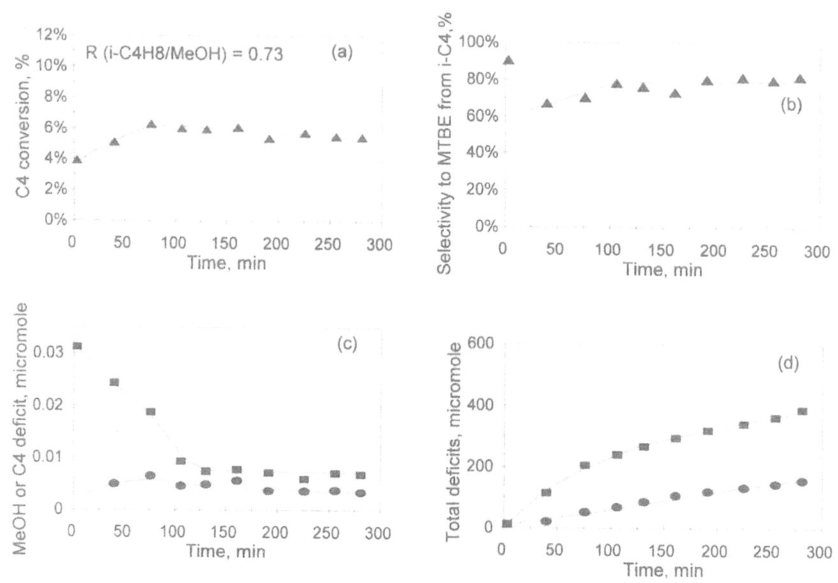

Figure 10 Typical examples of catalytic run at 40°C: (a) isobutene conversion; (b) selectivity; (c) deficit of methanol (■) and isobutene (●); (d) accumulated deficit of methanol (■) and isobutene (●) (run 44, $R = 0.73$).

signaled by darkening of the catalyst. The deficit of both methanol and isobutene was also partly created by volume sorption of MTBE. Such sorption, slower than that of methanol, was observed in separate experiments. By the integration of deficits determined from particular portions of gas taken for analysis one could obtain the total deficit (accumulated deficit) as a function of time. It should be observed here that Figure 10d shows that sorption of methanol at the conditions of catalytic reaction is distinctly slower than was observed during sorption experiments already described.

Besides MTBE, only small amounts of isobutene dimer were detected as the reaction products. It could be ascertained using mass spectrometry that the amount of another side product which might be expected, dimethyl ether, could not exceed 0.4%.

The results of two series of experiments aiming to determine reaction order with respect to methanol at 313 K at the initial and steady states are shown in Figure 11, where in double logarithmic scale the dependence of reaction rate on the partial pressure of methanol is presented [21]. From the slope of such linear plots, reaction orders with respect to methanol and

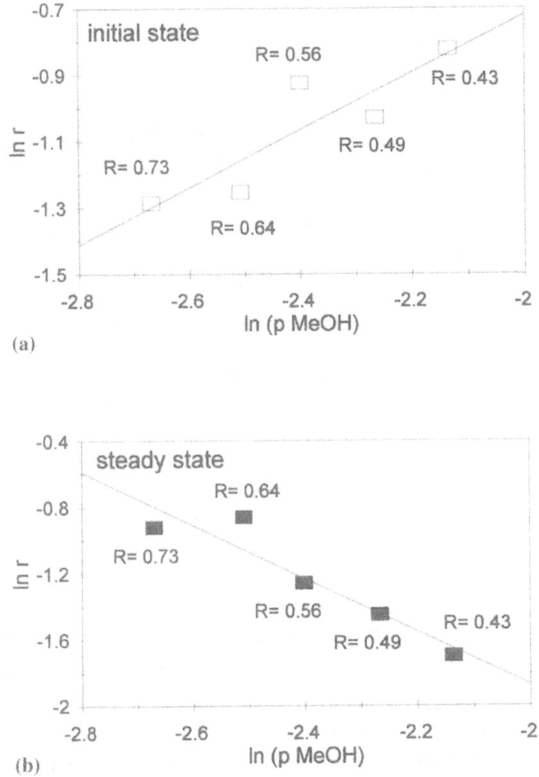

Figure 11 Logarithmic plots of rate of isobutene conversion at 40°C versus methanol pressure: (a) initial period of the run; (b) steady state of the reaction. (From Ref. 21.)

isobutene at 313 and 353 K were calculated; the data obtained are collected in Table 1 [21]. At all investigated conditions, reaction order with respect to isobutene, at initial as well as steady state, was positive. However, at 313 K, positive reaction order 0.86 with respect to methanol observed at the beginning of the reaction turned to negative − 1.61 at the steady state, thus showing not typical "unique behavior" observed earlier by Shikata et al. [13].

In looking for a reasonable explanation of the observed facts, it is necessary to propose a model of the catalytic system in which, depending on the conditions, reaction order with respect to one of the substrates, methanol, can change from positive to negative values. As already said, Shikata et al. [13] assumed that nonpolar isobutene can penetrate the bulk of $H_6P_2W_{18}O_{62}$ crystallites saturated with methanol and the catalytic reaction occurs in the pseudo-liquid phase. According to these authors, the

Table 1 Reaction Orders

Temperature		Reaction order: Value	Reaction order: Correlation coeff.
313 K	With respect to CH_3OH:		
	Initial	0.86 ± 0.26	0.7831
	Steady state	-1.61 ± 0.33	0.8856
	With respect to C_4H_8:		
	Initial	1.38 ± 0.19	0.9461
	Steady state	1.62 ± 0.28	0.9136
353 K	With respect to CH_3OH:		
	Initial	0.44 ± 0.06	0.9043
	Steady state	0.40 ± 0.06	0.9020
	With respect to C_4H_8:		
	Initial	0.97 ± 0.05	0.9853
	Steady state	0.94 ± 0.14	0.9027

Source: Ref. 21.

decrease of the reaction rate at high methanol pressure is due to the formation of inactive protonated methanol oligomers $(CH_3OH)_nH^+$ ($n > 3$). However, the experiments described above show that this model cannot be applied in our case, because no penetration of isobutene at the reaction conditions into the bulk of the $H_4SiW_{12}O_{40}$ catalyst was confirmed and no enhancement of isobutene sorption by presorption of methanol was stated. Hence, in the alternative model the catalytic reaction of "bulk type II," according to the classification of Misono, should be assumed, in which isobutene adsorbed at the surface reacts with methanol, penetrating the solid and forming pseudo-liquid phase. It should be recalled here that MTBE formation is classified in organic chemistry as an electrophilic addition, and it is generally accepted that a carbocation forming, e.g., on a Brønsted acid site is involved in the rate-determining step. In our case this will be a *t*-butyl carbenium ion forming at the surface with the participation of protons supplied from the bulk.

As already said, protons localized in the hydrogen bond between neighboring Keggin anions are more loosely bonded than those in protonated methanol monomers or larger protonated clusters. This is evidenced by relatively high negative enthalpy of formation of methoxonium ion ($\Delta H_1 = -58.9$ kJ/mol, $\Delta H_2 = -53.5$ kJ/mol. Hence it has been assumed in our model that there are loosely bonded protons ("free" protons) from $O—H^+—O$ hydrogen bonds which participate in the formation of carbenium ions at the surface.

As was shown in Figure 8, sorption of methanol by dehydrated $H_4SiW_{12}O_{40}$ is very rapid, and equilibrium, even at room temperature, is reached in a few minutes. However, under the conditions of catalytic reaction, a certain deficit of methanol in the mass balance indicated that volume sorption of methanol occurred more slowly. All these facts justify the assumption that at the reaction's steady state an equilibrium exists between methanol in gas phase and methanol in the next-to-surface layer of the cataylst and there is also fast equilibrium between free protons and methanol molecules in the surface layer. On the other hand, there is no equilibrium between methanol in the next-to-surface layer and in the bulk. A certain methanol concentration gradient enables slow penetration of methanol molecules from the surface layer to the bulk.

The proposed reaction scheme must comprise—at least—the following steps:

$$C_4H_{8(g)} + H^+_{(s)} \rightarrow C_4H^+_{9(\sigma)} \tag{5}$$

$$CH_3OH_{(g)} \rightarrow CH_3OH_{(s)} \tag{6}$$

$$C_4H^+_{9(\sigma)} + CH_3OH_{(s)} \rightarrow C_4H_9OCH_{3(g)} + H^+_{(s)} \tag{7}$$

$$nCH_3OH_{(s)} + H^+_{(s)} \rightleftarrows (CH_3OH)_nH^+_{(s)} \tag{8}$$

where g and s indicate molecules in the gas or solid phase, and σ a molecule adsorbed at the surface.

Equation (8) comprises in fact a series of equations with different integral values of n. In the following discussion the overall equilibrium formula will be used:

$$K_4 = \frac{[(CH_3OH)_nH^+_{(s)}]}{[CH_3OH_{(s)}]^n[H^+_{(s)}]} \tag{9}$$

where $[H^+_{(s)}]$ is the concentration of free protons in the solid, assuming that an integral value of n approximates the state in which a protonated cluster with n methanol molecules predominates and a fractional value of n indicates the predominance of two clusters differing in n by unity and present in comparable concentrations.

Let us now assume that reaction (7) is the rate-determining step. The reaction rate will be then expressed by the equation

$$r = k_3[C_4H^+_{9(\sigma)}][CH_3OH_{(s)}] \tag{10}$$

Assuming virtual equilibria of reactions (5) and (6), we obtain

$$K_1 = \frac{[C_4H^+_{9(\sigma)}]}{p_{C_4H_8}[H^+_{(s)}]} \tag{11}$$

$$K_2 = \frac{[CH_3OH_{(s)}]}{p_{CH_3OH}} \tag{12}$$

where $p_{C_4H_8}$ and p_{CH_3OH} are partial pressures of isobutene and methanol in the gas phase and then, from (10), (11), and (12),

$$r = k_3K_1K_2[H^+_{(s)}]p_{C_4H_8}p_{CH_3OH} \tag{13}$$

At the initial state of the reaction, when the concentration of methanol in the solid is small in comparison to the concentration of free protons, the latter can be assumed as constant and equal to the initial one $[H^+_{(s)}]_o$. Then reaction (13) turns into

$$r = \alpha p_{C_4H_8}p_{CH_3OH} \qquad \text{where} \qquad \alpha = k_3K_1K_2[H^+_{(s)}]_o \tag{14}$$

In fact, at 313 K the initial reaction order 0.86 with respect to methanol is near the expected value. On the other hand, the reaction order with respect to isobutene, 1.29, indicates deviation from the proposed model, suggesting parallel occurrence of a bimolecular reaction. This might be, e.g., the formation of isobutene dimer 2,5-dimethyl-3-hexene, which on further reaction with methanol would produce one molecule of MTBE and regenerate one molecule of isobutene.

At the steady state of catalytic reaction, the concentration of free protons at high enough partial pressure of methanol must markedly decrease owing to reaction (8). Its concentration evaluated from Eqs. (9) and (12), expressing the equilibrium of reaction (6), is given by

$$[H^+_{(s)}] = \frac{1}{K_4K_2^n}[(CH_3OH)_nH^+_{(s)}]p^{-n}_{CH_3OH} \tag{15}$$

and, hence,

$$r = \frac{k_3K_1}{K_2^{n-1}K_4}p_{C_4H_8}p^{1-n}_{CH_3OH}[(CH_3OH)_nH^+_{(s)}] \tag{16}$$

The concentration of the predominant cluster at low values of $[H^+_{(s)}]$ may be approximated by the equation $[(CH_3OH)_nH^+_{(s)}] \approx c_{CH_3OH}$, where c_{CH_3OH} is the total concentration of methanol in the next-to-surface layer (both concentrations being expressed as molecules of CH_3OH/KU). Then we transform Eq. (15) into

$$r = \frac{k_3 K_1}{K_2^{n-1} K_4} p_{C_4H_8} p_{CH_3OH}^{1-n} c_{CH_3OH} \tag{17}$$

At 313 K the dependence of c_{CH_3OH} on p_{CH_3OH} has been experimentally determined within methanol partial pressure 8.53 and 14.9 kPa (overlapping with the p_{CH_3OH} range in which determination of reaction order with respect to methanol was carried out). The sorption isotherm within this partial pressure limits has been expressed as

$$c_{CH_3OH}(CH_3OH\ \text{molecules/KU}) = 0.109 p_{CH_3OH}^{1.76} \tag{18}$$

Inserting this into Eq. (17) gives

$$r = \frac{0.109 k_3 K_1}{K_2^{n-1} K_4} p_{C_4H_8} p_{CH_3OH}^{1-n} p_{CH_3OH}^{1.76} = \beta p_{C_4H_8} p_{CH_3OH}^{2.76-n} \tag{19}$$

Hence at $n > 2.76$ a negative order of reaction with respect to methanol should be obtained. In fact, at 313 K it was -1.61, corresponding to $n \approx 4$, suggesting that at the steady state in the next-to-surface layer of the catalyst equilibrated with methanol in the gas phase during the catalytic reaction, a tetrameric species is the predominant cluster.

It should be observed here that assuming reaction (5), i.e., the formation of carbenium ion, as the rate-determining step led to an kinetic equation of analogous form to Eq. (19). However, it differed by the values of constants. The value n thus obtaining was somewhat lower ($n \approx 3$).

At 353 K there was not any change of reaction order as determined at the initial stage and at the steady state of the catalytic reaction. The equilibrium concentration of methanol is much lower than at 313 K. It corresponds to 0.33 CH_3OH/H^+ at 8.53 kPa and 0.46 CH_3OH/H^+ at 9.31 kPa. Hence, the same approximations can be introduced as those leading to Eq. (13), according to which both reaction orders should be equal to 1. In fact, reaction order with respect to isobutene is close to 1 (Table 1), but the order with respect to methanol is decreased to about 0.5, suggesting that at this temperature the reaction is diffusion-controlled. The case of reaction $A + B \rightarrow C$ on a porous catalyst in which only one reagent is diffusionally controlled and both exhibit reaction order 1 at the freely accessible catalyst surface has been discussed by Satterfield [32]. He has shown that on a porous catalyst reaction order of the component diffusionally limited remains unity, while the reaction order of the component diffusionally not limited decreases to 0.5. Hence, we can conclude that in the case of our reaction occurring on a porous catalyst at 353 K there are diffusion limitations of isobutene while there are not for methanol. This is in agreement with the fact that the diffusion coefficient (molecular or Knudsen) of isobutene must be lower than that of methanol, the molecular mass of which is almost twice lower. The conclusion that at 353 K reaction is diffusion-controlled is strongly supported by the estimation of the apparent activation energy, which was as low as about 25 kJ/mol.

The above study of MTBE formation over solid $H_4SiW_{12}O_{40}$ confirms the essential results of the earlier publications by Shikata and Misono [16]: the role of pseudo-liquid phase which, depending on the methanol concentration may be present in a catalytically active or catalytically non (or rather less) active state. The latter is the cause of the "unique dependence" of reaction rate on the methanol pressure. However, this study also stresses the role of protons, indicating them as the real catalytically active centers.

Most of the papers hitherto mentioned in this chapter dealt with research carried out with unsupported heteropolyacids as catalysts. Only in Refs. 8 and 19 was the use of carbon and silica as the supports mentioned. Until now no systematic studies on the selection of supports for MTBE catalytic synthesis on heteropolyacids have been published. However, some papers have described the behavior of HPA catalysts supported on typical supports. In particular, Baba et al. [33] described synthesis of a very active catalyst for gas-phase MTBE formation. It contained $H_4SiW_{12}O_{40}$ supported on Amberlyst-15 resin. At 323 K isobutene conversion on this catalyst was 42.0% and selectivity was 99.7%. At the same conditions, conversion on Amberlyst-15 was 11.0%; and on carbon-supported $H_4SiW_{12}O_{40}$ it was only 2.1%. A similar, although weaker, synergetic effect was observed if other heteropolyacids—$H_4SiW_{12}O_{40}$ and $H_3PMo_{12}O_{40}$—were used for the preparation of Amberlyst-15-supported catalysts. Looking for the explanation of this effect, the authors observed that it seems of the utmost importance for protons and heteropoly anions to exist in close proximity at the catalyst's surface. Protons originating from heteropolyacids naturally satisfy this condition. In the HPA/ion-exchanger system, protons originating from the ion exchanger can also interact with supported heteropoly anions, and they exhibit higher activity than those of resin not supporting HPA anions.

High catalytic activity and 100% selectivity to MTBE of carbon-supported $H_4SiW_{12}O_{40}$ and $H_3PW_{12}O_{40}$ catalysts was reported in Ref. 34. The authors claimed that, compared to the conventional commercial catalysts Amberlyst-15 resin and HZSM-5, the catalysts obtained by them proved to have much higher catalytic activity under lower temperature. However, no data supporting this conclusion were presented in their paper.

Active catalysts for ether formation (including MTBE) have been the object of several patents [35–37].

The behavior of polyaniline-supported $H_4SiW_{12}O_{40}$ in gas-phase MTBE synthesis was described in Ref. 38.

III. LIQUID-PHASE SYNTHESIS OF MTBE FROM ISOBUTENE AND METHANOL ON HPA-BASED CATALYSTS

Industrially, MTBE is produced by methanol-to-isobutene addition in the liquid phase [39]. This is evidently connected with the fact that much higher MTBE yield can be obtained in liquid, due to higher volume concentration of reagents and also to the increased activity coefficient of methanol in the liquid methanol–hydrocarbon mixture [40]. Liquid-phase synthesis of MTBE on heteropolyacids was the object of patents [6,41,42], but only few papers dealing with this system have been published.

Maksimov and Kozhevnikov [43] studied the influence of the composition and structure of heteropolyacids by measuring isobutene absorption in methanol at 315 K in a static system. The highest activity was obtained with Dawson-type heteropolyacid. The orders of activities,

$$H_6P_2W_{18}O_{62} > H_3PW_{12}O_{40} > H_4SiW_{12}O_{40} > H_5PW_{10}V_2O_{40} > H_7PW_8V_4O_{40} \tag{20}$$

$$H_6P_2Mo_{18}O_{62} > H_3PMo_{12}O_{40} > H_4PMo_{11}VO_{40} > H_8CeMo_{12}O_{42} \tag{21}$$

corresponded to the order of acid strength. No information concerning the selectivity was given in the paper.

Molnar et al. [44] described an experiment carried out in an autoclave at 358 K. The initial pressure of isobutene was 50 bar and yields of MTBE on $H_3PW_{12}O_{40}/SiO_2$ and $H_4SiW_{12}O_{40}/SiO_2$ equal to 57% and 70%, respectively, were obtain after 5 hr. High selectivities, equal to 90% and 95%, were simultaneously observed. On the other hand, in two experiments carried out with unsupported heteropoly salts, $Cs_{2.5}H_{0.5}PW_{12}O_{40}$ and $Cs_{3.5}H_{0.5}SiW_{12}O_{40}$, MTBE yield as high as 88% and 76% was accompanied by lower selectivity of 50% and 52%.

IV. MTBE SYNTHESIS FROM METHANOL AND ISOBUTANOL

It is anticipated by many authors that the supply of isobutene for MTBE production will not be sufficient to meet future demand. Among the proposed alternative solutions, synthesis of MTBE by etherification of a methanol–isobutanol mixture was studied on sulfated organic resins [45] as well as on mineral catalysts such as, e.g., zeolites, ZrO_2/SiO_2, H-mordenite, and SiO_2-Al_2O_3 [46]. In several papers this reaction was also studied on heteropolyacid-containing catalysts. Kim et al. [47] investigated the reaction

$$CH_3OH + i\text{-}C_4H_9OH \rightarrow C_4H_9\text{—}O\text{—}CH_3 + H_2O \qquad (22)$$

in gas phase at 353 K on $H_3PW_{12}O_{40}$ and a series of its salts with cations of Groups I and II, Mn^{II}, Fe^{II}, Co^{II}, Ni^{II}, and also Bi^{3+}. The main problem discussed in this paper was the effect of the catalyst's acidity on its catalytic properties. The acidity was characterized by TPD measurements of the sorption of pyridine and ammonia. No clear correlation was detected between the sorption results obtained by these two methods. Based on parallel studies of IR spectra of pyridine and ammonia, it was concluded by the authors that pyridine is mostly absorbed on acid sites in pseudo-liquid state. It is easily removed by evacuation at 353 K. On the other hand, ammonia was strongly adsorbed not only on the bulk acid sites but also coordinatively bonded on metal cations, and it remained on the adsorption sites even after evacuation at 473 K.

The correlation between the acidity expressed as the number of the pyridines per KU remaining after evacuation and *t*-butyl alcohol conversion indicated that a Brønsted acidity is necessary for etherification.

The highest *t*-butyl alcohol conversion (82%), with selectivity of 60.1%, exhibited pure $H_3PW_{12}O_{40}$. High conversion was also exhibited by $Pb_{1.5}PW_{12}O_{40}$ (90.5%), $Fe_{1.5}PW_{12}O_{40}$ (89.7%), $Mn_{1.5}PW_{12}O_{40}$ (95%), and $BiPW_{12}O_{40}$, but the selectivities reached by them were much lower and remained in the range 29–52%.

Etherification of methanol and *t*-butyl alcohol in gas phase at 373 K was also studied by Nowińska and Sopa [48] using samples containing

10–60 wt% of $H_3PW_{12}O_{40}$ and $H_6P_2W_{18}O_{62}$ supported on silica and γ-alumina. After impregnation of silica with small amounts of $H_3PW_{12}O_{40}$ (9 wt%), low conversion of *t*-butanol was observed. When HPA loading was increased to 23 wt% the conversion of *t*-butanol rose significantly, and for 44 wt% loading reached 85 mol%, close to that observed for unsupported HPA (87 mol%). The selectivity to MTBE was almost independent of HPA content (mostly 61–63%). Considering the fact that silica-supported samples exhibited surface areas of 90–150 m^2/g and the surface area of unsupported $H_3PW_{12}O_{40}$ is only about 6 m^2/g, the high conversion of *t*-butanol over unsupported $H_3PW_{12}O_{40}$ indicates that the etherification reaction proceeds not only on the external surface but also in the bulk of heteropolyacid crystallites.

An interesting result was obtained when TOF (turnover frequency number, calculated per one proton present in the catalyst's active mass) was plotted against the loading in the silica-supported catalysts. In the case of $H_6P_2W_{18}O_{62}/SiO_2$ and $H_3PW_{12}O_{40}/SiO_2$ the TOF increased steeply until approximately monolayer coverage was reached and than slowly decreased to about four times lower value in the former case and twice lower in the case of unsupported heteropolyacid. In all measurements, Dawson's-type HPA activity was nearly twice higher than that of Keggin-type HPA. On the other hand, γ-alumina was a rather poor support, and at low coverage with $H_3PW_{12}O_{40}$ conversions were nearly zero (below 20 wt% HPA) and than slowly increased, reaching at about 50 wt% loading an activity slightly higher than that for unsupported HPA.

Two papers have dealt with methanol–*t*-butanol etherification in the liquid phase. Yadav and Krithivasan [49] studied a series of mineral catalysts including Al^{3+}, ZrO_2, and Cr^{3+} supported on exchanged clay, HZSM-5, sulfated zirconia, montmorillonite, and $H_3PW_{12}O_{40}$ on different supports: silica, carbon, and K-10 Bavarian montmorillonite clay. In catalytic reaction carried out in an autoclave at 358 K, the best catalyst contained 20 wt% $H_3PW_{12}O_{40}$ on K-10 clay. It gave 71% conversion of *t*-butanol and 99% selectivity. The other highly active catalysts, Amberlyst-15 (conversion 74%) and HZSM-5 (77%), were less selective (86% and 52%, respectively).

Knifton and Edwards [50] carried out catalytic experiments at 373–453 K using a continuous plug-flow reactor at 20 bar back pressure. $H_3PMo_{12}O_{40}/TiO_2$, $H_3PW_{12}O_{40}/TiO_2$, $H_3PW_{12}O_{40}/SiO_2$, and $H_3PW_{12}O_{40}/Al_2O_3$ in the form of extrudates were used as the catalysts. No pronounced differences between supported HPA catalysts may be concluded basing on the data published in the paper. Depending on temperature conversion of *t*-butyl alcohol, up to 70–80% could be reached at selectivities of about 75–77%. Conversions of about 80% were also obtained when mineral acid-treated montmorillonite clays were used as the catalyst. It should be observed here that at high levels of *t*-butyl alcohol conversion (> 80%) and operating temperatures ≥ 433 K, phase separation of the desired MTBE plus isobutene products from aqueous methanol was observed.

Generally, the selectivity with respect to MTBE observed in the case of etherification of alcohols is definitely lower than that observed in the case of methanol-to-isobutene addition. This is due to the side reactions of alcohols occurring on acid catalysts leading to such products as dimethyl ether, di-*t*-butyl ether, di-isobutyl ether, dehydration of *t*-butyl alcohol to isobutene and subsequent isobutene isomerization, etc. [46].

V. CONCLUSIONS

Heteropolyacids and their salts exhibiting high Brønsted acidity are active and in many cases selective catalysts for the synthesis of MTBE in gas and liquid phase. Physical chemistry of MTBE catalytic synthesis from isobutene and methanol in the gas phase has been studied by many authors, who have shown that yield as high as 50 wt%, comparable to the equilibrium yield (the reaction is exothermic and the yield increases with decreasing temperature), may be obtained at temperatures as low as 313–333 K with good selectivity.

The course of liquid-phase synthesis of MTBE has been described by only a few authors. However, it is mentioned in patent literature.

Catalytic synthesis of MTBE by dehydration of a methanol–*t*-butanol mixture in both gas and liquid phases can also by carried out on heteropoly compounds. Generally, however, the selectivity is lower than in the case of methanol addition to isobutene.

REFERENCES

1. M. T. Pope. Heteropoly and Isopoly Oxometalates. Springer-Verlag, Berlin: 1983.
2. T. Okuhara, N. Mizuno, M. Misono. Adv Catal, 41: 113–252, 1996.
3. J. F. Knifton. U.S. Patent 4,827,048 to Texaco Chemical Company, 1989.
4. J. F. Knifton. U.S. Patent 5,157,161 to Texaco Chemical Company, 1992.
5. J. F. Knifton. U.S. Patent 5,099,072 to Texaco Chemical Company, 1992.
6. A. T. Guttmann, R. K. Grasselli. U.S. Patent 4,259,533, 1981.
7. A. Igarashi, T. Matsuda, Y. Ogino. Sekiyu Gakkaishi 22: 331–335, 1979, Citation according to CA 92:163517n, 1980.

8. Y. Ono, T. Baba. The catalytic behaviour of metal salts of heteropolyacids in the vapour-phase of methyl-tert-butyl ether. Proceedings of the 8th International Congress on Catalysis, Berlin, 1984, p. 405.
9. J.-H. Park, Y.-W. Yi. Kongop Hwahak 8:582–588, 1997. Citation according to CA 127:250302y, 1997.
10. K. Inoue. Japanese Patent 5,163,188 to Mitsui Toatsu Chem Inc., 1993.
11. Y. Mimura. 7,267,888 Japanese Patent to Cosmo Oil Co Ltd., 1995.
12. R. J. Taylor Jr., P.-S. E. Dai, J. F. Knifton, B. R. Martin. U.S. Patent US 5,637,778 to Texaco Chemical Company, 1997.
13. S. Shikata, T. Okuhara, M. Misono. J Mol Catal A: Chem, 100: 49–59, 1995.
14. J. Tejero, F. Cunill, J. F. Izquierdo. Ind Eng Chem Res, 27: 338–343, 1988.
15. K. Y. Lee, T. Arai, S. Nakata, S. Asaoka, T. Okuhara, M. Misono. J Am Chem Soc, 114: 2836–2842, 1992.
16. S. Shikata, M. Misono. Chem Commun, 1293–1294, 1998.
17. G. Baronetti, L. Briand, V. Sedran, H. Thomas. Appl Catal A: Gen, 172: 265–272, 1998.
18. S. Shikata, T. Okuhara, M. Misono, Sekiyu Gakkaishi 37: 632–635, 1994. Citation according to CA 121: 283238m, 1994.
19. S. Shikata, S. Nakata, T. Okuhara, M. Misono. J Catal, 166: 263–271, 1997.
20. A. Bielański, A. Malecka-Lubańska, A. Micek-Ilnicka, J. Poźniczek. Topics in Catalysis 11/12: 43–53, 2000.
21. A. Małecka, J. Poźniczek, A. Micek-Ilnicka, A. Bielański. J Mol Catal A: Chem, 138: 67–81, 1999.
22. A. Bielański, J. Datka, B. Gil, A. Małecka-Lubańska, A. Micek-Ilnicka. Catal Lett, 57: 61–64, 1999.
23. A. Bielański, A. Małecka, L. Kubelkova. J Chem Soc Faraday Trans, 85: 2847–2856, 1989.
24. A. Bielański, J. Datka, B. Gil, A. Małecka-Lubańska, A. Micek-Ilnicka. Phys, Chem Phys, 1: 2355–2360, 1999.
25. G. M. Brown, M.-R. Noe-Spirlet, W. R. Busing, H. A. Levy. Acta Crystalologr, B33: 1038–1046, 1977.
26. M.-R. Spirlet, W. R. Busing. Acta Crystallogr, B34: 907–910, 1978.
27. N. Mizuno, M. Misono. Chem Rev, 98: 199–217, 1998.
28. G. Mirth, J. Lercher, M. W. Anderson, J. Klinowski. J Chem Soc Faraday Trans, 86: 3039–3044, 1990.
29. J. G. Highfield, J. B. Moffat. J Catal, 95: 108–119, 1985.
30. A. Micek-Ilnicka, A. Bielański, M. Derewiński, J. Rakoczy. Reakt Kinet Catal Lett, 61: 33–41, 1997.
31. A. Bielański, A. Micek-Ilnicka, B. Gil, E. Szneler, E. Bielańska. Ann de Quimica Int Ed, 94: 268–273, 1998.
32. C. N. Satterfield. Mass Transfer in Heterogenous Catalysis (Russian edition). Moscow: Khimia, 1976, pp 137–140.
33. T. Baba, Y. Ono, T. Ishimoto, S. Moritaka, S. Tanooka. Bull Chem Soc Jpn, 58: 2155–2156, 1985.

34. W. L. Chu, X. G. Yang, X. K. Ye, Y. Wu. React Kinet Catal Lett, 62: 333–337, 1997.
35. Y. Mimura. Japanese Patent 7,265,711 to Cosmo Oil Co. Ltd., 1995.
36. Y. Mimura. Japanese Patent 7,265,712 to Cosmo Oil Co. Ltd., 1995.
37. J. J. Kim, W. Y. Lee, I. K. Song. U.S. Patent 5,227,141 to Korea Institute of Science and Technology, 1993.
38. A. Bielański, R. Dziembaj, A. Małecka-Lubańska, J. Poźniczek, M. Hasik, M. Drozdek. J Catal, 185: 363–370, 1999.
39. G. Trevale, G. F. Buzzi, Spanish Patent to Euteco Impianti SPA (IT); Petroflex Ind & Com S.A. (BR), EP 0,078,422, 1983.
40. R. Trotta, I. Miracca. Catal Today, 34: 447–455, 1997.
41. T. Murofushi, A. Aoshima, U.S. Patent 4,376,219 to Asahi Kasei Kogyo Kabushiki, 1982.
42. T. Murofushi, Japanese Patent 58,074,630 to Asahi Kasei Kogyo KK, 1983.
43. G. M. Maksimov, I. V. Kozhevnikov. React Kinet Catal Lett, 39: 317–322, 1989.
44. A. Molnar, C. Keresszegi, B. Török. Appl Catal A: Gen, 189: 217–224, 1999.
45. C. P. Nicolaides, C. J. Stotijn, E. R. A. van der Veen, M. S. Visser. Appl Catal A: Gen, 103: 223–232, 1993.
46. K. Klier, R. G. Herman, M. A. Johansson, O. C. Feeley. Proceedings of 203rd ACS National Meeting, San Francisco, 1992, vol 37 preprint 1, pp. 236–246.
47. J. S. Kim, J. M. Kim, G. Seo, N. C. Park, H. Niiyama. Appl Catal 37: 45–55, 1988.
48. K. Nowińska, M. Sopa. Supported heteropoly acids as catalysts for MTBE synthesis. Proceedings of 8th International Symposium on Heterogeneous Catalysis, Varna, 1996. Sofia: A. Andreev, et al. (eds), Heterogeneous Catalysis, 1996, pp 523–528.
49. G. D. Yadav, N. Krithivasan. J Chem Soc, Chem Commun: 203–204, 1995.
50. J. F. Knifton, J. C. Edwards. Appl Catal A: Gen 183: 1–13, 1999.

7
Ethanol-Based Oxygenates from Biomass

Farid Aiouache and Shigeo Goto
Nagoya University, Nagoya, Japan

I. INTRODUCTION

It is not in doubt that strategic interest over the last three decades has been given by researchers and legislators to the controversial concept of gasoline oxygenates. This growing concern has led to many passionate debates on issues and eventual solutions to moderate the large dependence on fossil resources, to promote underutilized agriculture facilities, and to reduce environmental degradation. Although it has created an energetic force that definitively leans toward ethanol-based fuel oxygenates, the Kyoto Protocol remains in a certain sense a "utopia" as a result of many technical and other economical reasons.

Oxygenates, such as methanol (MeOH) and its derivative ethers, and ethanol (EtOH) and its derivative alkyl ethers, are used largely as octane boosters, to promote more complete combustion of gasoline and to reduce carbon monoxide (CO) and volatile organic compounds (VOCs) emitted from the exhaust pipes. Among these oxygenates, methyl *tertiary*-butyl ether (MTBE), which has seen rapid growth in use during the last decade, is beginning to expose certain threats to human health safety beyond drinking water contamination and global warming issues. Consequently, legislation that would ban it is under consideration in many countries.

Ethanol and its derivative ethers are becoming the major option to MTBE phase-out. Combining the benefits of using agricultural and other cellulose wastes with cleaner oxygenates would be a promising alternative route [1].

Table 1 Blending Characteristics of Oxygenates

Property	MTBE	TAME	ETBE	EtOH	TAEE	TBA
Blending Rvp (psi)	8	2.5	4.4	18	1	10
Octane blending	110	105	112	115	109	100
Boiling point (K)	328	358	345	351	379	356
Oxygen content (wt)	18.2	15.7	15.7	34.7	13.8	21.6
Solubility in water (g/100 g water)	4.3	1.15	1.2	Infinite	1.03	NA
Atmospheric reactivity[a]	2.6	7.9	8.1	3.4	NA	1.1

NA = not available.
[a]Hydroxyl reaction rate coefficient: 10^{12} K-cm^3/mol-s (gas-phase reaction rate with hydroxyl radical related to methanol one).

Table 1 presents the main characteristics of some fuel oxygenates. Ethanol (EtOH) shows the highest blending Reid vapor pressure (bRvp), leading to high evaporative emissions and requiring refiners to reformulate gasoline by removing pentanes or butane cuts to meet bRvp gasoline limit standards. On the other hand, *tertiary*-amyl ethyl ether (TAEE), with a bRvp value of 1, might be suitable for hot places. All oxygenates present acceptable values of octane number; however, oxygen content ratio is favorable to ethanol. This means that blending gasoline with ethanol (oxygen content 34.7 wt%) up to 10–15 vol% is sufficient to meet minimum oxygen content requirements in gasoline, while ethers with values of oxygen content 13.8–18.2 wt% are required to add more volumes (20–30 vol%). In addition, ethanol is completely soluble in water; this ethanol hygroscopic character leads to phase separation of gasoline and makes it hard to handle by refiners during transportation or storage. MTBE's lower but non-negligible solubility value of 4.3 wt% is responsible for its phasing-out, while for other oxygenates such as ethyl *tertiary*-butyl ether (ETBE) the potential environmental impact is not well established. Furthermore, in the United States aldehyde emissions are under U.S. Environmental Protection Agency (EPA) regulations, and Table 1 indicates a low atmospheric reactivity of *tertiary*-butyl alcohol (TBA) through hydroxyl reaction rate parameter (gas-phase reaction rate with hydroxyl radical related to methanol one). However, the expected high solubility in water and relative high bRvp make the conversion of TBA into equivalent ethers more interesting.

Many of the subjects treated here were reviewed in more detail by other authors. Oxygenate synthesis was recently overviewed by Ancillotti et al. [2], where kinetics, thermodynamics, and catalytic aspects are discussed. Lynd et al. [3] outlined the recent information on cellulose-to-ethanol technology, while Taylor et al. [4] and Dautzenberg et al. [5]

provided an interesting commentary on potential use of hybrid processes applied largely to fuel oxygenates.

In this chapter we focus on the synthesis of ethanol-based oxygenates and particularly on the three probable widespread oxygenates of the future: ethanol (EtOH), ethyl *t*-butyl (ETBE), and *t*-amyl ethyl ether (TAEE). Their synthesis will be treated from the point of view of chemical reaction engineering, likely concerned with reactor design and process intensification. A later part describes catalyst selection (ion-exchange resins, zeolites, heteropoly acids) and overall integrated units in a single vessel [reactive distillation (RD), reactive extraction, reactive pervaporation]. We end with some comments on process retrofitting from methanol-based ethers units to ethanol-based ones.

Figure 1 shows the number of publications on oxygenates per year from 1994 to 2000. It is divided into four themes, that is, oxygenates synthesis, process design (separation and reactive distillation), phase equilibria (solid–liquid, liquid–liquid, and gas–liquid), and environmental release (toxicity and global warming). The dominant number of papers focused on environmental issues and human toxicity. This might be

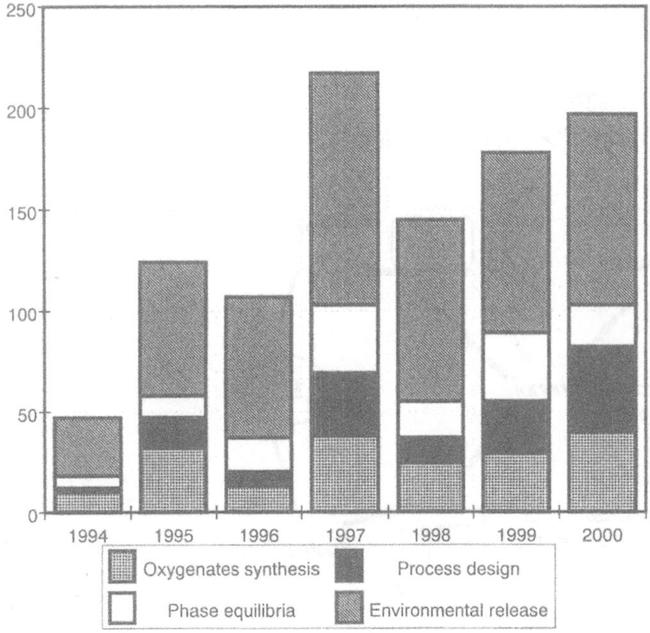

Figure 1 Publications on oxygenates (a total of 1126 papers). Patents are not listed.

explained by rapid increase of MTBE utilization during the 1990s, although mature information of MTBE effect on human health security was not completely established. The phase equilibria research trend was very similar to environmental release. This is a consequence of its application to water contamination, air pollution and global warming. Also to be noticed is the constant attention to oxygenates synthesis, although the leading role of MTBE is gradually being substituted by ethanol-based oxygenates.

II. ETHANOL-BASED OXYGENATE PRODUCTION NETWORK

A simplified representation of ethanol-based oxygenate production is indicated by Figure 2. From renewable resources, they can be produced through two routes:

Pyrolysis of biomass to tars followed by selective catalytic cracking of tars to C_2 hydrocarbons [6].

Fermentation of sugars contained in cellulose, hemicellulose, and carbohydrates materials [3].

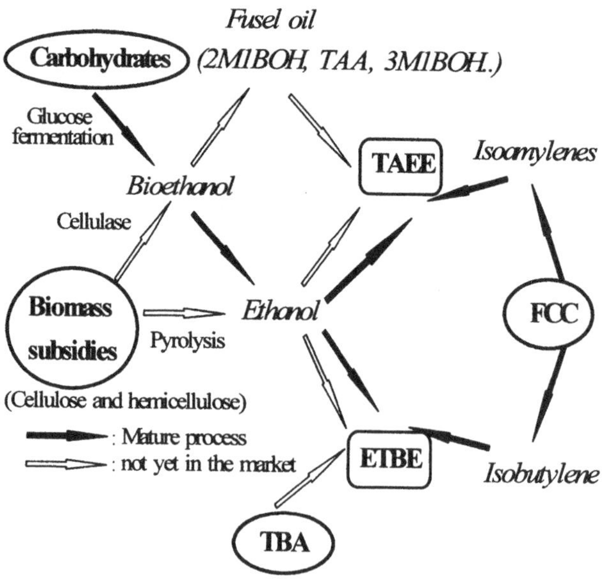

Figure 2 Overall pattern for ethanol-based ether production.

A. Bioethanol

The fermented products are currently produced by the conventional fermentation process of glucose found in carbohydrate feedstocks such as corn and beet. However, the future ethanol industry lies in cellulose and hemicellulose materials, largely abundant in wood and other agricultural wastes. This process, which is still not economical, converts cellulose to C_5 and C_6 carbon sugars through hydrolysis in acid medium. Despite tax exemptions adopted by many countries, both processes are still not competitive with the process from fossil resources.

The potential rapid growth of the process from biomass during this decade may come from the enzymatic hydrolysis of cellulose to sugars ready for fermentation. Then, actual ethanol costs from cellulose are expected to be reduced by one-third to one-fourth [7]. The worldwide potential of bioethanol is estimated to be around 2 million tons per year by 2020. One-fourth of the bioethanol process would originate from the mature sugar/starch crops process, and the remainder from the lignocellulosic biomass process (1).

The lead phase-out and energy crisis of the 1970s drove a renewed interest in ethanol as a sustainable resource and alkyl ether oxygenates as promising alternatives. Since the 1970s, Brazil and the United States, the main producers as shown in Figure 3, started boosting ethanol production through Proalcool programs and the National Energy Act, respectively. U.S. production has grown by 12% per year, while in Brazil the production jumped by 5 times during the 1980s, such that 20% of the cars today are running on 100% ethanol. On the other hand, Canada, the third largest producer of bioethanol, started production in the middle of the 1990s through fiscal incentives. Meanwhile, in Europe, where ETBE has had great

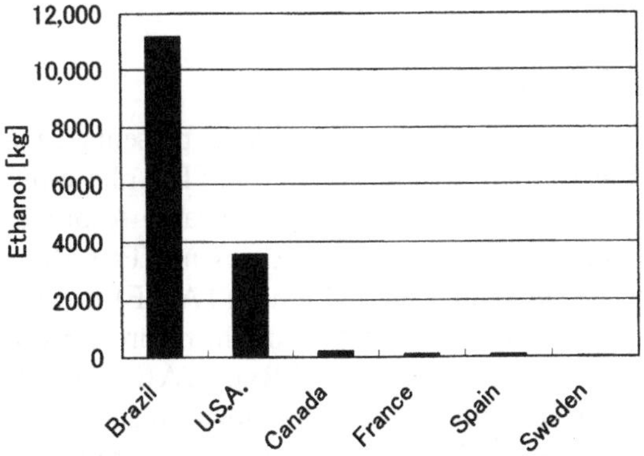

Figure 3 Main worldwide producers of bioethanol in 1998.

success, particularly in France and Spain, the majority of ethanol production is converted to ETBE.

Notwithstanding ethanol's oxygen content differential as illustrated in Table 1, MTBE has been used much more than ethanol. MTBE production was intensified during the 1990s due to favorable lower oil price during this decade and the high cost of ethanol production from the mature process, which is highly sensitive to climate fluctuations.

The main drawback of ethanol as an oxygenate for gasoline is its co-solubility effect in the ternary mixture, ethanol–water–gasoline. Hydrophobic compounds such as benzene, toluene, ethylbenzene, and xylene (BTEX) present higher solubilities in the aqueous phase with ethanol addition. Indeed, Herman et al. [8] noticed that an addition of 15 vol% of ethanol to a water–gasoline mixture increased BTEX concentrations by approximately 20–60%.

B. ETBE

ETBE is commercially manufactured by the reaction of isobutylene (IB) and EtOH on an acid ion-exchange resin. The first successful report of an industrial-scale ETBE synthesis setup came from France and Italy in 1992 (Oxy-Fuel) [9]. Besides isobutylene, *tert*-butyl alcohol (TBA), a major by-product of propylene oxide production from isobutane and propylene, can be employed instead of IB as a reactant [10]. This process, which is not yet on the market, is expected to be competitive when IB supply is limited.

High purification of ETBE from TBA would not be required to meet toxic emissions limits because TBA has low atmospheric reactivity (see Table 1) and low aldehyde emissions compared to ETBE.

C. TAEE Synthesis from Isoamylenes

Despite the fact that TAEE has low oxygen content, TAEE presents a real possibility to contribute to the oxygenate supply after MTBE phase-out. Indeed, it may be partially or totally produced from renewable resources. The large availability of C_5 olefins in fuel catalytic cracking (FCC) cuts (10–15 wt%), commonly used for *tert*-amyl methyl ether (TAME) production, can be advantageous for TAEE synthesis. TAEE might be preferred to other alcohols for hot places because of its lower bRvp. TAEE may be totally produced from renewable resources when ethanol is reacted with amyl alcohols largely present in fusel oil. The fusel oil is a by-product of bioethanol rectification.

III. ETHANOL-BASED ETHERS

The phasing-out of MTBE and predicted lower prices for ethanol will very likely lead to ethanol-based ethers as major gasoline oxygenates. Among these oxygenates, ETBE and TAEE are expected to be in widespread use.

A. ETBE (2-Ethoxy-2-methylpropane)

1. Some Industrial Processes for ETBE Manufacture

The commercial production of ETBE began in the 1990s [11], and commercial MTBE plants can be easily retrofitted to produce ETBE [12]. The industrial scheme for ETBE production consists of a primary reactor, a distillation or RD column, and an additional extractor column or membrane pervaporation for ETBE purification from nonreacted ethanol and other reaction by-products (isobutylene, diethyl ether, and *tert*-amyl alcohol). In the CDTECH and ETHERMAX process, isobutylene and ethanol streams are fed to a fixed-bed downflow adiabatic reactor. The equilibrium-converted reactor effluent is introduced into an RD column where the reaction continues. A top extractor is added to the RD column to remove ethanol with water. This scheme can provide overall isobutylene conversion of up to 95%. For the Institut Francais du Petrole (IFP) process, the reactants are converted at temperatures lower than 90°C and pressures lower than 2 MPa. Then the main effluents are purified for further applications or recycled. In the Snamprogetti and Philips etherification processes, ethers are produced by the addition of alcohol to reactive olefins in the presence of an ion-exchange resin at mild temperature and pressure. The feed passes through two reactors in series—an isothermal tubular reactor and an adiabatic drum reactor—and second reactor effluent goes to the product fractionation tower where the ether product leaves the bottom stream and hydrocarbon is recovered overhead [13].

2. ETBE Synthesis from TBA

Notwithstanding that ETBE was first synthesized over 70 years ago from ethanol and *tert*-butyl alcohol (TBA) in homogeneous phase and using sulfuric acid as catalyst [14], the major following research work used IB than TBA as starting reactant until early in the 1990s, as illustrated in Table 2. On the other hand, many works investigated both IB and TBA simultaneously with aqueous ethanol solutions. In this series, we cite works of Pescarollo et al. [25], Kochar et al. [26], and Jayadeokar et al. [17]. A subazeotrope 85/15 by weight of ethanol–water and isobutylene may produce ETBE accompanied with TBA from IB hydration. Christine et al. [21]

Table 2 ETBE Basic Research Work During the 1990s

Year	Reactant	Apparatus	Catalyst type	Mechanism	Kinetic expression	Er (kJ/mol)
1991 [15]	IB	Batch reactor	A15	E-R	Concentration basis	81.2
1992 [16]	IB	Differential tube reactor	A15	L-H	—	—
1992 [17]	IB	Semi-batch reactor	A15	E-R	Concentration basis	—
1993 [11]	IB	Batch reactor	Lewait K2630	—	—	—
1994 [18]	IB	Differential tube reactor	Lewait K2631	E-R	Activity basis	79.3
			Lewait K2631	L-H	Activity basis	133
1995 [19]	IB	Reaction calorimeter	Lewait K2631	E-R	Activity basis	86.5–89.2
1995 [20]	TBA	Semi-batch reactor	A15	L-H	Concentration basis	57.4
1995 [21]	TBA	Annular reactor	Sulfuric acid	—	—	—
1997 [23]	IB	Batch reactor	Lewait K2631	—	—	—
2000 [24]	TBA	Semi-batch reactor	A15, S-54, and D-72	L-H	Concentration basis	43.3–57.3

carried out this reaction in homogeneous media and under high temperatures and pressures (170°C and 3 MPa) to enhance the reaction rate by about 2 orders of magnitude above that realized commercially. Yin et al. [20] studied liquid-phase synthesis of ETBE from TBA and ethanol catalyzed by ion-exchange resin and heteropoly acid (HPA) at mild pressures and temperatures. Kifton et al. [27] also investigated different types of zeolite catalysts for direct synthesis of ETBE from TBA and EtOH. They obtained 40–70% yield and 65–95% selectivity in a temperature range of 40–140°C and a pressure range of 0.1–7 Mpa.

Chemistry. ETBE is obtained by the reaction of EtOH to IB or TBA. The reaction is reversible, moderately exothermic, and catalyzed by acid catalyst. Depending on the feed ratio, type of catalyst, and operating conditions side reactions are more or less present, such as isobutylene oligomerization to di-isobutene and TBA dehydration if it was used in lieu of IB.

The main reactions are represented by the following equations:

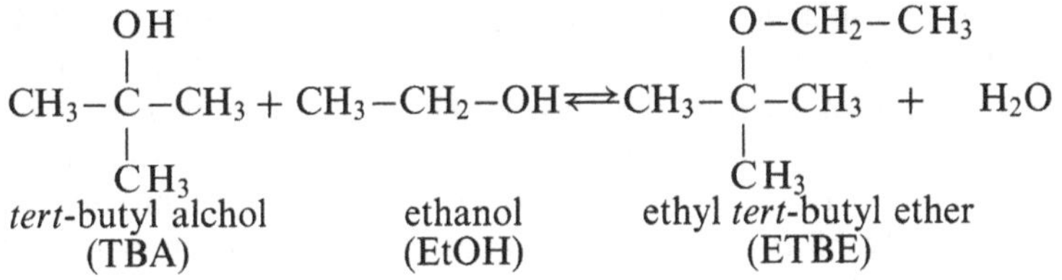

(1)

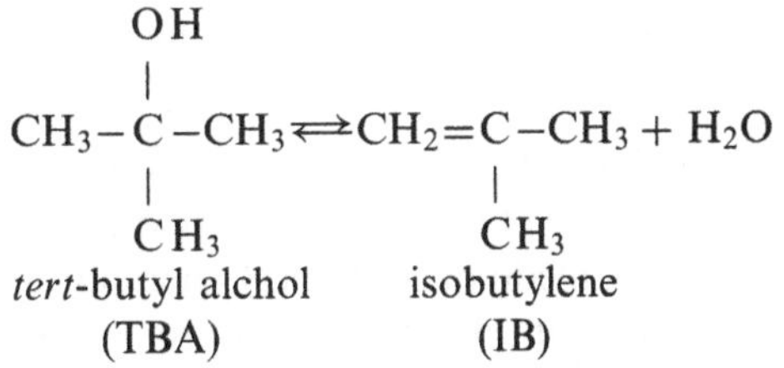

(2)

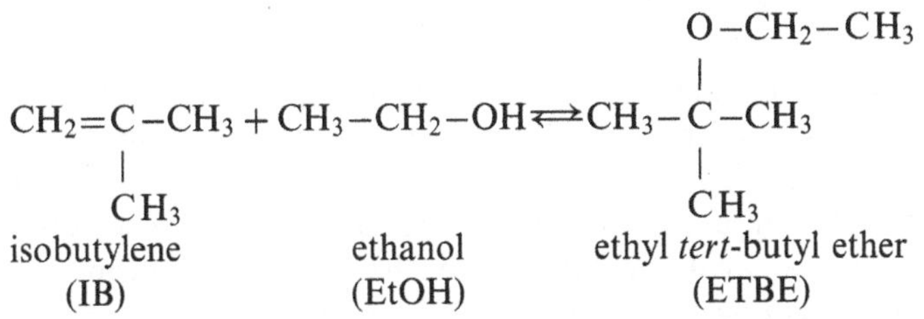

(3)

$$2CH_2{=}\underset{\displaystyle CH_3}{\underset{|}{C}}{-}CH_3 \rightleftharpoons \left[CH_2{=}\underset{\displaystyle CH_3}{\underset{|}{C}}{-}CH_3\right]_2 \qquad (4)$$

isobutylene (IB) diisobutene (DIB)

Reactivity of TBA with Ethanol in Various Acid Areas. As pointed out by Yin et al. [20], catalysts such as heteropoly acids (HPAs) and ion-exchange resins have been used to catalyze the synthesis of ETBE from TBA and ethanol in a batch reactor. Despite its high reactivity, acid resin showed less selectivity to ETBE and formed some by-products such as oligomers and diethyl ether. It was indicated that HPAs could be used advantageously in place of ion-exchange resins because the selectivity and activity of equivalent H^+ were generated. However, HPAs exhibited lower kinetic rates since they were more inhibited by water.

In addition, ion-exchange resins as catalysts for ether manufacture were found to display a higher reactivity of TBA with MeOH than EtOH, as shown in Figures 4a and 4b. Presenting a higher basicity, EtOH caused higher solvation of resin active sites SO_3H^+. These results were confirmed by several authors, including Ancillotti et al. [2] and Jayadeokar et al. [17], who studied the reaction of isobutylene with various light alcohols. In addition to solvability parameter, Gomez et al. [23] interpreted the lower rate for EtOH than MeOH by the preferential sorption of the respective alcohol in the bulk phase of resin. The higher the affinity of the alcohol for the resin, the higher the reaction rate will be. Expressing this affinity by solubility parameter as illustrated in Table 3, MeOH was the most sorbed component in the alcohol homolog series.

Reaction Mechanism. ETBE kinetics investigations were basically inspired by the many kinetic studies on MTBE synthesis. Matouq et al. [28] noticed that the acidity of the reaction medium might play an important role in the selectivity of TBA etherification. Indeed, for a series of alkali hydrogen sulfate acids ($MHSO_4$), where M is an alkali metal, potassium hydrogen sulfate, the weakest acid, produced the highest selectivity toward ETBE.

This may find interpretation through the global mechanism of this reaction. As shown in Figure 5, TBA is first protonated by a hydronium ion [21], followed by nucleophilic substitutions of protonated TBA by either a first-order (SN1) or a second-order (SN2) mechanism. The SN1 route proceeds under a weak acidity, while the SN2 mechanism generates isobutylene through heterolysis of protonated TBA.

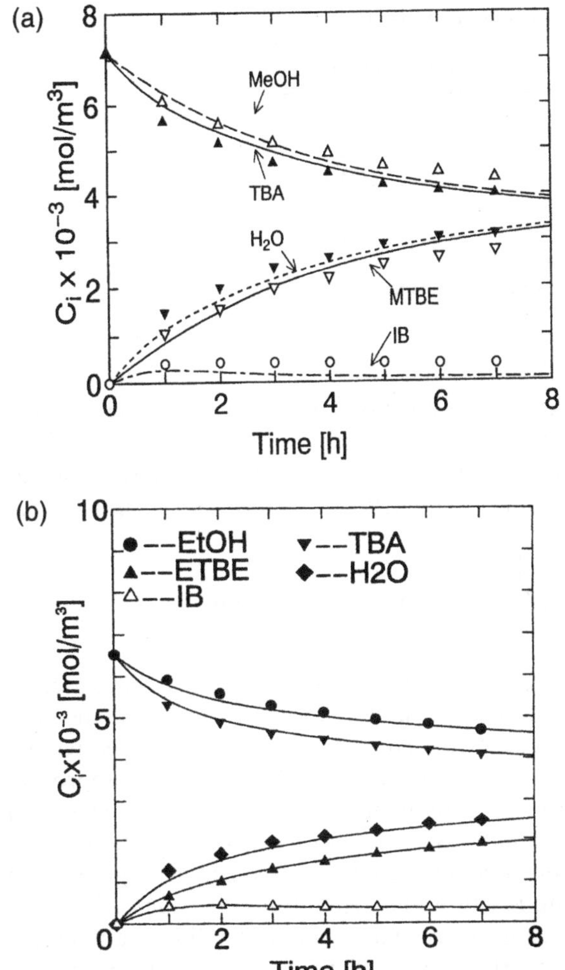

Figure 4 (a) Concentration profiles with time (A15 = 10 g, T = 323 K, $MtOH_0$ = TBA_0 = 1 mol). (b) Concentration profiles with time (A15 = 10 g, T = 323 K, $EtOH_0$ = TBA_0 = 1 mol).

Kinetics. In ion-exchange resins, the ETBE synthesis mechanism from TBA or IB was described through heterogeneous catalysis where sulfonic sites, SO_3H^+, are partially solvated by ethanol. Of the models tested, the Langmuir-Hinshelwood (L-H) model was more represented than the Eley-Rideal (E-R) one in the TBA-to-ETBE mechanism as illustrated in Table 2, while the E-R model was favored in the IB-to-ETBE mechanism. This can be explained by the more lipophilic character of TBA than IB

Table 3 Solubility Parameters of Some Polar Compounds

Compound	Solubility parameter[a]
Water	48
Methanol	29.2
Ethanol	26.4
tert-Butyl alcohol	23
tert-Amyl alcohol	20.4

[a]Total strength of intermolecular structure or cohesive pressure of component.
Source: Ref. 11.

toward sulfonate sites. Recent investigations in kinetic modeling of ether syntheses introduced some parameters such as Hildebrand solubility, nonuniform acid sites distribution in resin microgel phase [23], or interactions in the resin–liquid phase [29] to correct some controversial results illustrated in Table 2.

3. Combined Process for ETBE Production from TBA and Ethanol

Integration of two or more processes in a single unit started being used in the early 1970s with fuel catalytic cracking (FCC) units by simultaneously combining cracking reactions with catalyst movement. It was termed "process intensification." In comparison with the sequential approach, combining multiple in-situ operations improves energy efficiency and circumvents boundary conditions, i.e., azeotropic conditions and equilibrium-limited reactions encountered in the sequential approach. In these instances, compacting reaction and separation units such as an RD column and a membrane reactor offer under suitable conditions a potential decrease in equipment size, thermodynamic limitations, and improvement in process efficiency and safety.

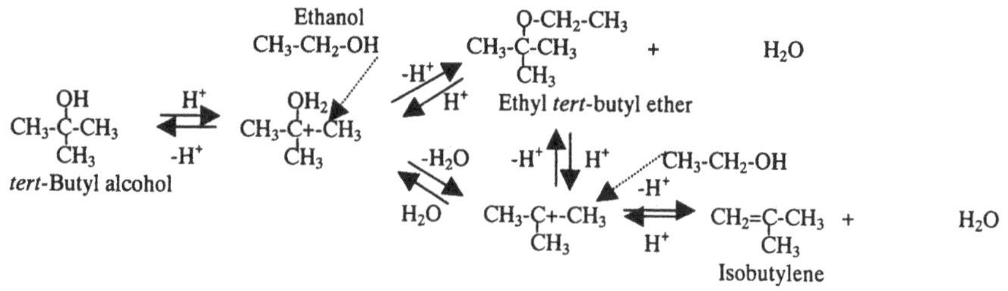

Figure 5 Proposed mechanism for ETBE formation from TBA.

Reactive Distillation for ETBE Synthesis. The major units of RD worldwide are applied to MTBE production where a heterogeneous structured catalyst is nested between stripping and enrichment sections of the distillation column. As stated by Taylor et al. [4], over the last decade many works have been published on the RD theme, mainly applied to MTBE production, while the first work on ETBE synthesis was pointed out by Sneeby et al. [12], who noticed that, due to phase behavior, any simple extrapolations of concepts from MTBE synthesis to ETBE synthesis may prove misleading.

As mentioned earlier, during this decade are expected many retrofitting operations of MTBE units to ETBE and other heavier ethers, leading to some changes in design aspects as well as in column control. As shown in Figure 6, under similar conditions of temperature and pressure, MTBE equilibrium constants as well as kinetics outrank those of ETBE. Consequently, in retrofitting an MTBE distillation column to ETBE one requires a higher Damkohler number (Da). This number evaluates the ratio between reaction time and distillation time inside the column as expressed by Eq. (1):

$$\mathrm{Da} = \frac{m_c k_{\mathrm{ref}}}{F} \qquad (5)$$

where m_c is the mass of catalyst, k_{ref} is the reaction rate constant at the reference temperature, and F is the feed flow rate.

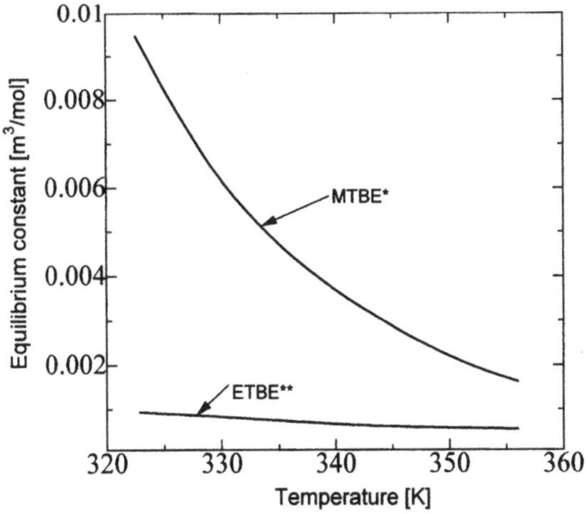

Figure 6 Equilibrium constant as a function of temperatures for MTBE and ETBE synthesis (*Matouq et al. [28]; **Yin et al. [20]).

On the other hand, the relative volatility of ETBE is lower than that of MTBE, as indicated in Table 1. Thus separate RD experiments in MTBE and ETBE synthesis results in ETBE being predominantly recovered, with the residue products in the ETBE distillation column, whereas MTBE is recovered with the distillate products.

Several authors have highlighted the decrease in both selectivity and yield of ETBE when an aqueous ethanol was used. For example, Cunill et al. [11] noticed a decrease of 7% in etherification selectivity when 3% water was added to an equimolar isobutylene–ethanol feed. For this reason, Quitain et al. [30] proposed an RD column for ETBE synthesis in diluted solution of bioethanol (2.5 mol%) in water. Then, the change in the selectivity as well the reduction in water inhibition effect was studied by continuous removal of water from the reactive section. Under the column configuration shown in Figure 7, ETBE at about 60 mol% could be obtained in the distillate and almost pure water in the residue. Concentration profiles of distillate and residue are shown in Figure 8. The conversion of TBA and the selectivity to ETBE were 99.9% and 35.9%, respectively.

Referring to the optimum conditions for continuous RD and liquid–liquid equilibria calculations, an overall process for direct production of ETBE and bioethanol was proposed and simulated using the Aspen Plus

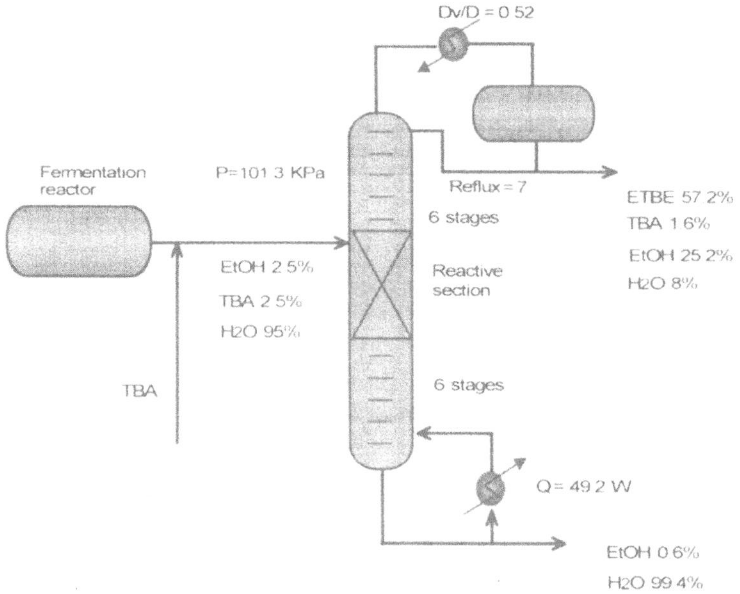

Figure 7 ETBE reactive distillation under standard conditions.

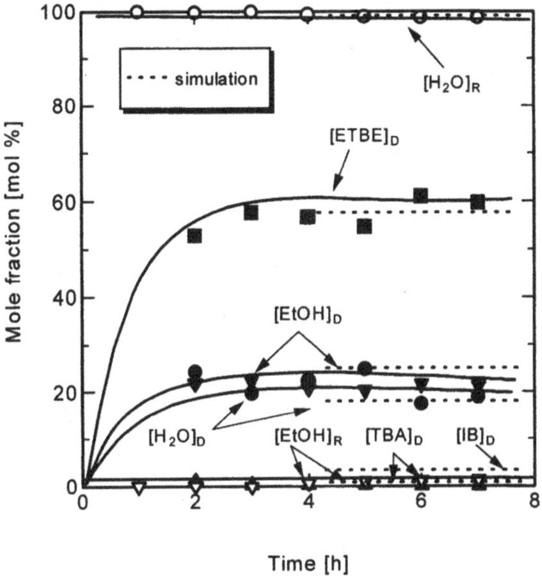

Figure 8 Concentration profiles of distillate and residue at standard operating conditions (total feed molar flow rate = 4.13×10^{-3} mol/s, reflux ratio = 0.7, catalyst = 0.1 kg, feed molar ratio = 1:1:38 TBA:EtOH:H_2O).

simulator. The column configuration was adopted from the work of Jacobs and Krishna [31]. The feed, consisting of bioethanol and TBA, was introduced below the reaction section of the column. The distillate was mixed with a small amount of residue to purify ETBE. The aqueous solution containing EtOH obtained from the purification process was recycled to the feed.

The isobutylene-rich gaseous product was used as a raw material for further etherification in a second RD column. At optimum conditions of feed composition, reflux ratio, and column pressure, the estimated value of ETBE composition was 94.3 mol% when it was purified with the residue. The IB-rich gas was converted to 88.8 mol% ETBE by using ethanol in the second reactive column. The global conversion of the process was 98.9 mol%, with a selectivity to ETBE of about 99.9%, as shown in Figure 9.

Reactive Distillation-Pervaporation.

Industrial processes for distillation-pervaporation unit. Contrary to the MTBE process, ETBE purification by conventional azeotropic distillation is poorly successful because of the presence of the azeotrope mixture EtOH–ETBE as indicated in Table 4. A number of processes, such as extraction of ethanol by water, distillation under high pressure, and

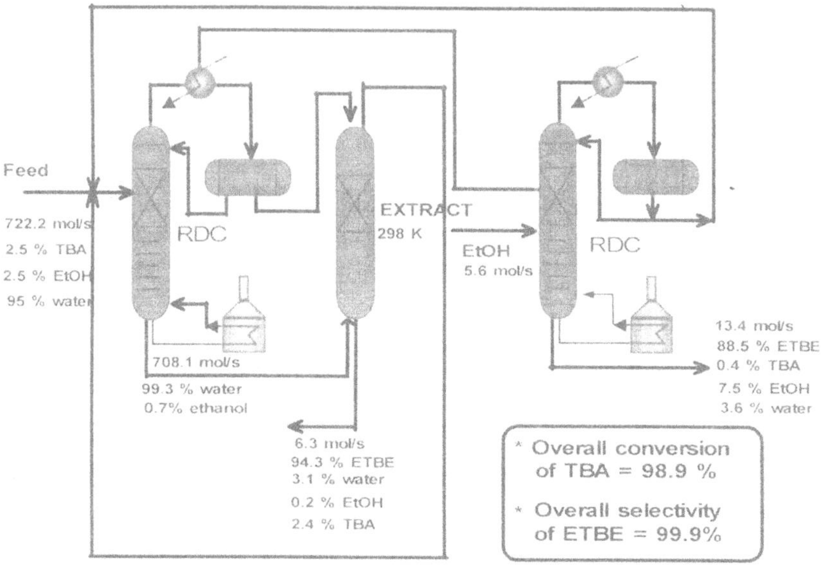

Figure 9 Overall process for ETBE synthesis from TBA and EtOH.

distillation-pervaporation, have been described for purifying the ETBE obtained from the bottom of azeotropic distillation.

Highlighting the last process, membrane technology has seen a lot of development during this last decade. In this case, the effluent from the reaction section passes over a pervaporation membrane which is selectively permeable to alcohol [32]. The retentate produced from the pervaporation step is thus depleted in alcohol. It is then distilled in a column, which produces ether from the bottom and the hydrocarbons present in the effluent from the reaction section overhead. However, this process could only produce ethanol-free ETBE if all the ethanol present in the effluent

Table 4 Azeotropic Conditions between Reaction Components at Atmospheric Pressure

	Composition (mol%)	Temperature (K)
ETBE–H_2O	73.4–26.6	338
ETBE–EtOH	62.9–37.1	340
EtOH–H_2O	90.3–9.7	352
TBA–H_2O	64.5–35.5	353
TBA–ETBE	25–75	351

from the reaction section at the pervaporation step could be extracted. However, like all other membrane processes, pervaporation alone cannot produce very pure products economically; thus, in this case, extraction of the last traces of ethanol from the retentate, even with a high-performance membrane, would mean the use of economically unrealistic membrane surfaces. In another process [32], the effluent from the reaction section is introduced directly into the azeotropic distillation column. A liquid fraction is extracted from this column as a side stream and then sent to a pervaporation step, which selectively extracts a portion of the alcohol. The alcohol-depleted retentate produced by this pervaporation step is then recycled to the azeotropic distillation column. ETBE from the bottom could be obtained from the distillation column with an ethanol content of less than 1% by weight, and even less than 0.1% by weight, provided that a membrane was used which could selectively extract ethanol from a mixture of EtOH/ETBE and hydrocarbons.

Reactive distillation (RD)-pervaporation. Information on RD-pervaporation is practically nonexistent in the open literature. A pervaporation configuration on an RD column would increase the reaction rate by lowering water inhibition influence, circumventing boundary azeotropic conditions between water and reaction mixture and reducing the consumption energy.

Forestiere et al. [33] claimed that a purity of ethanol in ETBE of less than 0.1% could be achieved when a permeation zone was added to RD. The mixture, extracted as a side stream from the RD column, was sent to a permeation zone. This side stream position was located in trays, which exhibited a substantially maximum composition of ethanol.

Yang et al. [34] carried out succinctly the etherification of TBA with MeOH and EtOH in a combined process of RD with pervaporation in order to separate water from the reaction mixture. Two types of membranes, polyvinyl alcohol (PVA 1001, Mitsui Engineering Co.) and micropore hollow-fiber membrane (Daicel Chemical Industry Co.) modules were used. The pervaporation cell unit was added to the bottom part of the column to remove water. Concentration profiles with and without pervaporation are shown in Figures 10a–10d. Higher reaction rates were observed since water mole fractions in the bottom were lowered by a significant amount when pervaporation was conducted. However, for the ETBE process, a reduction of water by 55% after 6 hr allowed an increase in ETBE purity of 40% in the top products. In both MTBE and ETBE combined processes, permeation selectivity toward water was not high, and the residual amount of water in the residue generated some azeotropic conditions between reaction mixtures, as shown in Figures 10b and 10d.

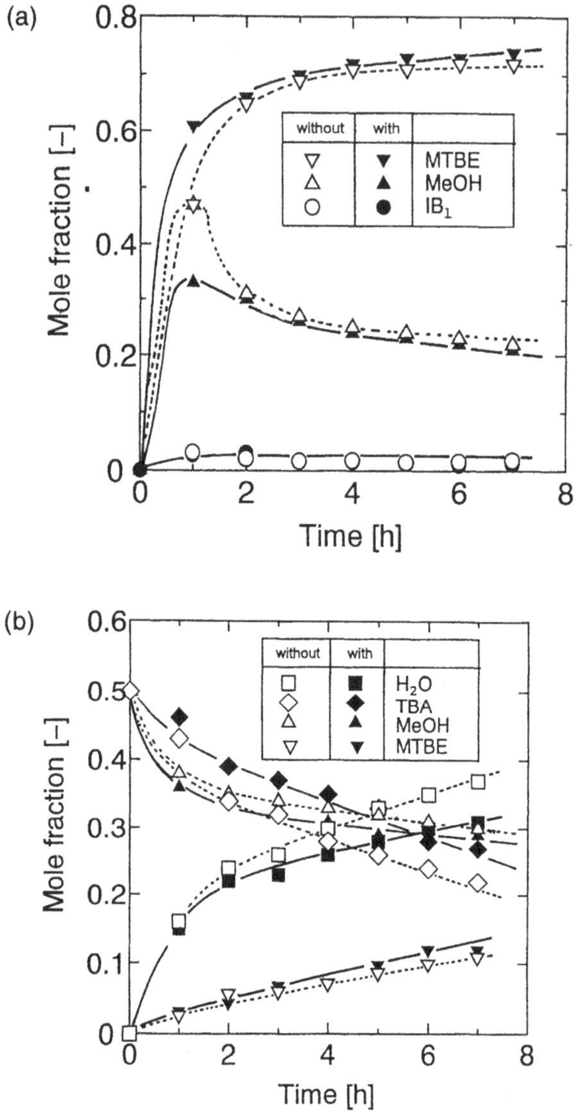

Figure 10 (a) Comparison between top products with and without pervaporation for MTBE synthesis. (b) Comparison between bottom products with and without pervaporation for MTBE synthesis. (c) Comparison between top products with and without pervaporation for ETBE synthesis. (d) Comparison between bottom products with and without pervaporation for ETBE synthesis.

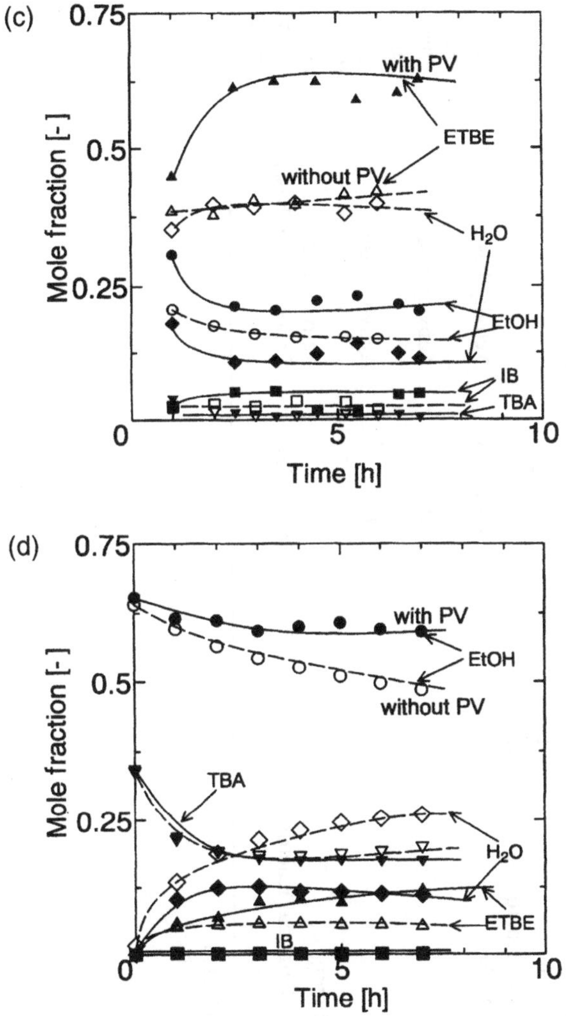

Figure 10 Continued.

B. TAEE (2-Ethoxy-2-methylbutane) Synthesis

1. TAEE Synthesis from Isoamylenes (IA)

Despite the growing interest in higher-ethanol-based ethers, few studies have been published regarding the formation of TAEE, as illustrated in Table 5. The Kraus group from Helsinki University of Technology has contributed greatly to TAEE synthesis understanding. The reactivity of isoamylenes

Table 5 TAEE Basic Research Work During the 1990s

	Reactant	Apparatus	Catalyst type	Mechanism	Kinetic expression	E_r (kJ/mol)
1993 [35]	IA	Plug/batch reactor	A15, Dowex 32	—	—	—
1994 [38]	IA	Batch	A16	—	—	—
1995 [37]	IA	Tube reactor	A15	—	—	—
1997 [36]	IA	CSTR	A16	Power law	Activity basis	47[a]
						69[b]
				E-R	Activity basis	79[a]
						98[b]
				L-H	Activity basis	90[a]
						108[b]
1998 [39]	IA + H_2O	CSTR	A16	L-H	Activity basis	117.7
1999 [40]	IA	Differential reactor	A15	Power law	Concentration basis	40.7[a]
						73.6[b]

[a]Reaction with 2-methyl-1-butene (2M1B).
[b]Reaction with 2-methyl-2-butene (2M2B).

with ethanol in ion-exchange resins was reported by Rihko et al. [35]. Similar to previous observations of ETBE and MTBE formation, ether was easily synthesized when isoamylenes were reacted with methanol and ethanol, respectively, for TAME and TAEE synthesis. A side isomerization reaction was observed from 2-methyl-1-butene (2M1B) to the less reactive isomer 2-methyl-2-butene (2M2B). The kinetic mechanism presented by Linnekoski et al. [36] followed the classical L-H model with separate surface acid sites for each reaction component.

Thermodynamic equilibria were studied by Kitchaiya et al. [37] and Rihko et al. [38]. Equilibrium constants and activities calculations were in good agreement with theoretical values. With aqueous ethanol, Jayadeokar et al. [17] and Linnekoski et al. [39] stated that an azeotropic mixture of water and ethanol leads to hydrate isoamylenes to *tert*-amyl alcohol as a side reaction. Compared to pure ethanol, a decrease of 50% in isoamylene conversion was observed.

The main reactions of TAEE synthesis from isoamylenes (2M1B and 2M2B) are represented by the following equations:

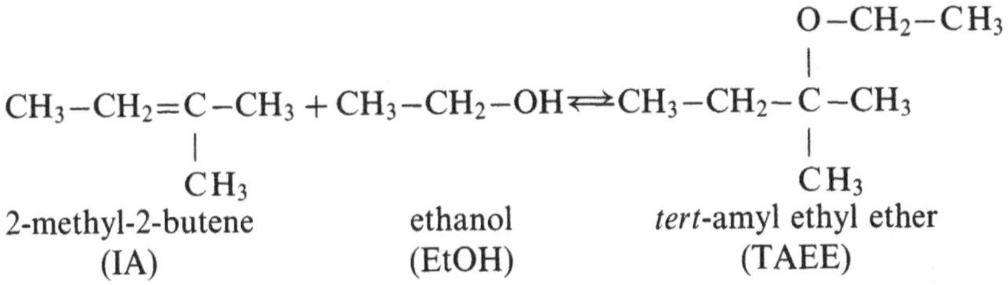

(6)

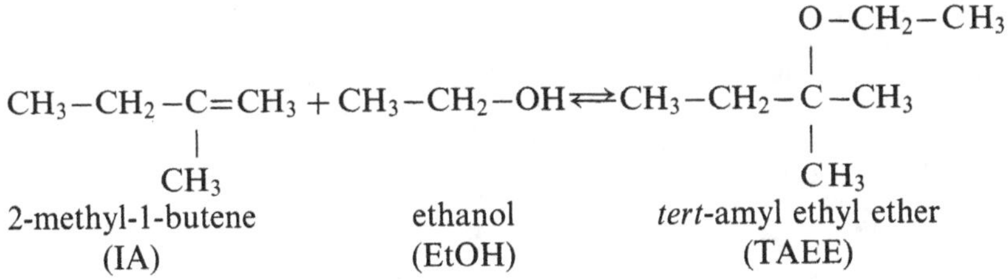

(7)

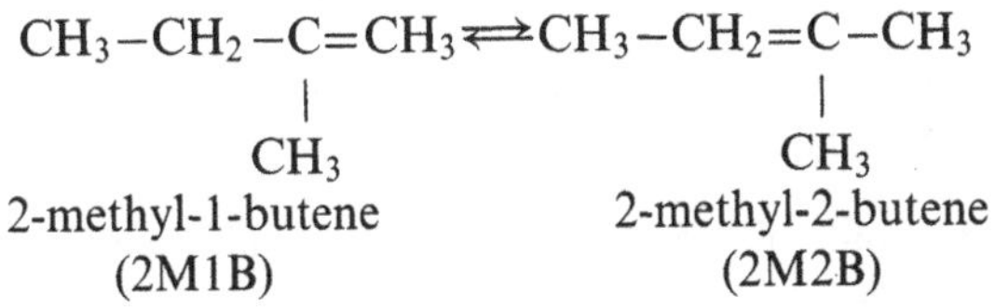

(8)

2. TAEE Synthesis from Fusel Oil

TAEE via 2-Methyl-1-butanol (2M1BOH). Besides fuel cracking additions, isoamylenes may be obtained through from several other fossil resources, which may not solve global warming worries. Goto et al. [41] proposed fusel oil, a by-product of bioethanol rectification, as interesting raw material for TAEE synthesis produced totally from renewable materials. TAEE was synthesized from 2-methyl 1-butanol (2M1BOH), one of the major components of fusel oil. Indeed, C_5 cracking cuts are used intensively in many processes, and their production may be limited in the future. Two RD columns in series were proposed, to carry out separately the dehydration to isoamylenes followed by the etherification part to TAEE. In both columns, ion-exchange resin A15 was used as a catalyst.

The low reaction rate of dehydration reaction is caused mainly by water inhibition effect. As shown Figure 11, a conversion of just 15% was observed over 60 hr. Then, in a conventional reactor, high values of catalyst loading or high reaction temperatures are required for high 2M1BOH conversions. Since the reaction temperature is limited by ion-exchange resin stability, RD was proposed to generate high conversions by continuous water removal from the reactive section. Then a trade-off between mass of catalyst and heat duty involved was generated. In this instance, a high mass of catalyst lead to water and isoamylenes with a high purity, respectively, in

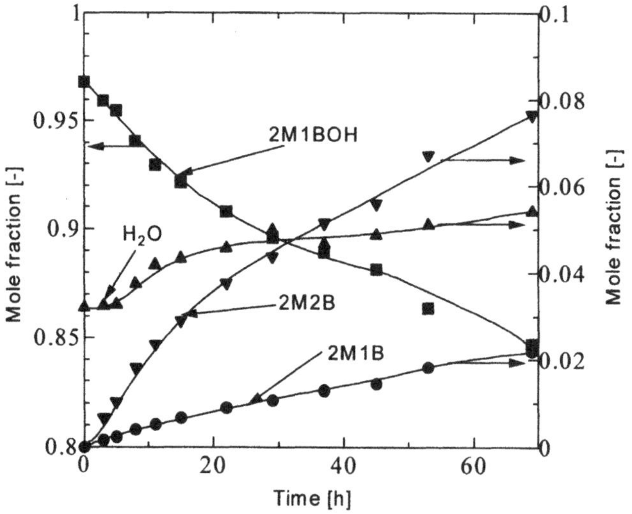

Figure 11 Mole fraction profile of 2M1BOH dehydration with time ($T = 368$ K, A15 = 20 g, $2M1BOH_0 = 1$ mol).

residue and distillate, while a high reboiler heat duty generates a high vapor flow and both products are received in the distillate. This compromise was circumvented when the stripping section of the dehydration column was eliminated. A complete conversion of 2M1BOH and pure isoamylenes was obtained in the top of the column under moderate conditions of reboiler heat duty and mass of catalyst. The low solubility of the organic phase in water suggested a top-column phase separator where only the organic phase was partially refluxed in the column. The configuration as well as the results are shown in Figure 12. The configuration of the etherification column required a similar configuration applied to the RD column for ETBE except that the rate for ETBE is about one order higher than the rate for TAEE. Then the desired equilibrium-controlled regime for TAEE needs higher values of Damkohler number. As a result, high conversions of isoamylenes were achieved when the reactive section was located close to the top of the column, where the forward reaction to TAEE synthesis is favored. Moreover, the long stripping section was exploited for better purification of TAEE from ethanol in the residue, partially penalized by some minimum azeotropes such as TAEE–TAA and TAEE–EtOH as illustrated in Table 6. A short rectifying section as well as high values of reflux ratios enabled

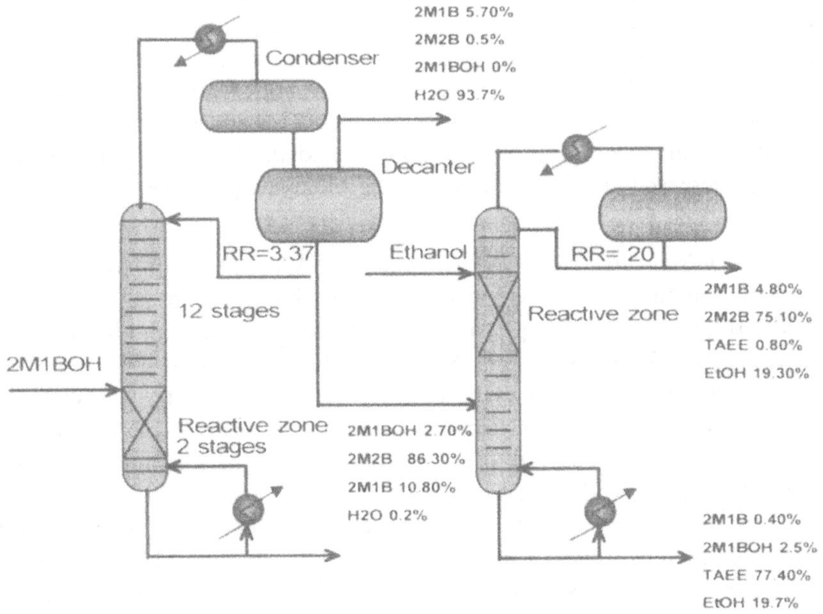

Figure 12 Overall scheme for TAEE synthesis.

Table 6 Azeotropic Conditions Between Reaction Components at Atmospheric Pressure

	Composition (mol%)	Temperature (K)
TAEE–H_2O	0.12–0.88	372
EtOH–H_2O	0.85–0.15	351
TAA–H_2O	0.40–0.60	361
2M2B–EtOH	0.88–0.12	312
TAA–TAEE[a]	0.55–0.45	372
EtOH–TAEE[a]	0.75–0.25	349
2M2B–H_2O–EtOH	0.90–0.05–0.05	310

[a]UNIFAC Dortmund prediction model.

ethanol to be mostly concentrated in the reactive section, leading to total conversion of isoamylenes.

TAEE via tert-Amyl Alcohol (TAA). Despite its low proportion in fusel oil, TAA is a side product in isoamylene etherification with bio-ethanol. TAA was also reacted with ethanol in a stirred batch reactor and an RD column. An ion-exchange resin (Amberlyst-15) was used as catalyst. This reaction is accompanied by side reactions such as the dehydration of TAA to isoamylenes and water and the etherification of isoamylenes to TAEE. Experimental data in a batch reactor has shown the strong inhibition effect of water on chemical kinetics rate. The Langmuir-Hinshelwood model fit the experimental data. This etherification is strongly limited by chemical equilibrium as illustrated in Figure 13. The yield of TAEE decreased from 66.3% to 53.8% when temperature was increased from 50 to 80°C.

Chemistry of TAA Etherification. TAA reactivity toward light alcohols, methanol and ethanol, in ion-exchange resin A15 has been successfully shown by Yang et al. [42] for TAME synthesis and by Aiouache et al. [43] for TAEE synthesis. Similar to TBA reactions, TAA reactivity with methanol was found to be more reactive than ethanol, as shown in Figures 14a and 14b. Under similar conditions, a decrease of 40% in TAA mole fraction was observed after 5 hr of TAME etherification time, while just 20% was noticed for TAEE synthesis.

In Figure 15, direct etherification of TAA to TAEE through nucleophilic substitution (SN1) is competed for by an (SN2) mechanism generating isoamylenes through heterolysis of protonated TAA. TAA is first protonated by a hydronium ion, followed by nucleophilic substitutions of protonated TAA by either a first-order (SN1) or a second-order (SN2)

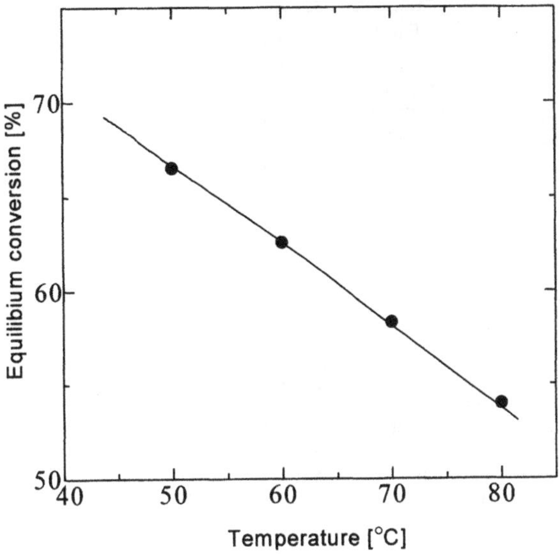

Figure 13 Equilibrium conversions as a function of temperatures for TAEE synthesis from TAA and EtOH.

mechanism. The SN1 route proceeds under weak acidity, while the SN2 mechanism generates isoamylenes through heterolysis of protonated TAA.

TAEE production through reactive distillation. Under the column conditions shown in Figure 16, mainly the forward reaction to TAEE is carried out in the top reactive stages by Le Chatelier's principle, while the bottom ones favors the decomposition of TAEE and TAA dehydration for isoamylene formation. Hence, reactive section close to the top of the column enables easy separation of the less volatile component, TAEE, from other reactants in the stripping section and maximizes ethanol, isobutylene, and TBA concentrations in the reactive section. A short rectifying section was used to purify the distillate, composed mainly of isoamylenes, TAA, and EtOH, from TAEE traces. A small excess of ethanol was used to achieve high conversions without isoamylene side-reaction drawbacks.

Figure 17 shows TAEE reaction rates as a function of conversion and temperature. The interesting rate at low conversion values motivated the use of a prereactor to achieve the majority of conversion. An equilibrium mixture obtained at 70°C from a batch prereactor with an initial composition in TAEE of 0.242 was introduced in the bottom part of the reaction zone (second stage). By using a high reflux ratio (RR) of 60, high

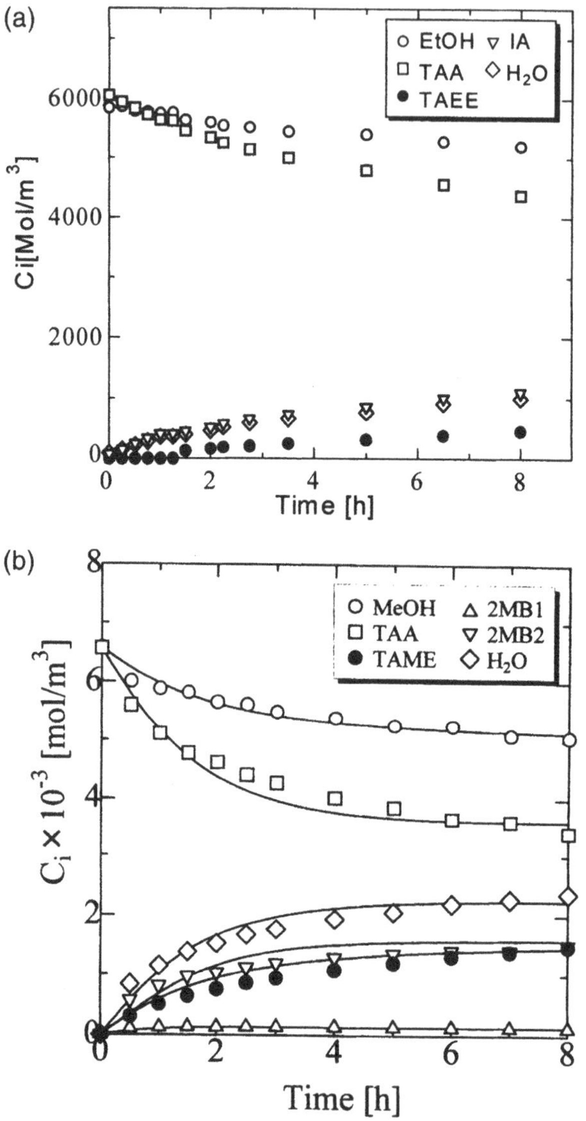

Figure 14 (a) Concentration profiles with time (A15 = 10 g, $T = 323$ K, $EtOH_0 = TAA_0 = 1.0$ mol). (b) Concentration profiles with time (A15 = 10 g, $T = 318$ K, $MeOH_0 = TAA_0 = 1.0$ mol).

conversion and selectivity to TAEE were obtained as shown in Figure 18. The additional RD column to the prereactor overcomes boundary limitations i.e., azeotropes and equilibrium conversions of TAA to TAEE (more than 83%) and purity of TAEE in the residue (93%).

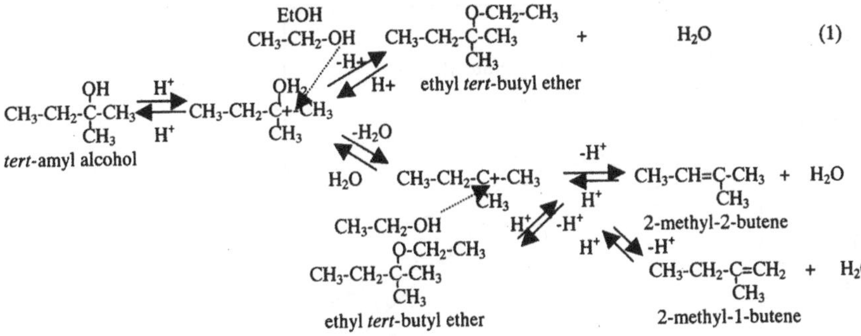

Figure 15 Proposed mechanism for TAEE synthesis from TAA and ethanol ($A15 = 10\,g$, $T = 323\,K$, $MeOH_0 = TAA_0 = 1.0\,mol$).

IV. CONCLUDING REMARKS

This overview has discussed two basic concepts:

1. The future rapid growth of ethanol-based oxygenates and particularly ETBE and TAEE to compensate phasing-out of MTBE
2. The application of "process intensification" to ethanol-based manufacture

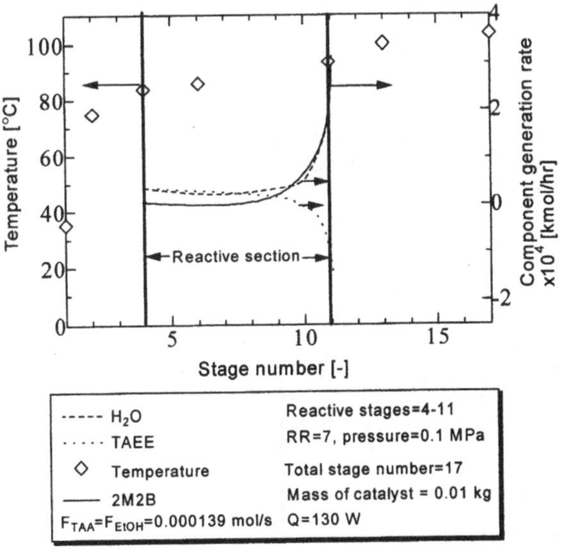

Figure 16 Temperature and rate of generation profiles in reactive distillation column.

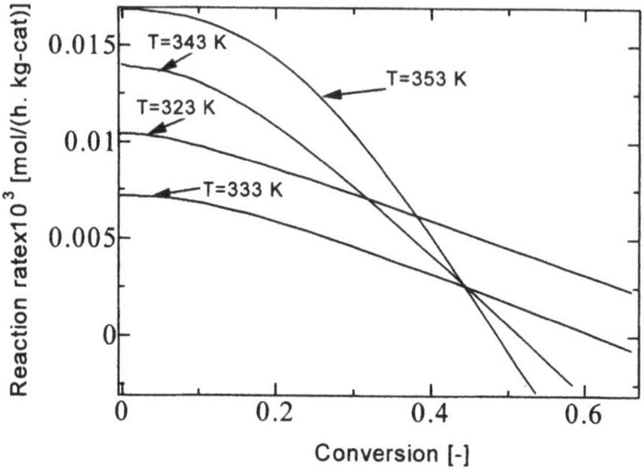

Figure 17 Reaction rate as a function of temperatures and conversions.

The following remarks are evidenced from this study.

1. Ethanol price may play a determinant factor for the future of ethanol-based ether process costs. The development of cellulose-to-ethanol processes will largely moderate actual carbohydrate process costs.

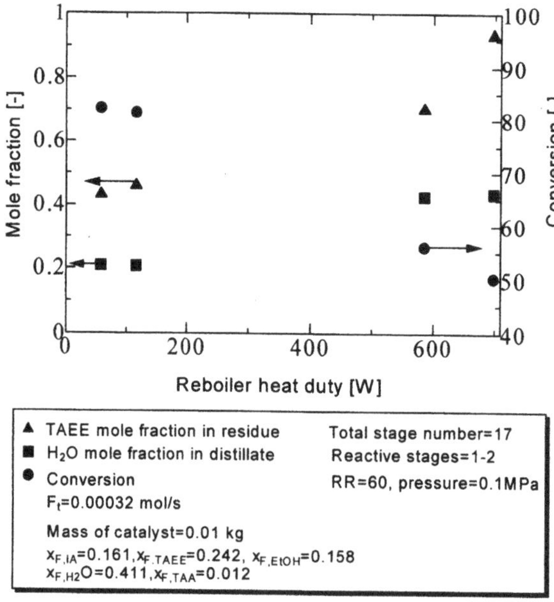

Figure 18 Effect of reboiler heat duty on conversion and purity of products.

2. Branched alcohols (TBA, 2M1BOH, and TAA) were used instead of traditional olefins for ether synthesis. Although traces of branched alcohols in gasoline would decrease gasoline reactivity during combustion, their utilization would cause a reduction in the buildup of greenhouse gas emissions.
3. Large retrofitting operations from methanol-based units to ethanol ones are foreseen during this decade. These operations will touch both sequential approaches to manufacture: reaction and separation and combined units, RD of MTBE. For example, more energy is consumed by ETBE units than MTBE ones. This is caused by lower reaction rates of ETBE, lower equilibrium conversions, and more azeotropes present in ETBE reaction mixtures. Furthermore, it was experimentally confirmed that ETBE units might show more instability through multiple steady-state control.
4. Reduction in energy consumption was achieved by in-situ purification of desired products during the reaction process. For example, a side extraction column for ETBE purification from EtOH could increase ETBE concentration in the distillate from 60% to 94.3%. On the other hand, a high-purified TAEE in an RD column could be achieved when a side membrane for water removal was combined with the RD column.

REFERENCES

1. Report to the Governor: Evaluation of biomass-to-ethanol fuel potential in California. PN 500-99-022, December 22, 1999.
2. F. Ancillotti, V. Fattore. Fuel Proc Technol, 57, 163–194, 1998.
3. R. L. Lynd. Ann Rev Energy Environ, 21, 403–465, 1996.
4. R. Taylor, R. Krishna. Chem Eng Sci, 55, 5183–5229, 2000.
5. F. M. Dautzenberg, M. Mukherjee. Chem Eng Sci, 56, 251–267, 2001.
6. J. P. Diebold, U.S. Patent 5,504,259, April 2, 1996.
7. National Renewable Energy Laboratory, Bioethanol Multi-year Technical Plan, Preliminary Draft, Golden, CO, July 1999.
8. S. E. Power, D. Rice, B. Dooher, P. J. J. Alvarez. Environ Sci Technol, 24–30, January, 1, 2001.
9. F. L. Plotter. Oxy-Fuel News 4: 6, May 11 1992.
10. C. Habenicht, L. C. Kam, M. J. Wilschut, M. J. Antal. Ind Eng Chem Res, 34: 3784–3792, 1995.
11. F. Cunill, M. Vila, J. F. Izquierdo, M. Iborra, J. Tejero. Ind Eng Chem Res, 32: 564–569, 1993.
12. M. G. Sneeby, M. O. Tade, R. Datta, T. N. Smith. Ind Eng Chem Res, 26: 1855–1869, 1997.

13. Hydrocarbon Proc, Nov, 114–116, 1996.
14. J. F. Norris, G. W. Rigby. J Am Chem Soc, 54: 2088–2100, 1932.
15. O. Francoise, F. C. Thyrion. Chem Eng Process, 30: 141–149,1991.
16. M. Iborra, J. F. Izquierdo, F. Cunill, J. Tejero. Ind Eng Chem Res, 31: 1840–1848, 1992.
17. S. S. Jayadeokar, M. M. Sharma. Chem Eng Sci, 47: 13: 3777–3784, 1992.
18. C. Fite, M. Iborra, J. Tejero, J. F. Izquierdo, F. Cunil. Ind Eng Chem Res, 33: 581–591, 1994.
19. L. Sola, M. A. Pericas. Ind Eng Chem Res, 34: 3718–3725,1995.
20. X. Yin, B. Yan, S. Goto. Int J Chem Kinet, 27: 1065–1074,1995.
21. C. Habenicht, L. C. Kam, M. J. Wilschut, M. J. Antal. Ind Eng Chem Res, 34: 3784–3792, 1995.
22. K. Sundmacher, R. Zhang, U. Hoffmann. Chem Eng Technol, 18: 269–277, 1995.
23. C. Gomez, F. Cunil, M. Iborra, F. Izquierdo, J. Tejero. Ind Eng Chem Res, 36: 4756–4762, 1997.
24. B. L. Yang, S. Yang, R. Yao. Reac-Funct Poly, 44: 167–175, 2000.
25. E. Pescarollo, F. Anchillotti, T. Floris. Chem Abstr, 96: 88291,1982.
26. N. K. Kochar, R. L. Marcell. Chem Abstr, 97: 112371, 1982.
27. J. F. Knifton, P. E. Dai. U.S. patent, 5,387,722, 1995.
28. M. Matouq, A. T. Quitain, K. Takahashi, S. Goto. Ind Eng Chem Res, 35: 982–984, 1996.
29. M. Mazzotti, B. Neri, D. Gelaso, A. Kruglov, M. Morbidelli. Ind EngChem Res, 36: 3–12, 1997.
30. A. T. Quitain, H. Itoh, S. Goto. J Chem Eng Japan, 32: 280–287,1999a.
31. R. Jacob, R. Krishna. Ind Eng Chem Res, 32: 1706–1709, 1993.
32. U.S. patent 4,774,365.
33. La forestiere, French patent FR-B-2 672 048.
34. B. Yang, S. Goto. Recent Res Dev Chem Eng, 1:27–39, 1997.
35. L. K. Rihko. Appl Catal A: Gen, 101: 283–295, 1993.
36. J. A. Linneshoski J, O. A. Krause. Ind Eng Chem Res, 36: 310–316, 1997.
37. P. Kitchaiya, R. Datta. Ind Eng Chem Res, 34: 1092–1101,1995.
38. L. K. Rihko, J. A. Linnekoski, A. O. I. Krause. J Chem Eng Data, 39: 700–704,1994.
39. J. A. Linnekoski, A. O. Krause, L. Struckmann. Appl Catal A: Gen, 170: 117–126, 1998.
40. N. Oktar, K. Murtezaoglu, G. Dogu, I. Gonderten, T. Dogu. J Chem Technol Biotechnol, 74: 155–161, 1999.
41. S. Goto, A. T. Quitain, F. Aiouache. Etherification of amyl alcohol in reactive distillation column Proceedings of CHISA 2000, Praha, 2000, Prague, Czechoslovakia, p 38.
42. B. L. Yang, M. Maeda, S. Goto. Int J Chem Kinet, 30: 137–143, 1998.
43. F. Aiouache, S. Goto. Synthesis of *tert*-amyl ethyl ether from *tert*-amyl alcohol and bioethanol in reactive distillation column. 6th World Congress of Chemical Engineering, Melbourne, Australia, 2001.

8

Catalytic Distillation Technology Applied to Ether Production

Aspi K. Kolah, Liisa K. Rihko-Struckmann, and Kai Sundmacher
Max Planck Institute for Dynamics of Complex Technical Systems, Magdeburg, Germany

I. REACTIVE DISTILLATION—PROCESS PRINCIPLES

Reactive distillation (RD) consists of an integrated multifunctional reactor in which a chemical reaction and distillation occur simultaneously in a single-unit operation apparatus. The reaction can be catalyzed either homogeneously or heterogeneously. The subset of heterogeneously catalyzed reactive distillation processes is commonly referred to as catalytic distillation processes. This technology of integrating reactors and separators has made enormous progress since it was first reported by Backhaus [1].

Excellent reviews on reactive distillation have been presented by Terrill et al. [2], Sharma [3], Doherty and Buzard [4], Stichlmair and Frey [5], Sakuth et al. [6], Taylor and Krishna [7], and Malone and Doherty [8]. Sundmacher et al. [9] have attempted to classify reactive distillation systems using various dimensionless numbers.

A. Motives for Application of Reactive Distillation

The motives for reactive distillation applications can be classified into two major groups as described below.

1. Reaction Problems

1. *Overcoming limitations of chemical equilibrium* to increase the conversion of desired product by simultaneously removing one of the products from the reaction system. This is the main objective in the use of catalytic distillation for methyl *tertiary*-butyl ether (MTBE) and more particularly for *tertiary*-amyl methyl ether (TAME) synthesis.
2. *Increasing the selectivity* of the desired products by removing the desired product from the reaction mixture in order to prevent it reacting further through consecutive reactions. However, in some cases a reduction of selectivity could be observed due to a localized unfavorable decrease in one of the components due to distillation, e.g., di-acetone alcohol production from acetone.
3. *The use of reaction enthalpy* of exothermic reactions for distillation, thereby avoiding unfavorable shifts in chemical reaction equilibrium, potential selectivity reduction, accelerated catalyst degradation, elimination of hot spots, etc. Endothermic reactions are not precluded from reactive distillation. Cumene synthesis from propene and benzene is an example of this use.

2. Separation Problems

1. *Separation of isomeric or close-boiling mixtures.* As an example, cyclohexane and cyclohexene, which have relative volatilities close to unity, thereby requiring a large number of distillation stages, can be conveniently separated using reactive distillation.
2. *Breaking of azeotropes.* Many conventional synthesis processes, such as those for MTBE and methyl acetate, have formation of azeotropes between various constituents of the process. These azeotropes are "reacted away," thus leading to considerably simpler phase equilibria for the system.
3. *High-purity separation*, such as hexamethylene diamine/water (in the Nylon-6,6 process), using adipic acid as the reactive entrainer.

B. Limitations and Disadvantages of Reactive Distillation Processes

1. Reaction systems intended to be used in a RD process should have optimal temperature and pressure range, compatible for both distillation and chemical reaction. In an ideal case the volatilities of the reactants and products are such that the concentration of

the reactants in liquid phase inside the reaction zone should be as high as possible and the concentration of the products should be minimum.

2. Longer development time for design, laboratory, and pilot testing due to complex nature of system. Moreover, the implementation of reactive distillation requires a very bold step from the management side, due to the presence of the omnipotent "it has always been done by this route" approach.
3. Cannot be applied for gas–liquid reactions, which require very high temperature and pressure.
4. Difficult to implement for very slow reactions due to very high residence time required, manifesting in very big column sizes and high liquid holdups.
5. The catalyst life should be long in the case of solid catalyzed reactive distillation processes in order to avoid frequent periodic shutdown of the plant for renewal of the catalyst.

C. Use of MTBE and Other Ethers as Octane Boosters

Methyl *tert*-butyl ether (MTBE, 2-methoxy-2-methylpropane) is a commonly used antiknock compound added to gasoline to increase its octane number. The implementation of the U.S. Clean Air Act 1970 and the Amendment of 1990 placed requirements on blending of MTBE in gasoline. This resulted in MTBE being rated as the chemical with the highest increasing production worldwide in the 1980s. In 1985, similar legislation implemented in the European Community led to increased application of octane boosters in Europe. In the United States, up to 16% MTBE is approved for blending in gasoline; for European markets the levels of MTBE usage can be found at the website in Ref. 10. Currently, intense controversy surrounds the use of MTBE as an octane booster due to its solubility and resulting contamination of groundwater, more information on which can be found in other chapters of this handbook.

Other well-known octane boosters include ethyl *tert*-butyl ether (ETBE, 2-ethoxy-2-methylpropane), *tert*-amyl methyl ether (TAME, 2-methoxy-2-methylbutane), *tert*-amyl ethyl ether (TAEE, 2-ethoxy-2-methylbutane), di-isopropyl ether [DIPE, 2-(1-methyl)ethoxypropane], isopropyl *tert*-butyl ether [IPTBE, 2-(1-methyl)ethoxy-2-methylpropane], etc. ETBE and TAEE have stimulated special interest, since they are synthesized using ethanol, which can be produced from renewable resources such as agricultural biomass waste. TAME and the other heavier ethers that are commercially manufactured are more environmentally friendly as antiknock components of gasoline, primarily due to their lower solubility in water (solubility in water of MTBE is 4.3 wt%, of ETBE is 1.2%, and of TAME is 2% [11]).

The blended octane number of MTBE is 109, its boiling point is 55°C, and it has a blend Reid vapor pressure of 8–10 psi [12]. MTBE when used above 11% provides the mandatory 2% oxygen in gasoline. The properties of TAME are blended octane number 104, boiling point 86°C, blended Reid vapor pressure 3–5 psi [12].

Excellent reviews on MTBE and other oxygenates are given by Ancillotti and Fattore [12], Trotta and Miracca [13], Peters et al. [14], and other chapters of this handbook.

MTBE is also used for production of high-purity isobutene, which is separated from a mixture of isobutene and other C_4 components. In this process MTBE is split into isobutene and methanol.

D. Thermodynamic Aspects

MTBE is prepared from the acid-catalyzed reaction of isobutene and methanol as shown below:

$$CH_3OH + CH_2 = \overset{\overset{CH_3}{|}}{C} - CH_3 \rightleftharpoons CH_2 - \underset{\underset{OCH_3}{|}}{\overset{\overset{CH_3}{|}}{C}} - CH_3$$

The reaction is weakly exothermic ($\Delta H = -37.7\,kJ/mol$) and equilibrium-limited. The temperature dependence of the chemical equilibrium constant of MTBE with respect to temperature is shown in Figure 1 [15]. The reaction synthesis of ETBE [16], TAME, and TAEE [17] are more equilibrium-controlled than that of MTBE, as also shown in Figure 1.

E. Kinetic Aspects

Kinetic models for MTBE synthesis have been reported by different workers over the years, a summary of which can be found in Ancillotti and Fattore [12]. Activity-based models are known to best describe the kinetics for the MTBE system.

A widely used activity-based Langmuir isotherm model was reported by Rehfinger and Hoffmann [15,18]. The experimental studies were performed in a continuously stirred tank reactor (CSTR). This model considers a two-step reaction: (1) sorption of reaction components on the active

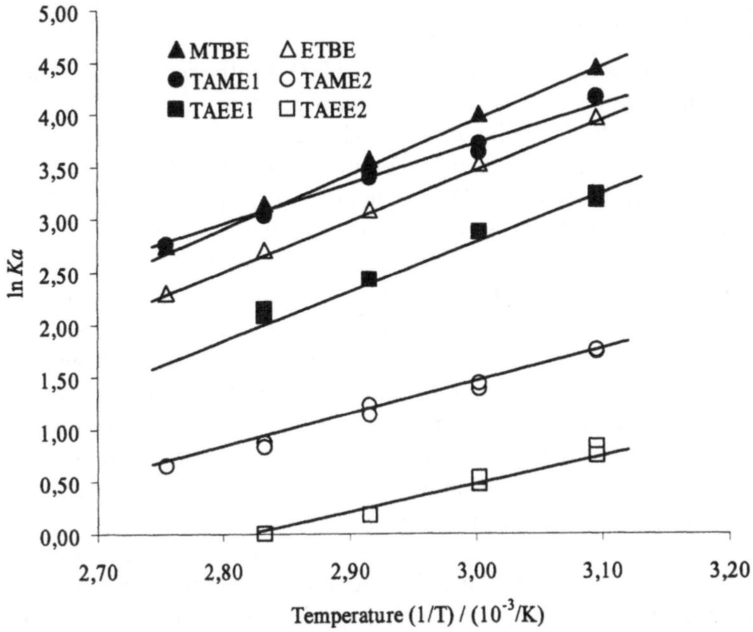

Figure 1 Chemical equilibrium constant for liquid-phase synthesis of MTBE, ETBE, TAME, and TAEE as a function of temperature: TAME1 for the reaction of 2-methyl-1-butene and methanol to TAME, TAME2 for the reaction of 2-methyl-2-butene and methanol to TAME, TAEE, respectively.

sites of the catalyst; and (2) chemical reaction of the sorbed molecules, which is the rate-controlling step. The rate of reaction is given by

$$r_{\mathrm{MTBE}} = k_{\mathrm{MTBE}}\ (T)\left(\frac{a_{\mathrm{IB}}}{a_{\mathrm{MeOH}}} - \frac{1}{K_{a,\mathrm{MTBE}(T)}}\frac{a_{\mathrm{MTBE}}}{a^2_{\mathrm{MeOH}}}\right) \tag{1}$$

where the terms are defined in the Notation section at the end of the chapter. The major side reactions in MTBE synthesis are dimerization of isobutene to form di-isobutene and methanol dehydration to dimethyl ether. *tert*-Butyl alcohol can form if there is water present in the process streams.

The kinetics for dimerization of isobutene to form di-isobutene are described by Haag [19]. The kinetics for methanol dehydration to form dimethyl ether have been given by Song et al. [20] considering case of a homogeneous catalysis. The kinetics of TAME synthesis have been reported by Rihko et al. [21].

F. Catalyst Selection

Any strong acidic catalyst can be used for MTBE synthesis, but industrially a macroreticular cation-exchange resin in hydrogen form (from sulfonic acid groups) is used. These acidic cation-exchange resins can be obtained commercially from a number of manufacturers, including Rohm and Haas (Amberlyst-15, Amberlyst-35), Dow Chemical Company (M-31, M-32), Bayer AG (K-2611, K-2631), Purolite (CT-175), etc.

A typical etherification catalyst such as Amberlyst-15 has a surface area of $45\,m^2/g$ and ion-exchange capacity of $4.8\,mEq\ H^+/g$ dry resin (Kunin et al. [22]). The morphology of these resins can be briefly described as structurally composed of small microgel particles (0.01–0.15 μm) forming clusters that are bonded at the interface. In brief, these resins are structurally composed of small microgel particles (0.01–15 μm) forming clusters which are bonded at the interface. Most of the active sulfonic acid groups are embedded within this gel phase. The porosity of these resins is from the void spaces present between and within these clusters. Both the gel phase and the pore phase are continuous inside these resins. In the etherification process, the polar alcohol causes the swelling of the polymer structure in the microgel particles. Solvation of the protons takes place, thereby affecting the strength of the acid site.

The sorption behavior of reaction components on catalytically active acid sites is described by the activity-based Langmuir sorption isotherms as shown below:

$$A_i + S \overset{K_{sj}}{\Leftrightarrow} A_iS \tag{2}$$

$$\theta_i = \frac{K_{s,i}a_i}{1+\sum_{j=1}^{N} K_{s,j}a_j}(i = 1, \ldots, N) \tag{3}$$

II. CATALYTIC PACKINGS—TYPES AND CHARACTERISTICS

Heterogeneous catalytic systems are preferred in reactive distillation systems due to the desire to eliminate catalyst separation and recycling problems associated with homogeneous catalysis, to avoid back reactions, and to exactly fix the position and height of the reaction zones.

Catalytic packings pose one of the most important challenges in the design and operation of catalytic distillation columns. This is because catalytic distillation is a three-phase operation in which vapor and liquid in countercurrent flow direction are in intimate contact with solid catalyst of size 1–3 mm. Larger catalyst size is not desired, since it would give rise to diffusion resistances.

To fulfill the requirements of an efficient catalyst, the catalytic packings should possess appropriate porosity and pore structure, large surface area, satisfactory activity and selectivity, easy catalyst installation and change-over, preferably online to avoid shutdown of the column, and long catalyst life.

The catalytic packings, in order to be effective for distillation, should exhibit low pressure drop through the packing bed, high separation efficiency (meaning a high number of theoretical plates per packed height), and good radial and axial dispersion of the liquid phase inside the catalyst structure. The external wetting efficiency should tend to unity, meaning good vapor–liquid contact, especially for fast equilibrium-controlled reactions, and high liquid residence time within the catalytic packings, important for slow reactions leading to the froth regime of operation.

In addition to these requirements, the catalytic packings should preferably have high mechanical, thermal, and chemical stability of the polymer structure, low osmotic swelling forces in order to limit breakages of the polymer structure, and above all, a low manufacturing cost.

Keeping in mind the above exacting requirements for catalytic packings, a general structure for catalytic packings has been described by what is known as the three levels of a porosity reactor by Krishna and Sie [23]:

1. Micropores inside the catalyst particles, providing the high surface area desired for reaction.
2. Millimeter-sized pores between the catalyst particles, held in a wire structure inside the catalytic pockets, providing channels for liquid flow down the bed by gravity.
3. Centimeter-sized channels between the various catalytic pockets, providing channels for the gas flow.

The open literature is replete with methods describing various techniques for immobilizing the heterogeneous catalyst in the distillation column, which are classified below:

1. Random packing
2. Structured packings broadly divided into the following subcategories:
    a. Catalytic trays
    b. Catalytic bales
    c. Structured supports
    d. Internally finned monoliths

Towler and Frey [24] have recently presented an excellent review of catalyst packing used in RD columns. The recent review by Taylor and Krishna [7] also discusses in detail various hardware aspects of catalytic packings. In addition to the catalytic packings described in this article, the reader is informed of the availability of propriety catalytic packings which are not described in this article.

Little study has been reported in the open literature to determine the effect of fluid dynamics in packed distillation column reactors. Fluid dynamics in catalytic distillation are very important since they affect not only the distillation but also the reaction in the column and are indispensable for scale up. Currently a consortium of European industries and academic institutions have been undertaking the INTINT program for development and hydrodynamic study of catalytic packings.

A. Random Packings

Random packings are often used for catalytic distillation columns, mainly due to the fact that they are a cheaper alternative compared to structured packings and they are also easier to handle. Some of them achieve higher separation efficiencies than structured packings. Various technologies have been adopted for the manufacture of random packings, summarized below.

1. Sintering of Polymers

In the sintering method, catalytically active material, e.g., a strong acid cation-exchange resin in bead form, is mixed with a thermoplastic powder in a rotating mixer. This mixture is packed in perforated plates of desired shape and molded in a furnace at high temperature. Details of various manufacturing technologies can be found in Fuchigami [25] and Spes [26]. This method suffers form the disadvantage that the macropores of the catalytically active ion-exchange resin can be blocked.

2. Melting of Polymers

Unfunctionalized polymer synthesized by using emulsion polymerization is mixed with thermoplastic material, molded with melting into the desired shape, and subsequently treated with activating agents such as sulfonic acid. This catalyst has a high activity and good heat resistance. Chaplits et al. [27] synthesized a catalyst having an ion-exchange capacity of 4.25 mEq/g. Smith [28] has described the manufacture of a catalyst by extruding the polymer followed by sulfonation. The ion-exchange capacity of this polymer is 5 mEq/g and is used for synthesis of MTBE in a laboratory column.

Flato and Hoffmann [29] used $6 \times 6\,mm^2$ raschig ring-shaped macroreticular ion-exchange catalyst to produce MTBE in a catalytic distillation column; the development and startup of the column is presented. The preparation of the catalyst is described in a patent by Gottlieb [30]. The catalyst has a porosity of 45% and an ion-exchange capacity of 4.54 mEq/g. Sundmacher and Hoffmann [31,32] also used this catalyst for synthesis of MTBE.

In the study of Flato and Hoffmann [29], the experimental column used for MTBE synthesis was 53 mm internal diameter (i.d.) and consisted of two sections 510 mm in height. The upper catalytic section was filled with a custom-prepared random catalytic packing molded in the shape of raschig rings called CVT rings of size $6 \times 6\,mm^2$. No prereactor was used. Up to 88% conversion of isobutene was obtained. Selectivity of MTBE obtained varied between 80% and 99.8%, increasing with pressure, increase in the methanol-to-isobutene molar feed ratio, and decrease in mole percentage of isobutene in the feed C_4 fraction. The purity of MTBE obtained as the bottom product from the reactive distillation column varied between 99.4% and >99.9%.

3. Block Polymerization

In the block polymerization method, polymerization is carried out for the monomers in the desired shape, e.g., in the form of raschig rings. This technique has been used by Gottlieb et al. [30] for preparation of the polymers. These packings have been evaluated for MTBE synthesis by Sundmacher and Hoffmann [34]. They have a good activity but suffer from very poor osmotic stability.

4. Precipitation Polymerization

To overcome the poor osmotic stability of the above block polymerization catalyst, catalyst synthesized by precipitation polymerization (PP) swells in the pore structure of a porous support and is thus suitable for catalytic distillation. Hoffmann et al. [35] and Kunz and Hoffmann [36] described the synthesis of random packings by precipitation polymerization on porous glass supports (GPP). The ion-exchange capacities of the GPP rings obtained was 0.9 mEq/g. The separation efficiency and pressure drops for these packings are shown in Figures 2 and 3, respectively. Some physical properties of these packings are presented in Table 1. These packings are ideal for use in catalysis of slow reactions.

These packings have a void fraction of 49%, catalyst loading of 51%, catalyst surface-to-volume ratio of $1129\,m^2/m^3$, and a packing surface-to-volume ratio of $576\,m^2/m^3$.

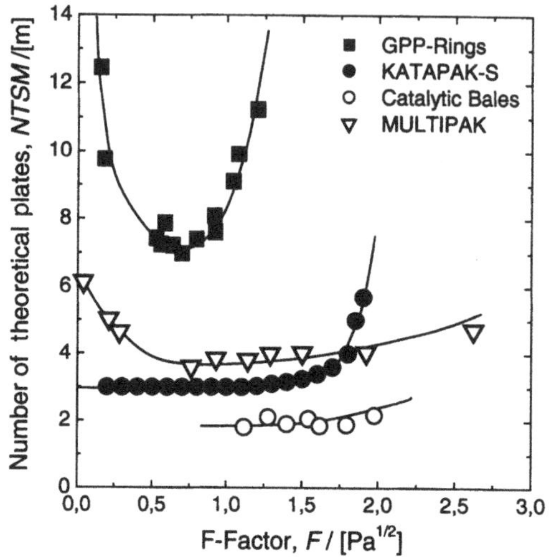

Figure 2 Separation efficiency for different packings used in reactive distillation.

These GPP rings have been used for MTBE synthesis by Sundmacher and Hoffmann [34] and for TAME synthesis by Sundmacher and Hoffmann [37]. The MTBE column configuration was the same as reported in the earlier section on melting of polymers.

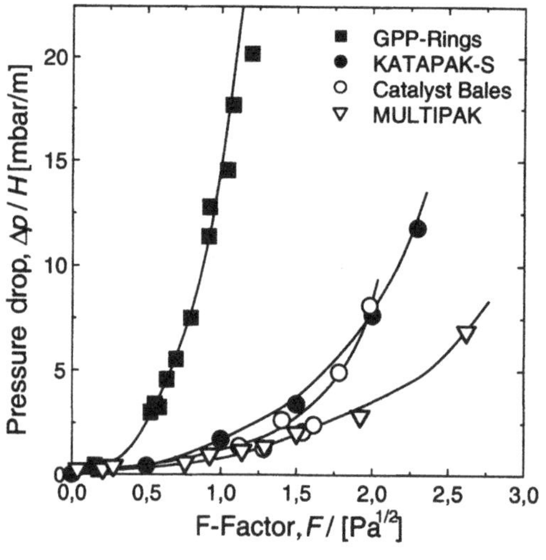

Figure 3 Pressure drop for different packings used in reactive distillation.

Table 1 Some Properties of Catalytic Packings

Description	Unit	Raschig ring	Catalyst bales	Katapak-S	Monolith
Void fraction	$[m^3/m_{col}^3]$	0.49	0.75[a]	0.75[a]	0.75[a]
Catalyst loading	$[m_{cat}^3/m_{col}^3]$	0.51	0.20[a]	0.20[a]	0.25[a]
Catalyst surface-to-volume ratio	$[m_{cat}^2/m_{cat}^3]$	1129	4000[a]	4000[a]	4000[a]
Packing surface-to-volume ratio	$[m_{cat}^2/m_{col}^3]$	576	800[a]	800[a]	1000[a]

[a]Lebens et al. [33].

B. Catalytic Distillation Trays

Another way of introducing catalyst in a catalytic distillation column is through use of catalytic distillation trays. One of the essential aspects for successful operation of catalytic distillation trays is the availability of sufficient quantity of catalyst for the reaction without a compromise in the pressure drop through the column. The design of catalytic trays for use in catalytic distillation processes is very different compared to that for use in normal distillation. Various different configurations of catalytic trays have been reported:

A catalyst bed is placed coaxially and alternating with at least one distillation tray. Vapor channels are provided on the catalyst tray, generally concentrically. Figure 4a shows a tray structure proposed by Jones [38]. Other similar tray arrangements have been patented for use in MTBE and TAME systems by Quang et al. [39], Nocca et al. [40,41], Sanfilippo et al. [42], etc. Configuration of novel catalyst-containing structures with separate distillation trays has been proposed by Yeoman et al. [43–45].

Pockets containing catalyst particles are placed adjacent to the downcomer in different configurations. The disadvantage of limited catalyst loading arises due to restricted space. Figure 4b shows the configuration proposed by Asselineau et al. [46] for MTBE synthesis, in which the liquid, after passing through the downcomer, goes onto the sieve plate where distillation takes place. An improvement in this technology in which the catalyst is placed concentric to a central liquid downcomer has been presented by Marion et al. [47].

Catalyst can also be placed within the downcomer's limited space as patented by Carland [48]. Here high catalyst loading cannot be achieved.

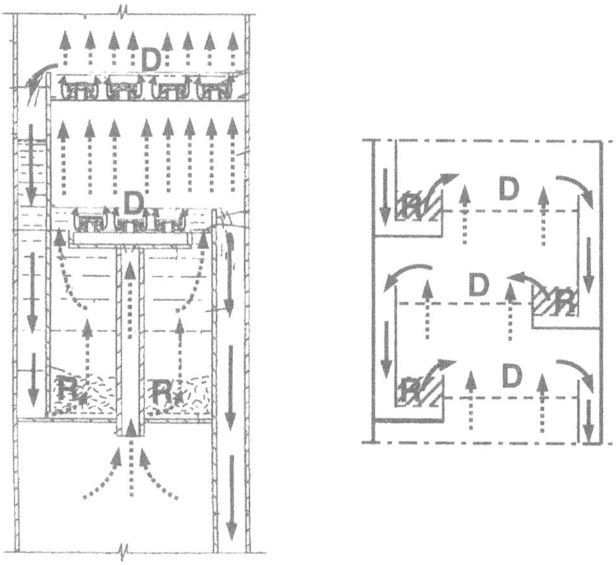

(a) (b)

Figure 4 Schematic diagram of catalytic distillation trays: (a) alternating catalyst bed and trays; (b) catalyst containing pockets adjacent to downcomer.

Other catalyst configurations include a vertical catalyst arrangement on catalytic trays in which the liquid flows along the catalyst envelopes (Jones [49]), cross flow of vapor and liquid (Sanfilippo et al. [42]), partial liquid flow through one or more catalyst bed placed vertically either on the side and/or the center, the downcomer having a liquid-permeable upper surface (Yeoman et al. [50]).

It is desirable to have configurations of catalytic trays in which the catalyst can be easily replaced, without a full shutdown of the column. Such arrangements can be found in patents by Carland [48] and by Jones [51], in which the reaction and distillation occur on the same tray, and by Yeoman et al. [43–45].

In most of the configurations described above, the catalyst is in a nonfluidized condition. There are cases where the catalyst can be in a fluidized condition, which can take place only when reaction and distillation occur on the same tray (Franklin [52]).

Studies are in progress at the University of Amsterdam exploring the design of tray columns and other catalytic packings using computational fluid dynamics (Krishna [53]).

C. Catalytic Bales

Catalytic bales packings, promoted by C. D. Tech (now ABB Lummus) and described by Smith [54–56], are widely used in industry for catalytic distillation processes. Catalyst particles are sewed in fiber pockets using a fiberglass cloth. The cloth is rolled up and wrapped with alternate wire gauze layers in cylindrical form to form the catalytic bales as shown in Figure 5. These are commonly referred to as "teabag" packings. These catalytic bales can be stapled onto trays of distillation columns piled one above another up to the desired height. Catalyst bales are cheap and widely used in commercial catalytic distillation columns. The catalyst can be replaced in the bale packings manually. These packings have a void fraction of 75%, catalyst loading of 20%, catalyst surface-to-volume ratio of $4000\,m^2/m^3$, and a packing surface-to-volume ratio of $800\,m^2/m^3$.

Separation efficiency and pressure drop characteristics for catalytic bales have been studied by Subawalla et al. [57] and are shown in Figures 2 and 3, respectively. It can be observed that the catalytic bales have poor mass transfer characteristics. Xu et al. [58,59] have studied pressure drop, holdup, and axial liquid dispersion. Zheng and Xu [60] have studied the mass transfer characteristics of bale packings.

Zheng and Xu [61] studied MTBE synthesis in a pilot-plant reactive distillation column without use of prereactors. A column of 0.95 m i.d. having a reactive catalytic section packed with catalytic bales, a nonreactive enriching section of 6 stages, and a nonreactive stripping section of 10 stages was used. Methanol is fed at the top of the catalytic section and pure isobutene at the bottom of the catalytic section. Ninety-one percent conversion of isobutene is achieved at an operating pressure of 950 kPa

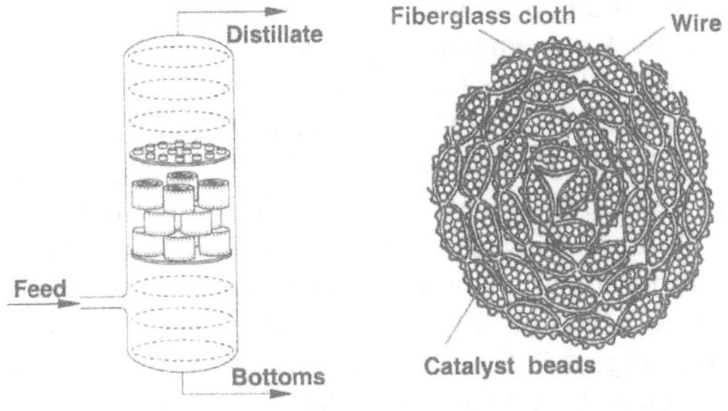

Figure 5 Schematic diagram of catalytic bales licensed by CR&L.

and a reflux ratio of 2. The temperature at the top of the column is 64.2°C, and it is 142.8°C at the bottom.

Other studies on MTBE synthesis using catalytic bales are described in detail by Smith [62–65], and by Bakshi and Hickey [66]. Gonzalez et al. [67] have used catalytic bales for synthesis of TAME.

Improvements in catalytic bales has been described by Johnson [68], in which deformable material is incorporated into the bundles to allow for swelling of the catalytic material. Crossland et al. [69] have also described improvements in bale packings.

D. Structured Packings

In structured packings the catalytic particles are embedded between two corrugated wire gauze sheets in zigzag form. These packings have the desired features of a catalyst support combined with effective distillation characteristics. They have good heat transfer characteristics, thus eliminating hot spots in the reactors. Structured packings for use in catalytic distillation reactors are commercially available from many companies for laboratory and industrial-scale columns.

Katapak-S and Katapak-SP are licensed by Sulzer Chemtech, Switzerland. These packings have a void fraction of 75%, catalyst loading of 20%, catalyst surface-to-volume ratio of 4000 m^2/m^3, and a packing surface-to-volume ratio of 800 m^2/m^3. Separation characteristics and pressure drop for Katapak-S have been studied by Moritz and Hasse [70] and by Moritz et al. [71] and are shown in Figures 2 and 3, respectively. Radial and axial dispersion of liqud in Katapak-S has been studied using computational fluid dynamics (CFD) by Baten et al. [72]. Technical details of these packings can be obtained from Stringaro [73], and their use in ether synthesis process from Shelden and Stringaro [74].

KataMax, licensed by Koch Engineering (USA), is widely used for MTBE production in the Huls-UOP process. Details of KataMax can be found in an article on MTBE by Pinjala et al. [75].

MultiPak (see Figure 6), licensed by the Julius Montz group of Germany, was developed in cooperation with the University of Dortmund. In Multipak packings the stacks are built of alternating layers of meshed catalyst envelopes and crimped structured distillation sheets. Multipak exhibits a slightly lower pressure drop, as shown in Figure 3. Separation characteristics of Multipak are shown in Figure 2. A major advantage of MultiPak is the identical geometric structure from lab to pilot scale.

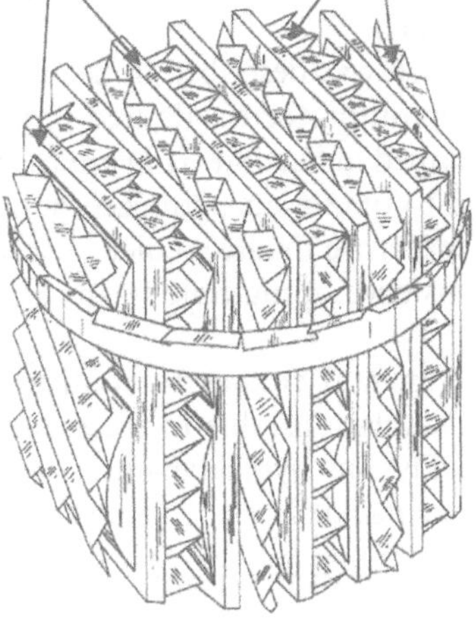

Figure 6 Schematic diagram of Katapak-S, a structured packing, licensed by Sulzer Chemtech.

Oudshoorn [76] presented a method for preparation of binderless films of catalytically active zeolite crystals on metal and ceramic. This support structure is used for the production of MTBE and ETBE.

Pinjala et al. [75] have described a pilot plant for MTBE synthesis using KataMax packings.

E. Internally Finned Monoliths

Use of internally finned monoliths in catalytic distillation is a relatively new concept. Details on internally finned monoliths can be obtained from Lebens et al. [77].

III. REACTIVE DISTILLATION SIMULATION/DESIGN AND VALIDATION

The steps involved in mathematical simulation and design analysis of a reactive distillation process are broadly classified as follows:

1. Preliminary feasibility study for a reaction system intended to be performed using reactive distillation using the concepts of residue curve maps, described in Section III.A.

2. Modeling of the reactive system using basic equilibrium stage models described in Section III.B using a simulation approach, or Section III.F using a design approach.
3. Selection of column hardware, which was described in Section II.
4. Determination of hardware configuration for the desired process.
5. Rigorous rate-based models, described in Section III.D are used to check the preliminary design and to obtain an optimized design.
6. Column dynamics and control aspects are finally studied; a special classification of the same, viz., existence of multiple steady states, is described in Section III.E.
7. It is important to perform experimental studies to confirm the predictions of the mathematical simulations, due to the existence of many "gray areas," especially in hydrodynamic aspects involved with use of heterogeneous packings.

A general guideline for designing catalytic distillation processes has been presented by Subawalla and Fair [78].

Modeling of reactive distillation systems fall into two categories, design and simulation. In a simulation approach the column parameters such as number of stages are specified along with the feed input and it is desired to determine the output compositions. In a design approach the input and output compositions are specified and the number of stages is required to be determined. These approaches demand the use of different types of calculations.

A. Thermodynamics of Reactive Distillation Processes

Calculation of simultaneous chemical reaction and phase equilibrium is essential for any theoretical study of reactive distillation systems, reviews of which are presented by Frey and Stichlmair [79] and by Seider and Widagdo [80]. Four possible modeling principles can be seen in the case of reactive distillation: both vapor–liquid equilibrium (VLE) and chemical equilibrium (CE) are assumed (equilibrium controlled reactions); VLE is not assumed but CE is assumed; VLE is assumed but CE is not assumed (equilibrium-controlled reactions); and both VLE and CE are not assumed. The concept of distillation line diagrams, commonly called residue curve maps (RCMs), is often used. In RCMs the change of liquid composition with time is plotted. Determination of RCMs, either experimentally or computationally, is invariably the first step in determining whether a given system is feasible for reactive distillation. The column behavior of a reactive distillation system can be predicted from the RCM.

Barbosa and Doherty [81–84] have studied RCMs for the first case in which both VLE and CE are reached. The authors make use of transformed variables defined by

$$X_i = \frac{(x_i/\nu_i) - (x_k/\nu_k)}{(\nu_k - \nu_T x_k)} \qquad Y_i = \frac{(y_i/\nu_i) - (y_k/\nu_k)}{(\nu_k - \nu_T y_k)} \tag{4}$$

where X_i and Y_i are the transformed liquid and vapor compositions, respectively. Knowing the actual compositions x_i and y_i, the transformed compositions X_i and Y_i can be easily calculated; the reverse calculations require search algorithms. The singular or azeotropic points are defined by

$$X_i = Y_i \tag{5}$$

These points are commonly called "reactive azeotropes," since these are points in a reactive distillation system where the concentrations of vapor and liquid phase cannot be changed due to distillation.

The advantage of using transformed variables is that X_i has the same value before and after chemical reaction. For a simple batch distillation the equations reduce to

$$\frac{dX_i}{d\tau} = X_i - Y_i \tag{6}$$

Integration of the above equations gives the RCM for the system.

Reactive and nonreactive RCMs are presented for the isobutene–methanol–MTBE systems in Figure 7. It can be observed from this figure that the azeotropes between isobutene–methanol, methanol–MTBE, and distillation boundaries "react away" for the reactive case, thereby greatly simplifying the phase equilibrium behavior. It is interesting to note that while the former is true, reactive azeotropes are found in the case of ideal reactive mixtures. Bifurcation analysis was performed by Okasinski and Doherty [85] to study the effect of the reaction equilibrium constant on the reactive azeotropes for the MTBE system.

RCMs for the MTBE system in the presence of inert component *n*-butene is shown by Ung and Doherty [86–88], Espinosa et al. [89–91], and Perez-Cisneros et al. [92]. The necessary and sufficient conditions for existence of reactive azeotropes is discussed.

RCMs for slow kinetically controlled reactions where CE is not reached but VLE is reached have been studied by Venimadhavan et al. [93] for homogeneously catalyzed MTBE systems and by Thiel et al. [94] for

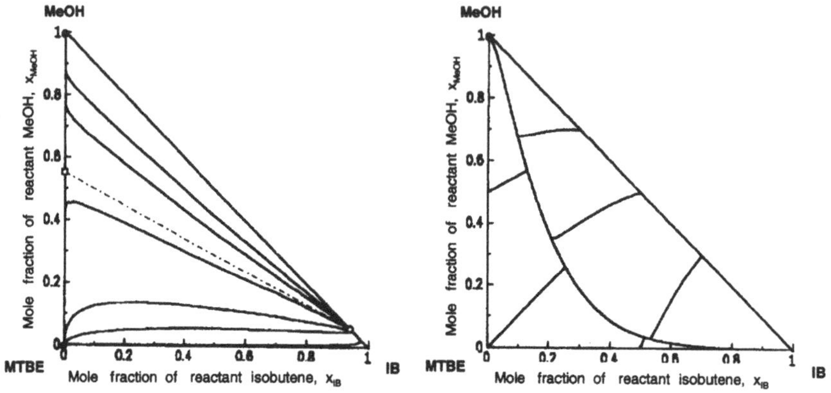

Figure 7 Reactive and nonreactive residue curve maps. (Adapted from Thiel et al. [94]. Reprinted with permision of Chemical Engineering Science. Copyright by Elsevier Science.)

heterogeneous catalyzed MTBE and TAME systems. For kinetically controlled reactions the Damkohler number is a very important parameter. The Damkohler number is defined as the ratio of the characteristic liquid residence time to the characteristic reaction time and is given as

$$\mathrm{Da} = \frac{H/F}{1/k} \tag{7}$$

Low Damkohler number indicates that phase equilibrium is not reached in the system, while high Damkohler number indicates that the system approaches chemical equilibrium.

Thiel et al. [94] show the effect of change of Damkohler number from 0 (nonreacting) up to 1, where the trajectories moves with increase in Damkohler number toward the curve of chemical equilibrium.

Figure 8 shows the RCMs for the MTBE system, where the homogeneously and heterogeneously catalyzed cases are compared at operating pressure of 0.8 MPa. It can be seen from this figure that heterogeneously catalyzed RCMs with initial mixtures containing less than 99% isobutene reach the stable node of pure methanol like the homogeneously catalyzed cases, although the shape of the curves are different. This is because the heterogeneously curves do not enrich as much methanol, due to the negative reaction order with respect to methanol, as the homogeneous ones. Moreover, the heterogeneously catalyzed RCMs with initial mixtures containing more than 99% isobutene show a second stable node.

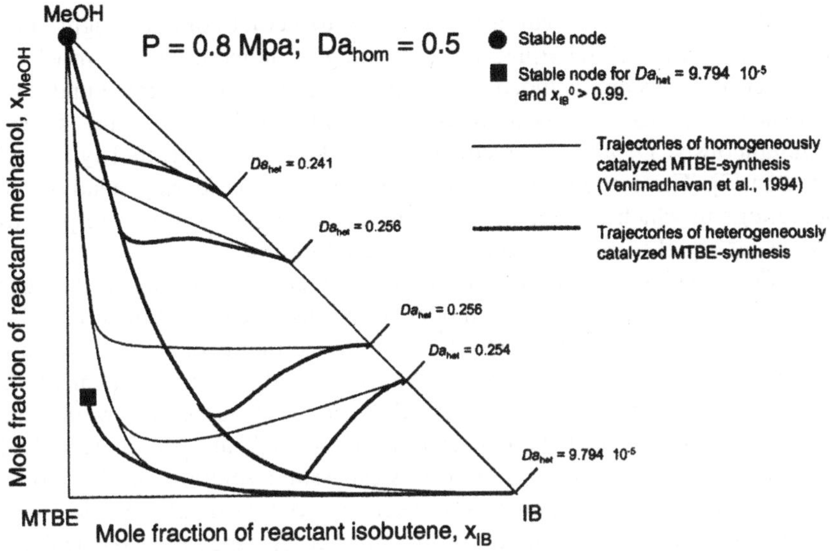

Figure 8 Homogeneously and heterogeneously catalyzed residue curve maps for the MTBE system. (Adapted from Thiel et al. [94]. Reprinted with permision of Chemical Engineering Science. Copyright by Elsevier Science.)

For the homogeneously catalyzed MTBE system, bifurcation analysis showed the existence of a critical Damkohler number for the disappearance of the distillation boundary (Venimadhavan et al. [95]).

For the MTBE system the attainable regions which can be obtained for different configurations of reactor and separator units using RCMs has been analyzed by Nisoli et al. [96].

A generalized reactive separation module for a flash-type operation in which the reaction, homogeneous or heterogeneous, and separation may occur in simultaneously or independently has been tested for the MTBE system by Gani et al. [97].

B. Simulation Using Equilibrium-Stage Models

Equilibrium-stage models have been well established over the years for use in conventional distillation. They have been widely used to model reactive distillation processes; see Taylor and Krishna [7] for a recent review. In this approach the distillation column is treated as a series of equilibrium stages in which the vapor and liquid streams leaving each stage are in thermodynamic equilibrium and finite reaction rates are accounted for in each reactive stage. Height equivalent to theoretical plate (HTU) is used to translate the number of equilibrium stages to actual column height. For these models the MESH equations, consisting of material balance in which the reaction terms are included, equilibrium relation, summation equations, and heat balance are written for the individual stages and solved simultaneously. Details of the equations can be obtained from Taylor and Krishna [7]. These equations can be solved either considering steady-state operations in which the derivative of the MESH equations with respect to time is set equal to zero, or unsteady-state operation.

For steady-state equilibrium-based models, various methodologies have been adapted to solve the model equations, such as, manual calculations performed tray to tray, short-cut methods, the theta method, tearing algorithms, relaxation methods, Newton's method, continuation methods such as homotopy, etc. For an excellent comprehensive review on this subject the reader is referred to Taylor and Krishna [7] and Sundmacher [98] for literature review upto 1994. For steady-state solution of the model equations an increasing number of workers in recent years have made use of commercially available simulation software such as Aspen Plus, particularly the RADFRAC inside-out algorithm (sequential module architecture), Pro/II, SpeedUp (equation-oriented architecture), etc.

A dynamic equilibrium model for MTBE synthesis was developed by Abufares and Douglas [99]. Steady-state simulations were performed using SpeedUp (equilibrium model) and the RADFRAC module of Aspen Plus (kinetic model). Figures for temperature profile and MTBE mole fraction adapted from their publication are shown in Figure 9 and compared with experimental results. Transient response of the column was studied for methanol feed and reflux ratio.

READYS, a dynamic simulator for the equilibrium-based models, is described by Ruiz et al. [100]. See Taylor and Krishna [7] for a general

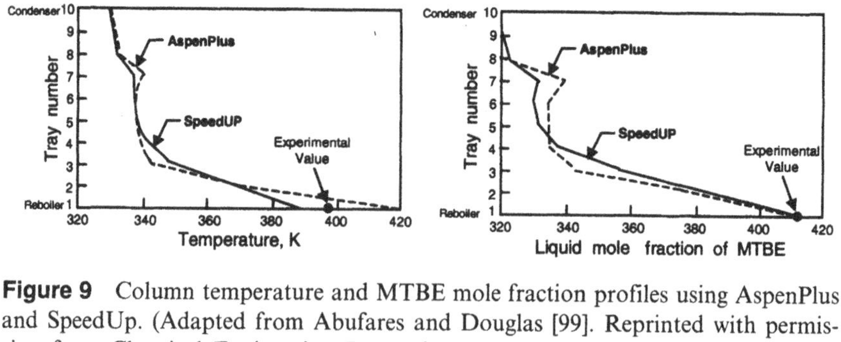

Figure 9 Column temperature and MTBE mole fraction profiles using AspenPlus and SpeedUp. (Adapted from Abufares and Douglas [99]. Reprinted with permission from Chemical Engineering Research and Design. Copyright by Institute of Chemical Engineers.)

overview of other approaches to dynamic simulation of reactive distillation systems using equilibrium models.

Sundmacher et al. [101] compared the use of a quasi-homogeneous equilibrium-based model and an equilibrium model incorporating mass transfer resistances inside the catalyst. Side reactions were considered for the second case. The quasi-homogeneous model underestimated the temperature in the reboiler because side reactions were not considered. The predictions using the later model are comparable to the one obtained using a fully rate-based model (Sundmacher et al. [102]), described in detail in a later section.

Influence of side reactions on the MTBE splitting system have been taken into consideration by Qi et al. [103] and Stein et al. [104]. MTBE splitting is a route for synthesis of high-purity isobutene. It is shown by these authors how the feed point and operating conditions can make a drastic change in obtaining either the desired component isobutene, or the undesired component diisobutene, when side reactions are considered.

More details of simulations using equilibrium-stage models for multiple steady states, a very interesting topic in ether synthesis, are given in Section III.F.

C. Physicochemical Analysis of Reactive Distillation Processes

In a homogeneously catalyzed reactive distillation process, diffusion is present only at the gas–liquid interface. Reactive mass transport occurs in the bulk liquid phase for slow reactions, in the liquid film for extremely fast reactions, and in both bulk the liquid and liquid film for intermediate cases. Transport phenomena diagrams for this case can be found in Sundmacher et al. [9].

To overcome the difficulties involved with use of multicomponent efficiencies, a nonequilibrium-stage model for distillation was developed by Krishnamurthy and Taylor [105] in which mass transfer is calculated using fundamental mass transfer models.

For modeling mass transfer in multicomponent systems, the Maxwell-Stephan theory has been used (Taylor and Krishna [106]). A simplified Maxwell-Stephan equation assumes the form

$$N_i = -c_t D_{i,eff} \frac{\partial x_i}{\partial t} \tag{8}$$

Information on solving the Maxwell-Stephan equations can be obtained from Frank et al. [107,108].

The generalized Fick's law can also be used for modeling of mass transfer in multicomponent systems, as given by

$$J_i = -c_t \sum_{k=1}^{c-1} D_{i,k} \frac{\partial x_i}{\partial z} \tag{9}$$

For solution of the Fick's law equations using numerical methods, refer to Kenig and Gorak [109].

For the heterogeneous catalyzed case in the presence of a catalytically active packing element, the reactive mass and energy transport most commonly takes place inside the microgel of the catalyst. Here nonreactive transport process in the two film layers between the gas–liquid and liquid–solid boundaries have to be accounted for. The local transport phenomena process for the heterogeneous case is shown in Figure 10.

Sundmacher and Hoffmann [31,32,34,110] have recommended the use of the Maxwell-Stephan equation for modeling of macrokinetic analysis of MTBE synthesis using a random catalyst. The catalyst effectiveness factor is determined and shows good agreement with experimental data obtained from a batch reactor. However, these authors ignore the Knudsen diffusion, which is considered in the dusty gas and dusty fluid models.

A detailed analysis of the simultaneous mass transport and reaction phenomena inside a porous catalyst was performed by Sundmacher

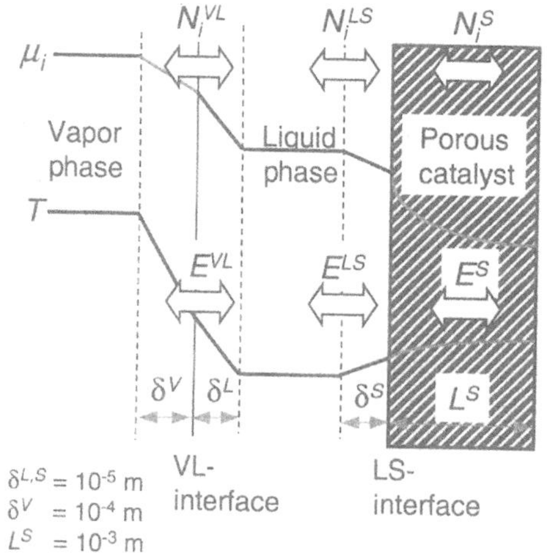

Figure 10 Transport phenomena in heterogeneously catalyzed reactive distillation process. (Adapted from Sundmacher et al. [9]. Reprinted with permission of Chemical Engineering Communication. Copyright by Gordon & Breach Publishing.)

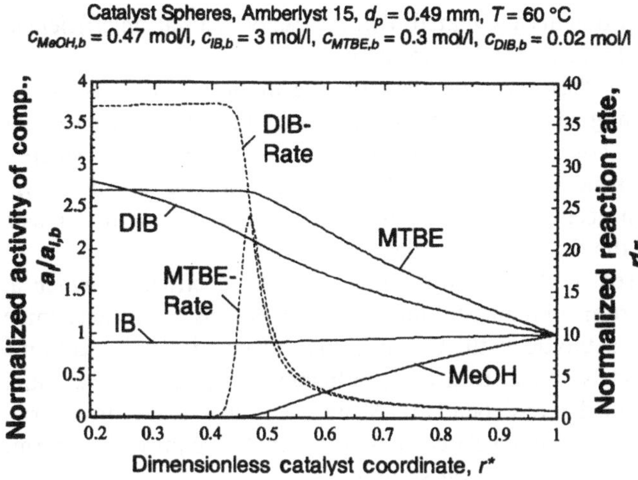

Figure 11 Simulated reaction rates and component activity profile within a catalyst particle for the MTBE system. (Adapted from Sundmacher et al. [101]. Reprinted with permission of Chemical Engineering Science. Copyright by Elsevier Science.)

et al. [101]. It was found that for low methanol bulk concentrations, the dimerization reaction of isobutene occurs in the inner core of the catalyst, where the MTBE formation rate vanishes as shown in Figure 11.

Uhde et al. [111] considered the side reaction of isobutene dimerization for modeling of activity and selectivity of ion-exchange catalysts. Good fit between experimental and simulated reaction rate profiles have been observed for both the main and side reactions.

The dusty gas model has been used for modeling multicomponent mass transfer in porous media where the mean free path length of the molecules is of order of the pore diameter (Maxwell [112]), Mason and Malinauskas [113]). The dusty fluid model developed by Krishna and Wesselingh [114] has been used for modeling of liquid-phase diffusion in porous media. Details of both these models can also be obtained from Higler et al. [115].

D. Simulation Using Rate-Based Models

For the modeling of slow reactions in reactive distillation it is necessary to use kinetic expressions which are dependent on the temperature and concentrations of the components involved. Normal distillation and reactive distillation processes almost never operate in complete equilibrium.

To overcome the problems associated with the use of distillation efficiencies, nonequilibrium rate-based models are used.

Rate-based models are dependent on the mode of catalysis, heterogeneous or homogeneous. Using rate-based models, heterogeneous reactions can be considered as either pseudo-homogeneous, where a lumped term is considered for catalyst diffusion and reaction, or using a rigorous approach accounting for catalyst diffusion and reaction or using the cell model or the dusty fluid model.

The rate-based models use finite mass transfer coefficients which are either calculated from fundamental mass transfer models or are determined experimentally for catalytic distillation packings, as mentioned in Section II. In rate-based models, mass transfer coefficients, interfacial area, liquid holdup, etc., are hardware-dependent. Moreover, the calculations in rate-based models for the driving force of mass transfer require the thermodynamic properties of the reaction system.

The commercial software Aspen Plus includes the RATEFRAC module for solving of rate-based reactive distillation problems. Pinjala et al. [75] used the RATEFRAC module to simulate a commercial Huels MTBE process packed with KataMax packings. Good agreement of the predictions with experimental pilot plant data was observed.

Zheng and Xu [61] used a MERQ rate-based model from the work of Taylor et al. [116] in which the vapor and liquid phases have individual material and energy balances. The model equations were solved using the Newton-Raphson method. Simulations were performed for a MTBE synthesis column and comparison of simulated results with experimental temperature profile of a pilot-plant MTBE column was carried out.

Sundmacher and Hoffmann [37] presented a detailed nonequilibrium model for packed columns, taking into account transport processes inside the porous catalytic packings. The porous catalytic packings are totally wetted internally and externally (Sundmacher [98]). Mass and energy transport occur across vapor–liquid and solid–liquid interfaces. The column is treated as a series of nonequilibrium stages. From the generalized Maxwell-Stefan equation, the vector of the mass flux density on the vapor side (n^V) and liquid side (n^L) of the vapor liquid interface is given by

$$n^V = c^V \beta^V_{\text{av}} k^V_{\text{av}} (x^V - x^{V1}) \qquad n^L = c^L \beta^L_{\text{av}} k^L_{\text{av}} (x^{L1} - x^L) \tag{10}$$

where β is the bootstrap matrix. For details on the bootstrap matrix the reader is referred to Taylor and Krishna [106] for further details.

The component mass balances are given by

Vapor phase	$M_{i,j}^{V} = V_{j+1}x_{i,j+1}^{V} - V_j x_{i,j}^{V} - (a_j\,\Delta V_j)n_{i,j}^{V} = 0$ (11)
Liquid phase	$M_{i,j}^{L} = L_{j-1}x_{i,j-1}^{L} - L_j x_{i,j}^{L} + F_j x_{i,j}^{F}$ $-(a_j\Delta V_j)n_{i,j}^{L} - \nu_i(\varepsilon_i^{S}\Delta V_j)\eta_j r_{V,j}^{L} = 0$ (12)
Vapor–liquid interface	$M_{i,j}^{I} = (a_j\Delta V_j)n_{i,j}^{V} - (a_j\Delta V_j)n_{i,j}^{L} = 0$ (13)

The mass transfer inside the catalyst affects the column performance and is called the catalyst effectiveness factor η, which is a function of the Thiele modulus and can vary along the axial coordinates. The steady-state solution is accomplished by solving the system of differential-algebraic (DA) equations using the solver LIMAX (Deuflhard et al. [117]). For experimental validation, a column packed with GPP rings was studied in Clausthal, and the results were simulated for MTBE synthesis using this model [37]. Parametric sensitivity of composition profiles across the column is determined for change in reflux ratio, feed ratio of reactants, and pressure. Effect of change in feed position was investigated with respect to the catalyst effectiveness factor. It can be observed from Figure 12 that the catalyst effectiveness factor can change dramatically if the feed strategy is improper. The simulations confirmed the experimental observations that high isobutene content in the C_4 feed and high reflux ratio led to more MTBE in the bottom product.

Most of the nonequilibrium models cannot account for flow patterns of vapor and liquid on tray columns or maldistribution in packed columns. Higler et al. [118,119] developed a nonequilibrium cell model to describe the mass transfer, reaction, and flow pattern on a catalytic distillation tray. On each tray the two-phase mixture is split up into a number of contacting cells with flexibility of interconnection, as shown in Figure 13. The connection pattern and backmixing characteristics can be chosen, thereby allowing study of the flow patterns and maldistributions. Mass transfer resistances are located in the vapor–liquid and liquid–solid films. The Maxwell-Stephan equations are used for calculation of mass transfer coefficients. Mass transfer inside the porous catalyst is determined using the dusty fluid model.

ChemSep, a nonequilibrium-based computer program model which can be used for modeling reactive distillation processes, is available for academic and commercial use from Clarkson University [120].

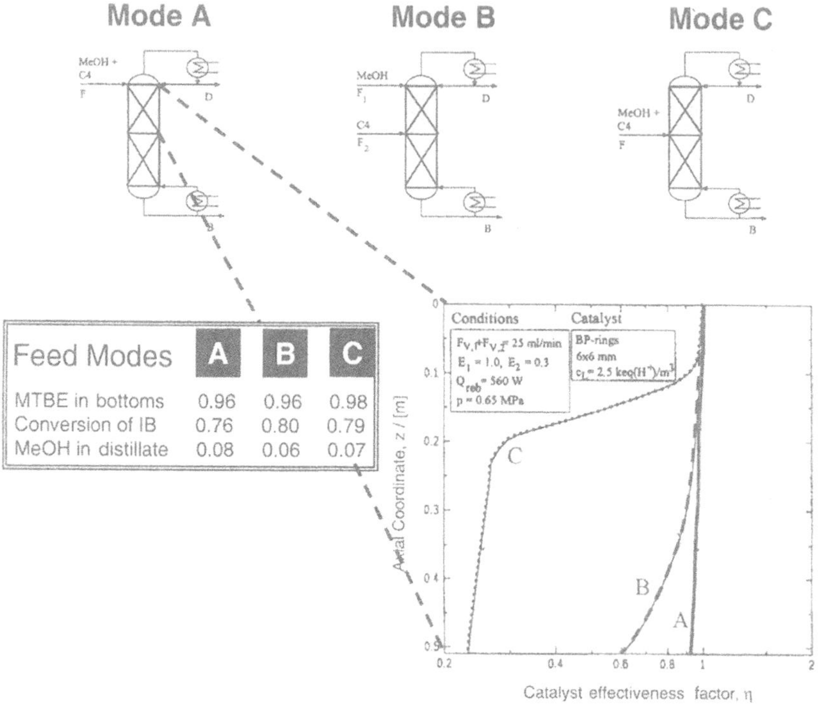

Figure 12 Simulated profiles of catalyst effectiveness factor for different feed modes. (Adapted from Sundmacher and Hoffmann [37]. Reprinted with permission of Chemical Engineering Science. Copyright by Elsevier Science.)

A dynamic nonequilibrium cell model has been described by Baur et al. [121], details of which are given in the section on multiplicities. Other details on rate-based models can also be obtained from the section on multiplicities.

It is appropriate to mention here that an integrated tool for the synthesis and design of catalytic distillation operations has been recently developed by a consortium of five industries and five universities (Kenig et al. [122]). The project was initiated by SUSTECH and supported by the European Union under the BRITE-EURAM program. The project was divided into two modules, SYNTHESISER and DESIGNER. SYNTHESIZER is a predictive tool which evaluates the feasibility of catalytic distillation usage for a desired system using different column configurations. DESIGNER is a process simulator used for modeling, design, sizing, etc., for the catalytic distillation process using rate-based models. A number of models have been used to describe the hydrodynamics and mass transfer coefficients. Newton's method or a relaxation method using LIMEX is used

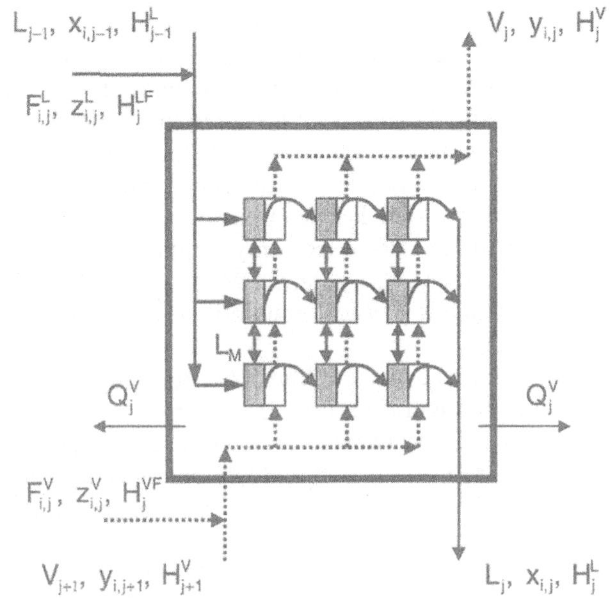

Figure 13 The nonequilibrium cell model. (Adapted from Higler et al. [118]. Reprinted with permission of American Institute of Chemical Engineers. Copyright 1999 AIChE. All rights reserved.)

for solving the model equations. The Kenig paper does not present any mathematical details. MTBE system was tested using this simulator and experimental results compared.

E. Multiple Steady States (MSS) in Reactive Distillation Columns

A column of a given design, exhibiting different product compositions at steady state for the same set of inputs, e.g., feed composition, feed position, feed flow rates, and operating parameters such as reflux ratio and distillate flow rate, is said to be displaying output multiplicities. Multiplicities can arise in reactive distillation due to distillation effects which are not relevant in the case of MTBE and TAME systems, reaction effects, or simultaneous reaction and distillation effects. An excellent review of MSS is given by Guttinger [123].

Pisarenko et al. [124] first reported multiple steady states in single-product reactive distillation columns. From 1993 on, a number of publications have appeared that have studied the presence of MSS in

MTBE and TAME catalytic distillation columns. Nijhuis et al. [125] used RADFRAC simulation, the equilibrium module of Aspen Plus, and showed that MSS can occur in the MTBE column (total 16 trays, of which 8 were reactive) by varying the methanol feed tray location between stages 9 and 11. A high isobutene conversion state accompanied by relatively low temperature in the catalytic section and a low isobutene conversion state accompanied by a higher temperature in the catalytic section was obtained. Fixing the methanol feed tray location but varying the reflux ratio or amount of catalyst also obtained MSS. The explanation for MSS is due to the simultaneous physical equilibrium and exothermic reaction.

Jacobs and Krishna [126] also used RADFRAC to study MSS in MTBE columns. The column configuration used is similar to the one used by Nijhuis et al. [125]. A detailed mechanistic explanation is provided by Hauan et al. [127,128] of why MSS occurs in the MTBE column considered by Jacobs and Krishna [126]. These authors suggested the explanation that in order to achieve a high isobutene conversion solution, the lower section of the column must contain sufficient MTBE so as to lift all the methanol into the catalytic zone through a minimum-boiling azeotrope. The reactive mixture in the lower part of the catalytic zone must be sufficiently diluted by *n*-butene to prevent MTBE decomposition before it escapes the catalytic zone. The bottom product will contain high levels of methanol if too little MTBE is present in the lower sections of the column or too little *n*-butene is present in the lower part of the catalytic zone, leading to a low-conversion solution. Hauan et al. [128] studied the effect of pulse change in the isobutene feed. The authors observed that a +4% change leads to sustained oscillations, and with +5% change the conversion jumps to the low-conversion steady state.

Schrans et al. [129] showed that MTBE synthesis can exhibit two stable and one unstable MSS using the column configuration studied by Jacobs and Krishna [126]. Dynamic simulations were used to determine the unstable MSS. Dynamic simulations showed oscillatory column behavior for a disturbance in C_4 feed composition. Eldarsi and Douglas [130] determined that the methanol-to-butene molar feed ratio is the main factor affecting the existence of MSS. Sneesby et al. [131] studied various aspects of input, output, and pseudo-multiplicities for MTBE and ETBE columns. Sneesby et al. [132] presented mechanistic interpretation of multiplicity in MTBE and ETBE columns.

Guttinger and Morari [133–135], considering infinite internal flow rates and an infinite number of stages, called the ∞/∞ method, determined that the MSS in the MTBE system mentioned above can be avoided by using control strategies such as bottom mass flow rate.

Mohl et al. [136] performed bifurcation analysis using continuation methods of the computer-aided modeling and simulation environment DIVA, which contains a powerful differential-algebraic equation solver especially suitable for large sparse systems with arbitrary structural properties (Kienle et al. [137]). They calculated the window of occurrence of MSS in MTBE and TAME systems as shown in Figure 14. In this study, multiplicities were found due to variation of real operating parameters such as reflux ratio and heating rate. It can be observed from the figure that the window of occurrence of MSS in MTBE systems is small and has to date never been experimentally verified. Compared to this, the multiplicity region is larger in the TAME system and has been experimentally verified.

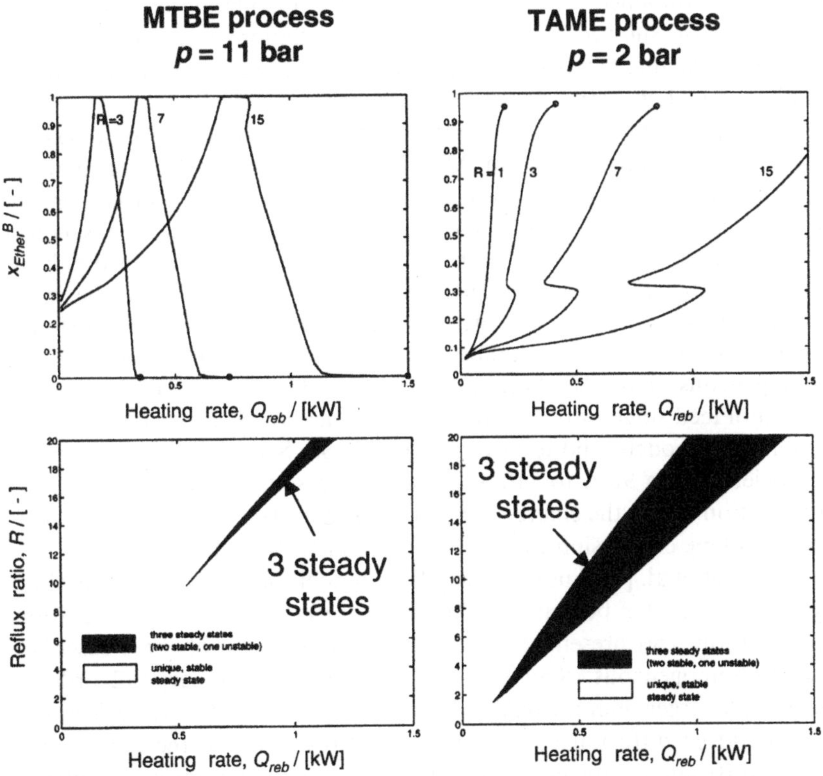

Figure 14 Window of multiple steady-state occurrence in MTBE and TAME systems. (Adapted from Mohl et al. [136]. Reprinted with permission of Chemical Engineering Science. Copyright by Elsevier Science.)

Bravo et al. [138] were the first to publish experimental verification of MSS in a pilot-plant TAME distillation column. Confirmation of MSS in TAME synthesis was later presented as well by Mohl et al. [139], Thiel et al. [140], Rapmund et al. [141], and Mohl et al. [136]. Figure 15a shows the experimental low and high conversion states of isobutene from Mohl et al. [136]. Figure 15b shows the experimental transient behavior of the TAME column, which, in response to a pulse injection of feed, changes from low conversion state to a high conversion state. It has been proved that the MSS for the TAME process are caused by kinetic instabilities using Hopf bifurcations.

Sundmacher and Hoffmann [142] were the first to report sustained spontaneous oscillations in an experimental MTBE column, which was packed using random glass-supported ion-exchange catalyst rings. Oscillations of reflux, boiling temperature, and pressure were observed. Surprising results were obtained, in which the nonreactive distillation of methanol and isobutene also showed oscillations. These oscillations were an outcome of the hydrodynamic behavior of the column, which was packed with random catalytic packings with a lower void fraction.

A nonequilibrium-stage model has been used by Higler et al. [143] to study the multiplicities in an MTBE column. MSS exist for both the nonequilibrium- and equilibrium-stage models using bottom product withdrawal rate as the operational specification. An interesting observation is that the "low conversion branch" of isobutene for the nonequilibrium model lies above that for the equilibrium model. This is because the reverse reaction is arrested due to incorporation of mass transfer resistances. Comparison of MSS for equilibrium and nonequilibrium models (pseudo-homogeneous) has also been presented by Baur et al. [144], using the methanol feed stage as the operational specification. They also observe that equilibrium models tend to "overpredict" MSS. Sensitivity of mass transfer coefficients from 90% to 110% of the base case is studied. An interesting observation is that the change of mass transfer coefficient to 110% decreases the isobutene conversion for the low-conversion branch.

Higler et al. [115] used a dusty fluid model for modeling of the MTBE system, using the bottom product withdrawal rate as the operational specification, and presented an interesting comparison of three cases, viz., pseudo-homogeneous, Knudsen diffusion coefficient is large (2 orders of magnitude larger than the binary diffusion coefficient, thereby neglecting it), and Knudsen diffusion coefficient is small (one-fifth of the lowest of the binary diffusion coefficients, therefore the column operates in the diffusion limited regime). Figure 16 shows the steady-state conversion of isobutene in the MTBE column along with the mass transfer rates inside the catalyst particles. It can be observed that the multiple steady states

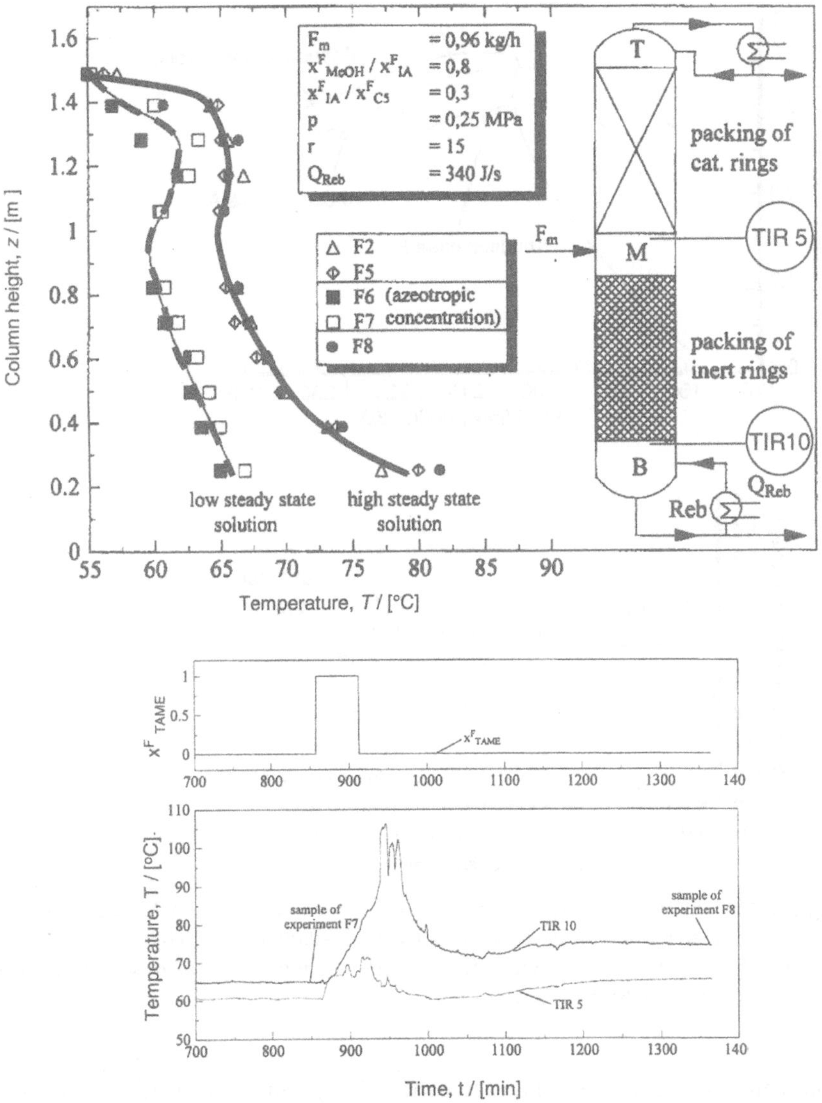

Figure 15 (a) Experimental and simulated composition profiles for TAME column displaying MSS. (b) Experimental transient behavior of TAME column. (Adapted from Mohl et al. [139].)

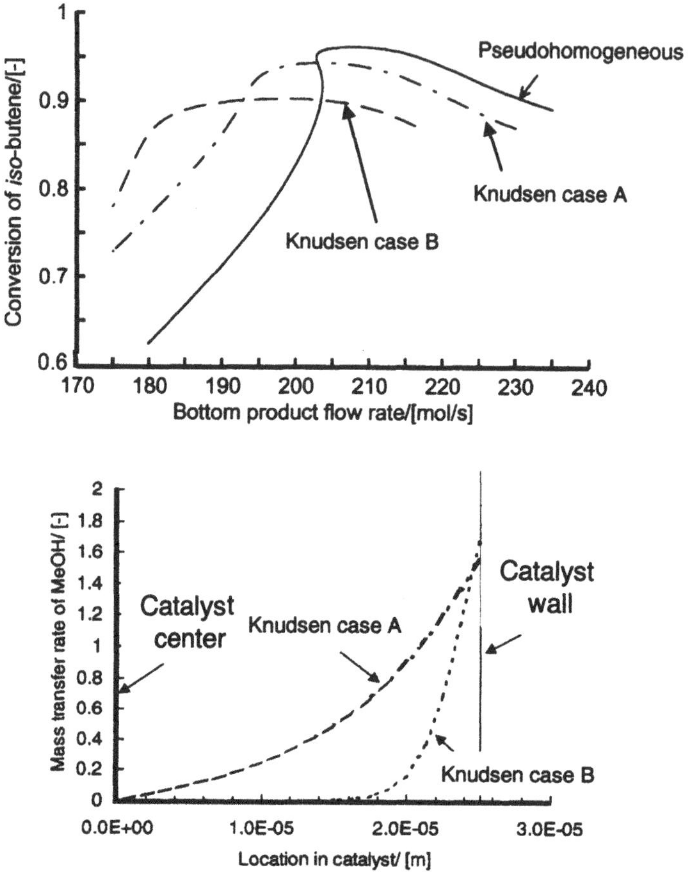

Figure 16 (a) Predicted steady-state conversion of isobutene in MTBE column. (b) Mass transfer rates inside catalyst particles. (Adapted from Higler et al. [115]. Reprinted with permission from Industrial Engineering Chemical Research. Copyright by American Chemical Society.)

completely disappear for the case of the dusty fluid model, leading to the conclusion that equilibrium-stage models apparently increase the window of occurrence of multiple steady states. The Knusden diffusion coefficient has a remarkable impact on the isobutene conversion. The TAME system was also studied for existence of multiple steady states using the equilibrium model, nonequilibrium pseudo-homogeneous model, and the dusty fluid model. Similar results were obtained for all three models, since for the TAME system the forward reaction rate constant is an order of magnitude lower than that of MTBE and hence is less sensitive to mass transfer resistances.

Baur et al. [144] developed a dynamic nonequilibrium cell model for modeling the MTBE formation process and compared the results with an equilibrium one. It was observed that the steady-state simulations using a 3×3 cell model (3 well-mixed cells in both gas and liquid phases) show similar multiplicities to an equilibrium process, though their dynamic responses to methanol feed flow perturbations are different.

For a review of studies on MSS for ETBE systems the reader is referred to Taylor and Krishna [7].

Input multiplicities is a virgin area, and virtually no information is yet available on this subject.

F. Design of Reactive Distillation Columns

Barbosa and Doherty [145,146] presented a design methodology for single- and double-feed reactive distillation columns using the concept of transformed variables presented earlier in this chapter. It is observed that the model equations are identical in form to a conventional distillation column. Using the concept of transformed variables, Espinosa et al. [147] designed a reactive distillation column for MTBE synthesis in the presence of an inert component.

Design methods for kinetically controlled systems using rate-based models have been reported by Buzad and Doherty [148] and by Doherty and Buzad [149] using the fixed-point approach. The fixed point in a composition profile diagram is a saddle-type point where the composition of the system is essentially the same for a large number of stages. The fixed point is a function of the system properties and is not dependent on the Damkohler number.

Mahajani and Kolah [150] developed a design method for packed catalytic distillation columns in which liquid-phase backmixing is totally absent in a three-component system. Mahajani [151] extended this method for multicomponent systems.

"Phenomena-based methods" have been used by Hauan and Lien [152] for the MTBE system, in which change in composition of a phase is given by the sum of the composition vectors describing mixing, separation, and reaction. The directions of these vectors indicate the feasibility of a particular subprocess, while their lengths indicate the extent of its efficiency. The equilibrium fixed point occurs when the reaction and separation vectors are zero; and when the sum of all three vectors is zero, though the individual vectors are nonzero, a kinetic azeotropic point occurs.

Graphic design methods of the Ponchon-Savarit type for conventional distillation have been developed for reactive distillation and tested for the MTBE system by Espinosa et al. [153]. Recently, Lee et al. [154–156] have presented graphic McCabe-Thiele and Ponchon-Savarit-type design methods for design of reactive distillation systems.

G. Optimization

Optimization of reactive distillation operations is an interesting methodology for studying the operational feasibility with respect to output product compositions and cost benefits of reactive distillation as compared with a conventional process. Optimization of reactive distillation systems was studied by Ciric and Gu [157] using an equilibrium-stage model for minimization of operating costs. Eldarsi and Douglas [158] studied the optimization of MTBE columns using Aspen Plus. Operating costs and profitability was studied taking into account the market price of MTBE, raw material costs, and utility costs. Recently, Qi and Sundmacher [159] performed optimization studies on the effect of catalyst distribution and feed strategies for reactive distillation columns.

IV. COMMERICIAL PROCESSES FOR ETHER SYNTHESIS

World production of MTBE was about 23 million tons in 2000 [160]. Two routes are followed for MTBE synthesis, one using fixed-bed reactor(s) and the second using adiabatic fixed-bed reactors followed by catalytic distillation columns. The bulk of the MTBE reaction takes place in the fixed-bed reactors, while the catalytic distillation column acts as a finishing reactor for final conversion. Both reactor configurations are used industrially for MTBE synthesis, while for TAME synthesis adiabatic fixed-bed reactors followed by catalytic distillation is necessarily used, due to the lower equilibrium conversion.

There are many companies worldwide that have ether synthesis plants and offer technologies for commercial licensing.

The ETHERMAX catalytic distillation process developed in a joint venture by Huls A.G., UOP, and Koch Engineering for MTBE is shown in Figure 17. A fixed-bed tubular reactor packed with commercially available cation-exchange resins is used as the primary reactor, where about 90% conversion of isobutene is achieved. This is followed by a catalytic distillation column where further conversion of isobutene takes place. The isobutene conversion after the catalytic distillation column is 99.9%. The nonreactive enriching section separates the inert components—mainly *n*-butane and unreacted methanol—from the reactants. The nonreactive stripping section separates the product MTBE from the reactants, a bottom composition contain >99% MTBE. The catalytic section of the column is packed with KataMax packing. The methanol is separated from the inert

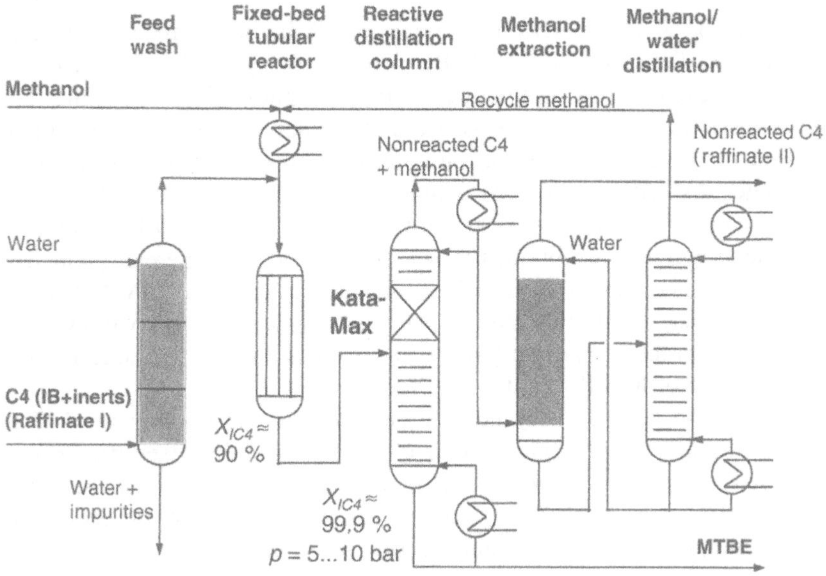

Figure 17 ETHERMAX process for MTBE synthesis.

hydrocarbons by liquid–liquid extraction with water and recycled back to the process as shown in the figure. More than 20 MTBE plants worldwide have been licensed to use this technology.

Institut Francais du Petrole (IFP) offers for commercial licensing two types of MTBE processes [161]. One version, called substoichiometric, achieves 94% conversion of isobutene using a prereactor and a catalytic distillation column without methanol recovery and recycle. About 97% MTBE is obtained as the bottom product from the catalytic distillation column. In this version the investment costs are minimized. In the second version, >99% of isobutene is converted and high-purity MTBE >98.5% is produced at the bottom of the catalytic distillation column. Here methanol recovery and recycle is done using extraction with water. The catalyst packings used are catalytic bales, which is a technology from CD Tech licensed to IFP. Another catalytic distillation technology for MTBE offered commercially by IFP is the CATACOL design [162]. This technology makes use of bulk cation-exchange resin as catalyst supported on catalytic trays around a central liquid downcomer. The first commercial MTBE plant licensed by IFP was commissioned in 1986. To date, IFP has about 10 MTBE units in operation and 4 under construction worldwide, having a combined capacity of 900,000 tons/year of MTBE.

CD Tech (now ABB Lummus) a U.S.-based company, offers for commercial licensing the CDMTBE process for synthesis of MTBE. The primary reactor is a boiling-point reactor packed with commercially available cation-exchange resins. Catalytic bales are used in the catalytic section of the catalytic distillation column. The bottom product from the catalytic distillation column contains typically 98.9% MTBE, 0.5% di-isobutene, 0.4% *tert*-butyl alcohol, <0.1% C_4's and <0.1% methanol. Isobutene conversion obtained is >97%. The first commercial plant from CD Tech was commissioned in 1989. To date more than 50 MTBE plants worldwide use technology from CD Tech, with capacities varying from 20,000 to 1 million tons/year [163]. CD Tech has perhaps the largest slice in the licensing of MTBE plants.

Snamprogetti of Italy licenses commercial technologies for MTBE, ETBE, and TAME using a conventional adiabatic fixed-bed reactor followed by a water-cooled tubular reactor (Miracca et al. [164,165]).

ARCO Chemical Technologies, the first and one of the largest oxygenate producers in the United States, produces oxygenates using fixed-bed adiabatic reactors.

Hao et al. [166] have studied the synthesis of MTBE in a pilot catafractionation tower of 100 mm i.d. without prereactor. Catalytic distillation trays are used, and 99.8% conversion of isobutene is achievable with a bottom product having 98.5% MTBE at an operating pressure is 0.9 MPa. This configuration was tested on a commercial MTBE installation at Luoyang Refinery, China, in 1990 for a production capacity of 2500 tons/year, achieving 95% conversion of isobutene without a prereactor. The bottom product contains 99% MTBE. Currently, five MTBE plants in China use this technology.

Fortum Oil and Gas has developed the NExTAME process for TAME synthesis [167]. In this process the concept of pre- and side reactors is used, coupled around a noncatalytic distillation column.

V. CONCLUSIONS AND FUTURE TRENDS

It is hoped that this review chapter, addressing the key issues of catalytic distillation for MTBE and other ether synthesis, will go a long way in bringing out the criticalities involved in use of catalytic distillation. Catalytic distillation is by itself a complex process at both the design and operation stages. MTBE and other ether production compose the largest commercial application of catalytic distillation. Reactive distillation will play a pivotal role in the design of reactors in the current millennium. Remember Krishna's message: "More killer applications are needed."

Even though sophisticated rate-based models are available for the simulation of catalytic distillation processes, the open literature severely lacks information on hydrodynamics and mass transfer aspects of catalytic distillation packings. A lot of research is desired in this area to bridge the current gap experienced in implementation of these models. The current trends, with use of computational fluid dynamics for the purpose, is a step in the right direction.

MTBE, the darling of the chemical industry of the 1980s with double-digit growth, is recently experiencing a downtrend with almost no or negative growth due to the controversies regarding "clean air but dirty water." Usage of MTBE, especially in the United States, is under discussion after leaking storage tanks led to high levels of detectable MTBE in groundwater in California and other states of the United States. MTBE will be banned in California and some other U.S. states by 2002, and the U.S. Environmental Protection Agency (EPA) is considering further restrictions. The scenario for use of alternative ethers such as ETBE, TAME, etc., is not yet very clear.

There is currently worldwide interest concerning reformulating gasoline through the alkylate route. Branched *iso*-octane, commonly called alkylate, can be synthesized through the dimerization of isobutene, yielding di-isobutene, which is subsequently hydrogenated. The days when di-isobutene was an undesired by-product in the MTBE process are over, and a lot of research is catalyzed toward studying the process for its synthesis [168]. Fortum Gas and Oil has launched a commercial process for isobutene dimerization [169]. A review of isobutene dimerization was recently presented by Kolah et al. [170].

NOTATION

a	liquid-phase activity coefficient, dimensionless
c_t	total concentration, mol/sec
$\mathrm{D_a}$	Damkohler number
$\mathrm{D}_{i,\mathrm{eff}}$	effective Fick diffusivity, m^2/sec
$D_{i,k}$	Maxwell-Stefan diffusivity, m^2/sec
F	feed flow rate, mol/sec
H	liquid holdup per stage, moles
J_i	molar diffusion flux of species i
k	reaction rate constant, mol/[sec-eq(H^+)]
Ka	chemical equilibrium constant based on activities, dimensionless
$K_{s,i}$	sorption equilibrium constant of component i
L_j	liquid flow rate from column section j, mol/sec

N	number of components
N_i	molar flux of species i, mol/m^2-sec
r	reaction rate, mol/[sec-eq(H^+)]
t	time, sec
V_j	vapor flow rate from column section j, mol/sec
ΔV_j	volume of column section j, m^3
X_i	transformed liquid composition defined by Eq. (4)
x_i	liquid mole fraction of component i
Y_i	transformed vapor composition defined by Eq. (4)
y_i	vapor mole fraction of component i
β	bootstrap matrix
ν_k	stoichiometric coefficient of component k
ν_T	$\sum_{1=1}^{c} \nu_i$
τ	time variable

Superscripts and Subscripts

k	reference component

Abbreviations

GPP	glass-supported precipitated polymer
IB	isobutene
MeOH	methanol
MTBE	methyl *tert*-butyl ether
MSS	multiple steady states

REFERENCES

1. A. A. Backhaus. U.S. Patent 1400849, 1921.
2. D. Terrill, L. Sylvestre, M. Doherty. Ind Eng Chem Process Des Dev, 24:1062–1071, 1985.
3. M. M. Sharma. J Sep Process Technol, 6:9–16, 1985.
4. M. Doherty, G. Buzad. Trans Inst Chem Eng, 70: 448–458, 1992.
5. J. G. Stichlmair, T. Frey. Chem Eng Technol, 22:95–103, 1999.
6. M. Sakuth, D. Reusch, R. Janowsky. In Ullmann's Encyclopedia of Industrial Chemistry, 6th ed. Wiley–VCH Verlag GmbH, Weinheim, 1999.
7. R. Taylor, R. Krishna. Chem Eng Sci, 55:5183–5229, 2000.
8. M. F. Malone, M. F. Doherty. Ind Eng Chem Res, 39: 3953–3957, 2000.
9. K. Sundmacher, L. Rihko, U. Hoffmann. Chem Eng Commun, 127: 151–167, 1994.
10. Web site: http://www.cleanfuels.net/doc/intl/doc/worldox/OfaFebruary2000.html.

11. W. J. Piel. Fuel Reformulation, March/April 1994, p 28.
12. F. Ancillotti, V. Fattore. Fuel Process Technol, 57: 163–194, 1998.
13. R. Trotta, I. Miracca. Catal Today, 34: 447–455, 1997.
14. U. Peters, F. Nierlich, E. Schulte-Korne, M. Sakuth. In Ullmann's Encyclopedia of Industrial Chemistry, 6th ed., Wiley–VCH Verlag GmbH, Weinheim, 1999.
15. A. Rehfinger, U. Hoffmann. Chem Eng Sci, 45: 1605–1617, 1990.
16. Francoisse, Th. Chem Eng Process, 30: 141–149, 1991.
17. L. K. Rihko, J. A. Linnekoski, O. I. Krause. J Chem Eng Data, 39: 700–704, 1999.
18. A. Rehfinger, U. Hoffmann. Chem Eng Sci, 45:1619–1626, 1990.
19. W. O. Haag. Chem Eng Prog Symp Ser, 63: 140–147, 1967.
20. W. Song, G. Venimadhavan, J. M. Mannig, M. F. Malone, M. F. Doherty. Ind Eng Chem Res, 37: 1917–1928, 1998.
21. L. K. Rihko, P. K. Kiviranta-Pääkkönen, O. I. Krause. Ind Eng Chem Res, 36: 614–621, 1997.
22. R. Kunin, E. F. Meitzner, J. A. Oline, S. A. Frisch. Ind Eng Chem Prod Res Dev, 1: 140, 1962.
23. R. Krishna, S. T. Sie. Chem Eng Sci, 49: 4029–4065, 1994.
24. G. P. Towler, S. J. Frey. In S. Kulprathipanja ed., Reactive Distillation Process, chap 2, Taylor & Francis, New York, 2002.
25. Y. Fuchigami. J Chem Eng Japan, 3:354–359, 1990.
26. H. Spes. German Patent 1,1285,170, 1969.
27. D. N. Chaplits, V. P. Kazakov, E. G. Lazariants, V. F. Chebotaev, M. I. Balashov, L. A. Serafimov. U.S. Patent 3965039, 1976.
28. L. A. Smith Jr. U.S. Patent 4250052, 1981.
29. J. Flato, U. Hoffmann. Chem Eng Technol, 15: 193–201, 1992.
30. K. Gottlieb, W. Graf, K. Schädlich, U. Hoffmann, A. Rehfinger, J. Flato. German Patent 3930515, 1989.
31. K. Sundmacher, U. Hoffmann. Chem Eng Sci, 47: 2733–2738, 1992.
32. K. Sundmacher, U. Hoffmann. Chem Eng Technol, 16: 279–289, 1993.
33. P. J. M. Lebens, F. Kapteijn, T. N. Sie, J. A. Moulijn. Chem Eng Sci 54: 1359–1365, 1997.
34. K. Sundmacher, U. Hoffmann. Chem Eng Sci, 49: 4443–4464, 1994.
35. U. Hoffmann, U. Kunz, H. Bruderreck, K. Gottlieb, K. Schädlick, S. Becker. German Patent 4234779(3), 1992.
36. U. Kunz, U. Hoffmann. Stud Surf Sci Catal, 91: 299–309, 1995.
37. K. Sundmacher, U. Hoffmann. Chem Eng Sci, 51: 2359–2368, 1996.
38. E. M. Jones Jr. U.S. Patent 5130102, 1992.
39. D. V. Quang, P. Amigues, J. F. Gaillard, J. Leonard, J. L. Nocca. U.S. Patent 4847430, 1989.
40. J. L. Nocca, J. Leonard, J. F. Gaillard, P. Amigues. U.S. Patent 4847431, 1989.
41. J. L. Nocca, J. Leonard, J. F. Gaillard, P. Amigues. U.S. Patent 5013407, 1991.

42. D. Sanfilippo, M. Lupieri, F. Ancillotti. U.S. Patent 5493059, 1996.
43. N. Yeoman, R. Pinaire, M. A. Ulowetz, O. J. Berven, T. P. Nace, D. A. Furse. U.S. Patent 5447609, 1995.
44. N. Yeoman, R. Pinaire, M. A. Ulowetz, T. P. Nace, D. A. Furse. U.S. Patent 5454913, 1995.
45. N. Yeoman, R. Pinaire, M. A. Ulowetz, O. J. Berven, T. P. Nace, D. A. Furse. U.S. Patent 5496446, 1996.
46. L. Asselineau, P. Mikitenko, J. C. Viltard. M. Zuliani. U.S. Patent 5368691, 1994.
47. M. C. Marion, J. C. Vitard, P. Travers, I. Harter, A. Forestiere. U.S. Patent 5776320, 1996.
48. R. J. Carland. U.S. Patent 5308451, 1994.
49. E. M. Jones Jr. U.S. Patent 4536373, 1985.
50. N. Yeoman, R. Pinaire, M. A. Ulowetz, T. P. Nace, D. A. Furse. U.S. Patent 5593548, 1997.
51. E. M. Jones Jr. U.S. Patent 5133942, 1992.
52. F. C. Franklin. U.S. Patent 4471154, 1984.
53. R. Krishna. In K. Sundmacher, A. Kienle eds., Reactive Distillation Status and Future Directions, chap. 7, Wiley-VCH Verlag GmbH, Weinheim, 2003.
54. L. A. Smith, Jr. U.S. Patent 424530, 1980.
55. L. A. Smith Jr. U.S. Patent 4443559, 1984.
56. L. A. Smith Jr. European Patent 476938, 1991.
57. H. Subawalla, J. C. Gonzalez, A. F. Seibert, J. R. Fair. Ind Eng Chem Res, 36: 3821–3832, 1997.
58. X. Xu, Z. Zhao, S. Tian. Chem Eng Res Des, Trans IChemE, 77(Part A): 625–629, 1997.
59. X. Xu, Z. Zhao, S. Tian. Chem Eng Res Des, Trans IChemE, 77(Part A): 16–20, 1999.
60. Y. Zheng, X. Xu. Chem Eng Res Des, 70(A5): 459–464, 1992.
61. Y. Zheng, X. Xu. Chem Eng Res Des, 70(A5): 465–470, 1992.
62. L. A. Smith Jr. U.S. Patent 4307254, 1981.
63. L. A. Smith Jr. U.S. Patent 4336407, 1982.
64. L. A. Smith Jr. U.S. Patent 5118873, 1992.
65. L. A. Smith Jr. U.S. Patent 5120403, 1992.
66. A. Bakshi, T. P. Hickey. U.S. Patent 5919989, 1999.
67. J. C. Gonzalez, H. Subawalla, J. R. Fair. Ind Eng Chem Res, 36: 3845–3853, 1997.
68. K. H. Johnson. U.S. Patent 5189001, 1993.
69. C. S. Crossland, G. R. Gildert, D. Hearn. U.S. Patent 5431890, 1995.
70. P. Moritz, H. Hasse. Chem Eng Sci, 54: 1367–1374, 1999.
71. P. Moritz, B. Bessling, G. Schembecker. Chem Ing Tech, 71:131–135, 1999.
72. J. M. Baten, J. Ellenberger, R. Krishna. Chem Eng Sci, 56: 813–821, 2001.
73. J. P. Stringaro. U.S. Patent 5470542, 1995.
74. R. Shelden, J. P. Stringaro. European Patent 396650, 1992.

75. V. Pinjala, J. L. DeGarmo, M. A. Ulowetz. Rate-based modelling of reactive distillation systems. Proceedings of AIChE 1992 Annual Meeting Session 3: "Distillation with Reaction," Florida, Nov 1992.
76. O. L. Oudshoorm, M. Janissen, W. E. J. van Kooten, J. C. Jansen, H. van Bekkum, C. M. van den Bleek, H. P. A. Calis. Chem Eng Sci, 54: 1413–1418, 1999.
77. P. J. M. Lebens, F. Kapteijn, S. T. Sie, J. A. Moulijn. Chem Eng Sci, 54: 1359–1363, 1999.
78. H. Subawalla, J. R. Fair. Ind Eng Chem Res, 38: 3696–3709, 1999.
79. T. Frey, J. Stichlmair. Chem Eng Res Des, Trans. IChemE, 77(Part A): 613–618, 1999.
80. W. D. Seider, S. Widagdo. Fluid Phase Equilibria 123: 283–303, 1996.
81. D. Barbosa, M. F. Doherty. Proc R Soc Lond, A413: 443–458, 1987a.
82. D. Barbosa, M. F. Doherty. Proc R Soc Lond, A413: 459–464, 1987b.
83. D. Barbosa, M. F. Doherty. Chem Eng Sci, 43: 529–540, 1988a.
84. D. Barbosa, M. F. Doherty. Chem Eng Sci, 43: 541–550, 1988b.
85. M. J. Okasinski, M. F. Doherty. AIChE J, 43: 2227–2238, 1997.
86. S. Ung, M. F. Doherty. Chem Eng Sci, 50:23–48, 1995.
87. S. Ung, M. F. Doherty. Ind Eng Chem Res, 34: 2555–2565, 1995.
88. S. Ung, M. F. Doherty. Ind Eng Chem Res, 34: 3195–3202, 1995.
89. J. Espinosa, P. A. Aguirre, G. A. Perez. Ind Eng Chem Res, 34: 853–861, 1995.
90. J. Espinosa, P. A. Aguirre, G. A. Perez. Chem Eng Sci, 50: 469–484, 1995.
91. J. Espinosa, P. A. Aguirre, G. A. Perez. Ind Eng Chem Res, 35: 4537–4549, 1996.
92. E. S. Perez-Cisneros, R. Gami, M. L. Michelsen. Chem Eng Sci, 52: 527–543, 1997.
93. G. Venimadhavan, G. Buzad, M. F. Doherty, M. F. Malone. AIChE J, 40: 1814–1824, 1994.
94. C. Thiel, K. Sundmacher, U. Hoffmann. Chem Eng Sci, 52: 993–1005, 1997.
95. G. Venimadhavan, M. F. Malone, M. F. Doherty. AIChE J, 45: 546–556, 1999.
96. A. Nisoli, M. F. Malone, M. F. Doherty. AIChE J, 43: 374–387, 1997.
97. R. Gani, T. S. Jepsen, E. S. Perez-Cisneros. Comput Chem Eng, 22: S363–S370, 1998.
98. K. Sundmacher. Reaktivdestillation mit katalytischen Fuellkoerperpackungen—Ein neuer Process zur Herstellund der Kraftstoffkomponente MTBE. PhD thesis, Universitt Clausthal, 1995.
99. A. A. Abufares, P. L. Douglas. Chem Eng Res Des, Trans IChemE, 73 (Part A): 3–12, 1995.
100. C. A. Ruiz, M. S. Basualdo, N. J. Scenna. Chem Eng Res Des, Trans IChemE, 73(Part A): 363–378, 1995.
101. K. Sundmacher, G. Uhde, U. Hoffmann. Chem Eng Sci, 54: 2839–2847, 1999.
102. K. Sundmacher, M. Gravekarstens, P. Rapmund, C. Thiel, U. Hoffmann. Intensified production of fuel ethers in reactive distillation columns—systematic model-based process analysis. AIDC Conf Ser, Vol 2: 215–221, 1997.

103. Z. Qi, K. Sundmacher, E. Stein, A. Kienle, A. Kolah. Sep Purif Technol, 26: 147–163, 2002.
104. E. Stein, A. Kienle, K. Sundmacher. Chem Eng, 107: 68–72, 2000.
105. R. Krishnamurthy, R. Taylor. AIChE J, 31:449, 1985.
106. R. Taylor, R. Krishna. Multicomponent Mass Transfer. New York, Wiley, 1993.
107. M. J. W. Frank, J. A. M. Kuipers, G. Versteeg, W. P. M. van Swaaij. Chem Eng Sci, 50: 1645–1659, 1995.
108. M. J. W. Frank, J. A. M. Kuipers, R. Krishna, W. P. M. van Swaaij. Chem Eng Sci, 50: 1661–1671, 1995.
109. E. Y. Kenig, A. Gorak. Chem Eng Process, 34: 97–103, 1995.
110. K. Sundmacher, U. Hoffmann. Chem Eng Sci, 49: 3077–3089, 1994.
111. G. Uhde, K. Sundmacher, U. Hoffmann. Chem Eng Technol, 22: 33–37, 1999.
112. J. C. Maxwell. Phil Trans R Soc, 157: 49, 1866.
113. E. A. Mason, A. P. Malinauskas. Gas Transport in Porous Media: The Dusty Gas Model. Amsterdam, Elsevier, 1983.
114. R. Krishna, J. A. Wesselingh. Chem Eng Sci, 52: 861–911, 1997.
115. A. Higler, R. Krishna, R. Taylor. Ind Eng Chem Res, 39: 1596–1607, 2000.
116. R. Taylor, M. F. Powers, M. Lao, A. Arehole. The development of a nonequilibrium model for computer simulation of multicomponent distillation and absorption operation. IChemE. Symp Ser, No 104, Distillation and Absorption, Brighton, UK, B321, 1987.
117. P. Deuflhard, E. Hairer, J. Zugek. Numer Math, 51: 501–516, 1987.
118. A. Higler, R. Krishna, R. Taylor. AIChE J, 45: 2357–2370, 1999.
119. A. Higler, R. Krishna, R. Taylor. Ind Eng Chem Res, 38: 3988–3999, 1999.
120. Web site: http://www.clarkson.edu/~chengweb/faculty/taylor/chemsep/chemsep.html.
121. R. Baur, R. Taylor, R. Krishna. Chem Eng Sci, 55: 6139–6154, 2000.
122. E. Kenig, K. Jakobsson, P. Banik, J. Aittamaa, A. Gorak, M. Koskinen, P. Wettmann. Chem Eng Sci, 54: 1347–1352, 1999.
123. T. E. Guttinger. Multiple steady states in azeotropic and reactive distillation. Dissertation ETH 12720, Swiss Federal Institute of Technology (ETH), Zurich.
124. Y. Pisarenko, O. Epifanova, L. Serafimov. Theor Found Chem Eng, 21: 281–286, 1987.
125. S. A. Nijhuis, F. P. J. M. Kerkhof, A. N. S. Mak. Ind Eng Chem Res, 32: 2767–2774, 1993.
126. R. Jacobs, R. Krishna. Ind Eng Chem Res, 32: 1706–1709, 1993.
127. S. Hauan, T. Hertzberg, K. M. Lien. Comput Chem Eng, 19: S327–S332, 1995.
128. S. Hauan, T. Hertzberg, K. M. Lien. Comput Chem Eng, 21: 1117–1124, 1997.
129. S. Schrans, S. D. Wolf, R. Baur. Comput Chem Eng, 20:S1619–1624, 1996.
130. H. D. Eldarsi, P. L. Douglas. Chem Eng Res Des, Trans IChemE, 76(Part A): 509–516, 1998.

131. M. G. Sneesby, M. O. Tade, T. N. Smith. Chem Eng Res Des, Trans IChemE, 76(Part A): 525–531, 1998a.
132. M. G. Sneesby, M. O. Tade, T. N. Smith. Ind Eng Chem Res, 37: 4424–4433, 1998b.
133. T. E. Guttinger, M. Morari. Comput Chem Eng, 21: S995–S1000, 1997.
134. T. E. Guttinger, M. Morari. Ind Chem Eng Res, 38: 1633–1648, 1999.
135. T. E. Guttinger, M. Morari. Ind Chem Eng Res, 38: 1649–1665, 1999.
136. K. D. Mohl, A. Kienle, E. D. Gilles, P. Rapmund, K. Sundmacher, U. Hoffmann. Chem Eng Sci, 54: 1029–1043, 1999.
137. A. Kienle, G. Lauschke, G. V. Gehrke, E. D. Gilles. Chem Eng Sci, 50: 2361–2375, 1995.
138. J. L. Bravo, A. Pyhalahti, H. Jarvelin. Ind Eng Chem Res, 32: 2220–2225, 1993.
139. K. Mohl, A. Kienle, E. Gilles, P. Rapmund, K. Sundmacher, U. Hoffmann. Comput Chem Eng, 21: S989–S994, 1997.
140. C. Thiel, P. Rapmund, K. Sundmacher, U. Hoffmann, K. D. Mohl, A. Kienle, E. D. Gilles. Proceedings of the First European Congress on Chemical Engineering, Florence, 2: 1423–1426. May 4–7, 1997.
141. P. Rapmund, K. Sundmacher, U. Hoffmann. Chem Eng Technol, 21: 136–139, 1998.
142. K. Sundmacher, U. Hoffmann. Chem Eng J, 57: 219–228, 1995.
143. A. P. Higler, R. Taylor, R. Krishna. Chem Eng Sci, 54: 1389–1395, 1999.
144. R. Baur, A. P. Higler, R. Taylor, R. Krishna, Chem Eng J, 76: 33–47, 2000.
145. D. Barbosa, M. F. Doherty. Chem Eng Sci, 43: 1523–1537, 1988.
146. D. Barbosa, M. F. Doherty. Chem Eng Sci, 43: 2377–2389, 1988.
147. J. Espinosa, P. Aguirre, G. Perez. Ind Eng Chem Res, 35: 4537–4549, 1996.
148. G. Buzad, M. F. Doherty. Chem Eng Sci, 49: 1947–1963, 1994.
149. M. F. Doherty, G. Buzad. Comput Chem Eng, 18: S1–S13, 1994.
150. S. M. Mahajani, A. K. Kolah. Ind Eng Chem Res, 35: 4587–4596, 1996.
151. S. M. Mahajani. Chem Eng Sci, 54: 1425–1430, 1999.
152. S. Hauan, K. M. Lien. Chem Eng Res Des, Trans IChemE, 76(Part A): 396–407, 1998.
153. J. Espinosa, N. Scenna, G. A. Perez. Chem Eng Commun, 119: 109–124, 1993.
154. J. W. Lee, S. Hauan, K. M. Lien, A. W. Westerberg. P Roy Soc Lond A Mat, 456: 1953–1964, 2000.
155. J. W. Lee, S. Hauan, K. M. Lien, A. W. Westerberg. P Roy Soc Lond A Mat, 456: 1965–1978, 2000.
156. J. W. Lee, S. Hauan, A. W. Westerberg. AIChE J, 46: 1218–1233, 2000.
157. A. R. Ciric, D. Gu. AIChE J, 40: 1479–1487, 1994.
158. H. S. Eldarsi, P. L. Douglas. Chem Eng Res Des, Trans IChemE, 76 (Part A): 517–524, 1998.
159. Z. Qi, K. Sundmacher. Proceedings of 2nd International Symposium on Multifunctional Reactors, Nurenberg, 2001.
160. Anon. Oil & Gas J, Oct 9, 2000, p 52.

161. A. Hennico, J. Leonard, J. A. Chodorge, J. L. Nocca, IFP. Etherification Technology. The MTBE and TAME Processes, 1991.
162. J. L. Nocca and J. A. Chodorge, Catacol a Low Cost Reactive Distillation Technology for Etherification, IFP France.
163. Catalytic Distillation Technologies. Ether Units in Operation. CDTECH, Houston, TX.
164. I. Miracca, L. Tagliabue and R. Trotta. Chem Eng Sci, 51:2349–2358, 1996.
165. Web site: http://www.snamprogetti.it/inglese/about/about_technologies_propr.htm.
166. X. R. Hao, J. S. Wang, Z. R. Yang, J. Bao. Chem Eng J Biochem Eng, 56 (2): 11–18, 1995.
167. M. Koskinen , P. Ylinen, H. Järvelin, P. Lindqvist, Hydrocarbon Eng, Dec/Jan 1998/1999: 28–34.
168. G. Parkinson, Chem Eng, Jan 2001, 27–33.
169. Web site: http://www.fortum.com/sitemap.asp.
170. A. K. Kolah, Z. Qi, S. M. Mahajani. Chem Innov, 31(3): 15–21, 2001.

9
Commercial Production of Ethers

Harri Järvelin*
Fortum Oil and Gas Oy, Fortum, Finland

I. INTRODUCTION

Etherification processes are relatively new additions to refinery configuration. The first methyl *tertiary*-butyl ether (MTBE) units were built in the 1970s, when phase-out of lead-containing octane enhancers started. The big expansion of etherification units took place in the late 1980s and early 1990s, when the U.S. Clean Air Act Amedment (CAAA) was passed. U.S. legislation mandated use of oxygentaes in reformulated fuel gasoline (RFG) in areas where specified air quality targets were not met. As a result, several new MTBE and *tertiary*-amyl methyl ether (TAME) plants were built and MTBE became a worldwide-traded chemical. Ethanol is another oxygenate which can be used in RFG, but due to higher production costs than ethanol, ethers are preferred almost everywhere.

World MTBE consumption is about 500,000 bbl/day, or 20,000,000 metric tons per year. The United States is by far the most important market for ethers. Half of the world production capacity is located in the United States, and more than half of the world consumption is there. Most of the U.S. MTBE and TAME is made on the Gulf Coast. California and East Coast states are the biggest users of ethers.

Other areas of interest are Europe, Far East Asia, the Middle East, and Latin America. Several big MTBE plants are located in the Middle East due to the vast oil and gas reserves. Europe and Far East Asia have lots of captive MTBE and TAME production and comsumption.

**Current affiliation*: Alberta Envirofuels Inc., Edmonton, Alberta, Canada.

MTBE is by far the most important ether used in gasoline. World TAME capacity is about one-tenth of MTBE capacity. Ethyl *tertiary*-butyl ether (ETBE) is produced only in two plants in France.

II. CATALYSTS AND REACTIONS

MTBE is produced by reacting methanol and isobutylene in the presence of cationic ion-exchange resin [1]. In principle, any acidic catalyst, solid or liquid, can be used, but due to easy availability and relatively low cost, ion-exchange resins are almost entirely applied in etherification processes. Zeolite catalysts have also been considered in this application, but for cost/benefit reasons they are not used. The major disadvantage of ion-exchange resin catalysts is that they are difficult to regenerate and the cost of regeneration is too high for commercial interest. In-situ regeneration is practically impossible, and when resin catalyst is regenerated ex situ, a producer must wait until regeneration is completed if to reuse the catalyst. That would mean significant production losses.

The resin catalyst which is typically used in commercial etherification processes is a co-polymer of sulfonated divinylbenzene and styrene [2]. The same types of resins are commonly used in water purification processes. Sulfonic acid is the active component of the catalyst. The catalyst donates a hydrogen ion to isobutylene. The carbonium ion formed reacts with methanol or other alcohol to form an ether compound. A simplified reaction of isobutylene and methanol to MTBE is shown in Figure 1.

When carbonium ions are formed, it is also possible that two isobutylene molecules react with each other to form isobutylene dimer. Di-isobutylene is a good gasoline component. MTBE producers are not too concerned about this side reaction as long as only small amounts of di-isobutylene are formed. The best way to control this side reaction is to have excess amounts of methanol present in the etherification reactors. If the methanol/isobutylene ratio drops to less than stoichiometric level, di-isobutylene formation increases rapidly and simultaneously isobutylene trimers and tetramers are also formed.

$$(CH_3)_2\,C\,(CH_2) + (CH_3)\,OH \overset{H^+}{\Leftrightarrow} (CH_3)_3\,C\,O\,(CH_3)$$

Isobutylene Methanol MTBE

Figure 1 MTBE reaction mechanism.

$(CH_3)_2C(CH_2) + (CH_3)_2C(CH_2) \Leftrightarrow (CH_3)_3C(CH)C(CH_3)_2$

Isobutylene Isobutylene Di-isobutylene

$(CH_3)OH + (CH_3)OH \Leftrightarrow (CH_3)O(CH_3) + H_2O$

Methanol Methanol Dimethyl ether Water

$(CH_3)_2C(CH_2) + H_2O \Leftrightarrow (CH_3)_3C(OH)$

Isobutylene Water Tertiary butyl alcohol

Figure 2 Side reactions occurring along with MTBE formation.

Trace amounts of dimethyl ether (DME) and *tertiary*-butyl alcohol (TBA) are also formed in MTBE units. Figure 2 illustrates the main side reactions of commercial MTBE reactors. Dimethyl ether is produced when two methanol molecules react with each other. DME does not end up in the MTBE product, but it must be separated from the unreacted C4 stream before that stream can be directed to further refinery units. In world-scale MTBE units, DME and traces of other oxygenates must be separated from unreacted C4 stream before that stream can be circulated to the dehydrogenation unit. Corresponding compounds are formed in ETBE and TAME units. Figure 3 shows the desired reactions, while Figures 4 and 5 illustrate the typical side reactions of these processes. Some TAME

$(CH_3)_2C(CH_2) + (CH_3)(CH_2)OH \overset{H^+}{\Leftrightarrow} (CH_3)_3CO(CH_2)(CH_3)$

Isobutylene Ethanol ETBE

$(CH_3)_2C(CH)(CH_3) + (CH_3)OH \overset{H^+}{\Leftrightarrow} (CH_3)(CH_2)(CH_3)_2CO(CH_3)$

Isoamylene Methanol TAME

Figure 3 TAME and ETBE reaction mechanisms.

$$(CH_3)_2C(CH)(CH_3) + (CH_3)_2C(CH)(CH_3) \Leftrightarrow C_{10}H_{20}$$

Isoamylene Isoamylene Di-isoamylene

$$(CH_3)_2C(CH)(CH_3) + H_2O \Leftrightarrow (CH_3)(CH_2)C(CH_3)_2(OH)$$

Isoamylene Water Tertiary amyl alcohol

Figure 4 Typical side products in TAME process are DME, di-isoamylene (DIA), and *tertiary*-amyl alcohol (TAOH). DME reaction is shown in Figure 2.

producers also manufacture heavier ethers than TAME. These ethers can provide extra revenue to these producers. The reaction mechanism between heavier olefins and methanol is analogous to the MTBE and TAME reactions.

A water molecule is also formed in the DME reaction. This water, and water from the methanol makeup (methanol used in MTBE processes is typically 99.5% pure, the rest being water), react with isobutylene and form *tertiary*-butyl alcohol (TBA). TBA has good gasoline properties and is left in the MTBE product.

Each commercial MTBE unit is designed to produce a maximum amount of MTBE, and as little as possible of di-isobutylene, TBA, or DME. Catalyst and especially processing conditions dictate how well this target is reached. The MTBE reaction is an equilibrium reaction in which, on one hand, low temperatures favor MTBE equilibrium concentration, and on the other hand, high temperatures increase the rate of reaction. The typical temperature range in commercial MTBE reactors is from 40 to 80°C. It is beneficial to operate the first reactors at a higher temperature and the last reactors at a lower temperature. This arrangement ensures high MTBE reaction rate at the beginning, and then by lowering the temperature, the MTBE producer can reach a high yield of MTBE due to more favorable reaction equilibrium concentration. Obviously, cooling is required between

$$(CH_3)(CH_2)OH + (CH_3)(CH_2)OH \Leftrightarrow (CH_3)(CH_2)O(CH_3)(CH_2) + H_2O$$

Ethanol Ethanol Diethyl ether Water

Figure 5 Typical side products in ETBE process are diethyl ether (DEE), di-isobutylene (DIB), and *tertiary*-butyl alcohol (TBA). DIB and TBA reactions are shown in Figure 2.

reactors. If isobutylene concentration in the feedstock is high, the cooling requirement can be severe because of the exothermic nature of the reaction.

Operating pressure is not as important as temperature. Typically, pressure is set to ensure that the process operates in a liquid phase. The disadvantage of setting the pressure too high is higher capital cost due to more expensive equipment.

The role of the catalyst is to donate hydrogen ions to carbonium ion formation. Especially at the start of run conditions, the type of catalyst is rarely significant. However, some catalysts have more active sites than others and thus the producer can choose to buy a more active catalyst, which is typically more expensive, and try to extend the life of the catalyst. Table 1 summarizes the most important catalyst properties. MTBE catalyst pricing is typically based on hydrogen ion concentration, which is believed to provide either longer catalyst life or better MTBE yield. However, the catalyst cost/benefit optimum is always site-specific. If a producer has smallish reactors, it is probably beneficial to buy more active and more expensive catalyst, which allows higher MTBE yield. If feedstock impurities are the problem, more active catalyst may provide longer catalyst life. On the other hand, if a producer has large reactors and is not limited by the catalyst life, investing in expensive catalyst may be a waste of money. However, commercial feasibility depends on the actual size and type of reactors, feedstock impurities, the price of catalyst, operational procedures, gasoline specifications, and market conditions.

The structure of the catalyst is often overlooked when selecting MTBE catalyst [3]. Pore volume (also used as porosity) and average pore diameter illustrate the accessibility of reactants to the active sites of the catalyst. There is no use for high activity if those active sites are not accessible by reactants. However, the higher the pore volume, the poorer the catalyst strength, thus resulting in catalyst fines, which can cause pressure buildup or even block equipment and cause substantial losses due to extra downtime.

Table 1 Physical Properties of Ion-Exchange Resin Catalysts Used in Commercial MTBE Processes

Catalyst property	Typical value, range
Acidity (hygrogen ion concentration)	4.7–5.5 mEq/g
Surface area	25–50 m^2/g
Pore volume	300–700 L/kg
Average pore diameter	300–800 Å
Particle size, diameter	0.5–0.7 mm
Water content	50–60%

Pressure drop is something each MTBE producer watches carefully. Since ion-exchange resin has a small particle diameter already, even a relatively small amount of catalyst fines can cause serious operating problems by blocking screens that support the catalyst bed.

MTBE catalyst, unlike almost any other catalyst, contains a significant amount of water when delivered to the MTBE producer. Water is needed to keep iso-exchange resin together. Dry catalyst turns very easily to dust. MTBE producers must carefully calculate how much of the "catalyst volume" is water and how much is catalyst. It is possible that lower-priced catalyst is actually more expensive if calculated on dry catalyst basis. Dry ion-exchange resin catalyst can also be bought, but it is rather expensive. Because catalyst is almost always loaded in wet conditions, one has to take into account that the catalyst will shrink when methanol and hydrocarbons replace water. Specific catalyst loading and unloading procedures have been developed for etherification processes [4].

The operating temperature in MTBE reactors is typically limited to 90°C. Higher temperatures, with increased rate of desulfonation, lead to loss of catalyst activity. Some newer and more expensive catalysts can be operated safely at up to 120°C.

III. FEEDSTOCKS AND FACILITIES

Feedstock to etherification units can come from three different sources:

1. Captive refinery units get C4 and C5 olefinic feedstreams from fluidized catalytic cracking (FCC) units. C4s are the top product of the debutanizer distillation column. C5s originate from FCC gasoline distillation. The product from refinery MTBE units is almost always used locally as a gasoline-blending component [5].
2. Steam crackers can produce highly olefinic C4 stream commonly called raffinate I. It can be used as a feedstock to the MTBE unit after converting di-olefins to olefins [6].
3. World-scale MTBE units get their feedstock from gas fields [7,8]. Butanes are separated from natural gas and from lighter hydrocarbons. This condensed gas is often called field butane. Isobutane concentration varies, typically between 30% and 40%, the rest being *n*-butane and small quantities of pentanes. Some world-scale units are designed to use concentrated isobutane. However, this feedstock is available only in the U.S. Gulf Coast area. There are also a few world-scale MTBE units which utilize by-product from propylene oxide production plants.

Captive refinery units are spread all over the world. Almost all refiners who have FCC units can build an MTBE unit and improve the refining margin. Most of the MTBE units which are based on raffinate I feedstock are located in Far East Asia or in Europe, where steam cracker feedstock is heavier and favors formation of C4 olefins. World-scale units are located in areas where huge amounts of natural gas are produced and thus condensed gas, i.e., field butane, is readily available at a reasonable price. These areas include the U.S. Gulf Coast, the Middle East, Far East Asia, and western Canada. Product from these large facilities is practically the only chemical-grade MTBE that is available in the global market.

Figure 6 illustrates a typical integration of refinery MTBE/ETBE and TAME units. Depending on technology, it is possible to combine MTBE and TAME units in a single unit, but that has not been done on a commercial scale so far. Some TAME units are currently producing ethers from C6 and C7 tertiary olefins. Ethers, FCC gasoline, and alkylate are globally important gasoline components. However, refineries are always producing other gasoline components.

The C4 fraction from a typical FCC unit contains about 20% isobutylene. Other C4 hydrocarbons and trace amounts of C3's and C5's do not react in MTBE/ETBE processes in any significant amount. Isoamylenes, present in the C4 stream, react with alcohol and form TAME or

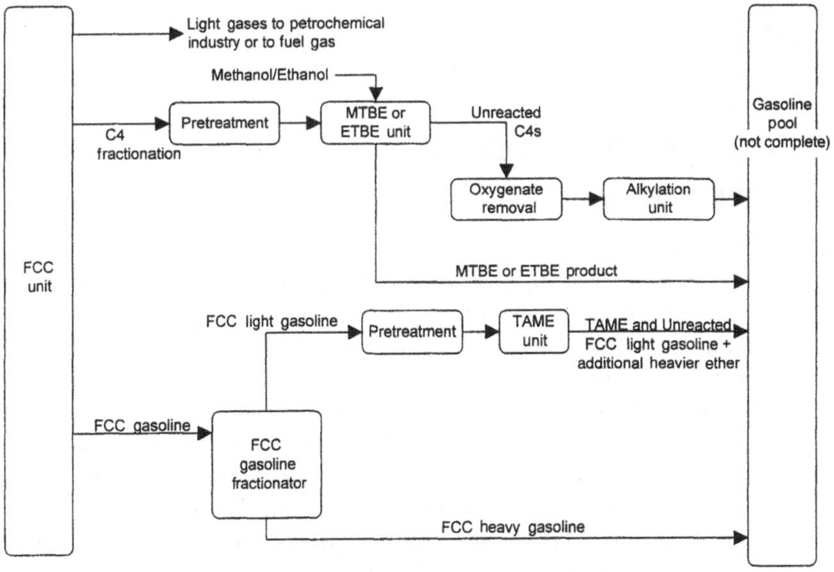

Figure 6 MTBE/ETBE and TAME processes in refinery connection.

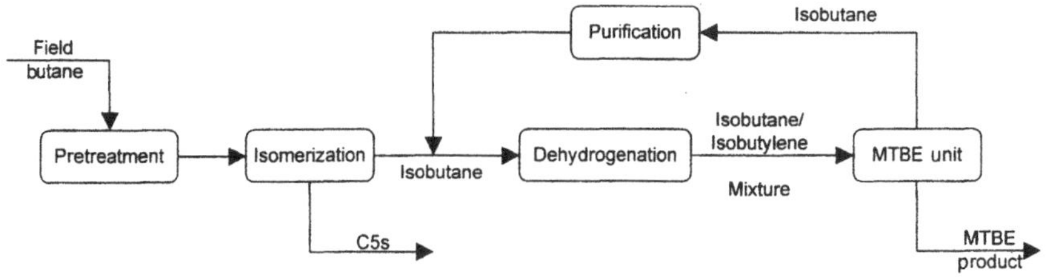

Figure 7 World-scale MTBE plant configuration.

tertiary-amyl ethyl ether (TAEE), depending on the alcohol used. The FCC C4 stream is a relatively diluted mixture of isobutylene and is relatively easy to process. Temperature control seldom creates many problems. There are several options available for temperature control in commercial fixed-bed reactors: recirculation of reactor effluent, isothermal reactors, or boiling-point reactors.

Figure 7 illustrates a typical configuration for a world-scale MTBE unit. So far MTBE is the only ether which is manufactured in these large units. The name "world-scale" is used because these are stand-alone units, with a typical capacity of 10 times that of a typical refinery MTBE unit. World-scale plants are significantly more expensive than refinery units, due to required isomerization and dehydrogenation units.

Isobutylene concentration in MTBE unit feedstream in world-scale units is around 50%. Units that utilize by-product from propylene oxide plants can have very concentrated isobutylene feedstock.

IV. COMMERCIAL PROCESSES

A. MTBE Processes

The first MTBE units were built in the late 1970s and early 1980s. Figures 8 and 9 illustrate a conventional unit, which has typically two or three prereactors, a product distillation, and a methanol recovery section. These units were able to achieve up to 96% isobutylene conversion. This reflects the equilibrium concentration of isobutylene and MTBE. Since then, the basic process concept has not changed. In the late 1980s catalytic distillation technology saw its breakthrough and had an impact on MTBE technology. By assembling the catalyst in the product distillation column, it was possible to reach 99+% isobutylene conversion. This technology took advantage of the low MTBE concentration in the product distillation column above the feedpoint. However, most MTBE units located in refineries are still based on the conventional technology. In many cases, refiners cannot justify

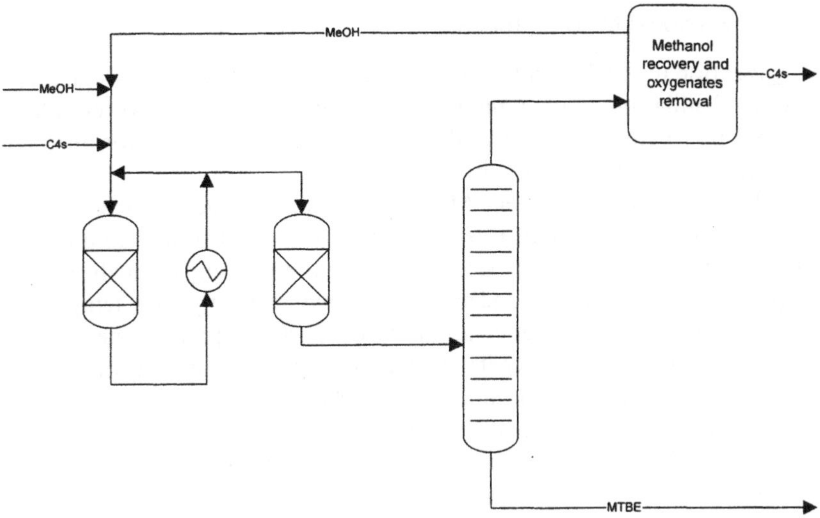

Figure 8 Conventional MTBE process.

higher investment costs, since unreacted isobutylene is typically fed to alkylation units where isobutylene reacts with isobutane and forms alkylates, branched C8 hydrocarbons, which have good gasoline properties. World-scale plants, on the other hand, benefit remarkably by utilizing this advanced technology. Since they have to circulate unreacted isobutylene, the lower conversion means lower production rate and lower revenue.

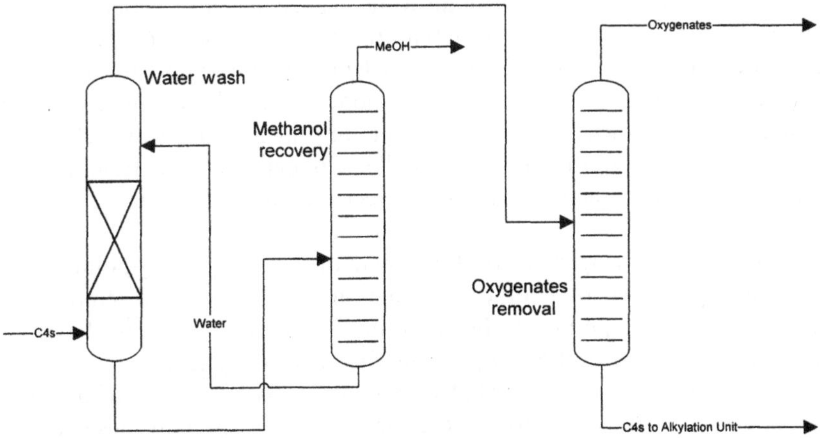

Figure 9 Alcohol recovery and oxygenates removal systems.

1. Conventional MTBE Process

A conventional MTBE unit contains a set of MTBE synthesis reactors, a product distillation, and a methanol recovery section. However, almost always a pretreatment section and an oxygenate removal unit for unreacted C4 hydrocarbon are also required.

The feedstock to refinery MTBE units is a C4 fraction from an FCC unit. Isobutylene concentration in this stream is only about 20%. Etherification reaction is very selective and thus remaining C4 hydrocarbons are practically inert. Low concentration allows relatively easy temperature control, but obviously, higher amounts of inert compounds mean higher investment and operating costs. Higher isobutylene concentration in the world-scale units and in the units which take their feedstock from steam crackers must be taken into account when selecting reactor type. The more concentrated feedstock requires more attention to the reactor temperature control system.

Various reactor types are available for MTBE synthesis, all of which emphasize temperature control and MTBE yield.

Most commonly, cooled reactor effluent is circulated back to the reactor. This is a very simple method, but due to circulation, equipment is bigger than the once-through systems. These reactors operate in adiabatic conditions, meaning higher temperatures at the end of the reactor compared to the front part of the reactor. Since reaction equilibrium favors MTBE at lower temperature, this is a disadvantage. However, it is very common to have two or three fixed-bed reactors in series. This allows the MTBE producer to maintain lower temperatures in the second and third reactors and obtain high MTBE yield.

Tubular reactors with cooling water jackets are also widely used [9]. These reactors are more expensive than fixed-bed reactors, but they operate in isothermal conditions and allow excellent reaction temperature control. They can be operated in once-through mode, which allows smaller reactor volume. Water-cooled tubular reactors are in principle like heat exchangers, which have catalyst in the tube bundle. Changing of the resin catalyst requires a special tool, and may take longer than a catalyst change-out of a fixed-bed reactor.

Some designs operate fixed-bed reactors at the boiling-point condition. Heat produced by exothermic reaction cannot raise bed temperature above the boiling point of the mixture. These reactors must have gas removal capabilities and gas/liquid interface control systems.

The material used in MTBE reactors is commonly carbon steel, except for the catalyst support structure, which is specially designed for this purpose because it has to hold very small particle size resin. Failure of the support structure can bring the etherification unit down for several days.

Reactor effluent consists of MTBE, unreacted methanol, formed by-products, some unreacted isobutylene, and inert compounds. This mixture is directed to the product distillation tower where MTBE is separated from all other components, pumped to product storage tanks, and then blended into gasoline or, in the case of world-scale units, sold in the world market.

Methanol and C4 hydrocarbons form an azetropic mixture in which methanol concentration is 4%. However, methanol must be separated from hydrocarbons since the following units cannot tolerate methanol. Also, it is always beneficial to recycle methanol and thus decrease the need for fresh makeup methanol. Methanol is separated from C4 hydrocarbons in an extraction column. Methanol and hydrocarbons are fed to the bottom part of the extractor and water to the top section. The column is typically filled with random packing or sieve trays, which enhance methanol mass transfer from the hydrocarbon phase to the water phase. Practically methanol-free hydrocarbon stream is obtained above the water feedpoint, and water/methanol mixture is the bottom product. The methanol is separated from the water by distillation and led back to the MTBE reactor, and the water is directed to the extractor column.

The C4 hydrocarbon stream from the extraction column is, in refinery application, directed to the alkylation unit, where olefins react with isobutane and form alkylate. However, the alkylation unit cannot tolerate even a trace amount of oxygenates, which would result in excess acid consumption. Thus C4 hydrocarbons have to go through one more purification section. Typically, remaining oxygenates are removed by distillation, but adsorption can also be used if the oxygenate concentration is small enough. In world-scale units, unreacted C4 hydrocarbons are circulated back to the dehydrogenation process. The degree of oxygenate removal depends on the type of process.

2. Reactive Distillation in MTBE Process

Etherification processes, especially MTBE process, are among the few applications in which reaction synthesis and separation can be combined. Reactive distillation (sometimes referred to as catalytic distillation) can be used successfully in processes in which selectivity and conversion are high enough that an additional separation step is not required. The first MTBE units based on reactive distillation technology were designed in the 1980s. However, it has not been possible to eliminate fixed-bed prereactors. By adding some catalyst to the product distillation column,

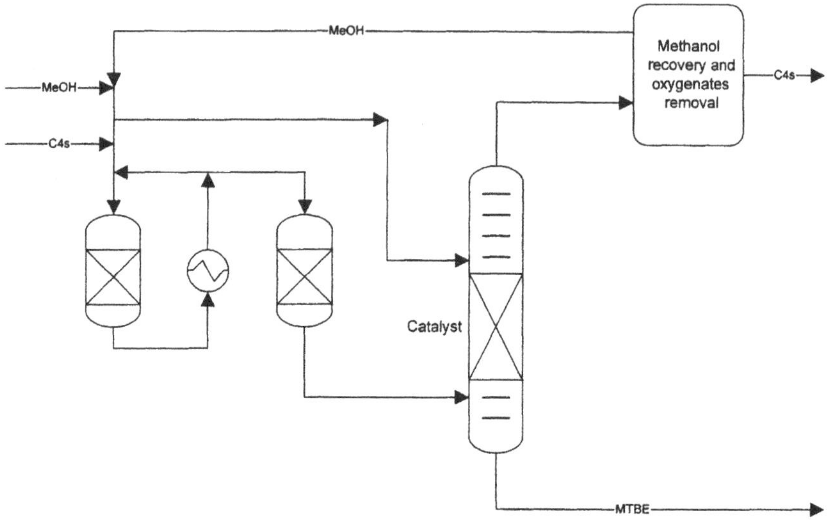

Figure 10 Reactive distillation MTBE process.

producers have been able to achieve 99+% isobutylene conversion, combined with very high selectivity. The reactive distillation process is shown in Figure 10.

There are several tools and processes available to design an MTBE process. One should always look at the total investment and operational costs when selecting a process. It is not wise to look at the reaction section separately from the product and by-product purification. Typically, reactors used in the chemical process industry have high investment costs due to the catalyst cost. The distillation section quite often has higher operational costs due to steam reboiler and cooling requirements. The MTBE process is slightly different, since catalyst cost is not significant. However, the conventional MTBE process suffers from an equilibrium limitation, which can be overcome by utilizing reactive distillation technology. If the rate of production is high enough to justify extra investment, the question is whether capital funds are available for investment, assuming there is a market need for the product. World-scale units can benefit most from higher isobutylene conversion achieved by reactive distillation.

In addition to higher conversion, reactive distillation also provides energy savings to MTBE manufacturers. This is realized since reaction between isobutylene and methanol is exothermic and thus releases energy to the reactor. In fixed-bed reactors this heat must be removed. However, when reaction takes place inside the distillation column, reaction heat reduces reboiler duty and thus steam consumption. In addition to the excellent heat integration, reactive distillation eliminates catalyst-bed hot spots, which may occur in fixed-bed reactors due to flow maldistribution.

The disadvantage of reactive distillation is the extra cost when catalyst is added to a distillation column. Obviously, the column has to be designed specifically for this service. It is longer than the conventional MTBE/C4 separation column and its support structure is different from the normal distillation column.

Catalyst placement can be done in many different ways. Catalyst can be placed on the top of a tray or trays, in the downcomers, in specific wire mesh structure, or it can be added to structured packing. Catalyst can be in one or several sections. The gas phase can either go through the catalyst zone or it may bypass the catalyst zone. Currently there are three different catalyst packing methods used in MTBE processes. All these processes are proprietary and are covered by several patents.

Resin catalyst can be placed in the pockets of a fiberglass cloth belt, which is rolled up using a spacer of mesh knitted steel wire and forms big bales, which are then placed in the distillation column [10]. Loading of these bales can take a long time and therefore reactive distillation should not be used if catalyst contamination is a problem. This system assures good contact between isobutylene and methanol and results in good conversion. It is not a very effective distillation device, but in this application this is not critical.

Catalyst can also be placed on the support tray and force a liquid phase to go through the catalyst zone. This system simulates a process which has a fixed-bed reactor inside a distillation column. Gas phase can be directed through the catalyst bed or the gas phase may bypass the catalyst. In this case, it is actually wrong to use the term "reactive distillation," since there is no distillation occurring in the catalyst zone. Loading and changing of the catalyst requires specific design and operating procedures.

Catalyst can also be placed between two layers of structured packing metal gauze [11]. This system has good distillation capabilities. The catalyst volume usually is smaller compared to the other two methods, and this can lead to shorter operational periods between structure change-out.

Operating pressure and volume of the catalyst dictate the optimum location of the reaction zone. Generally, operating pressure sets the temperature in the MTBE distillation column. The catalyst section operates at a steady temperature since the heat produced by MTBE formation is consumed by boiling off components near the reaction zone. Reaction zone temperature is naturally kept close to the boiling point. If air coolers are used as overhead condensers, the operating pressure is set low, which enhances cooling during warm weather periods. Large quantities of catalyst can be placed toward the top section of the fractionator, where the temperature is lower and MTBE formation is slower but equilibrium is more favorable. Placing the catalyst at the point where methanol and isobutylene concentration is at its highest can minimize the catalyst volume. Also, this arrangement ensures high selectivity since, if the operator is not able to maintain desired methanol/isobutylene ratio in the reaction zone, either DME or di-isobutylene formation will increase.

Reactive distillation allows higher isobutylene conversion in the MTBE process; however, it does not provide any advantage for purification of the unreacted C4 hydrocarbon stream. Extraction with water/methanol separation distillation is required as well as an oxygenates removal section.

Reactive distillation technology has been widely used in MTBE production. It can also be used in manufacturing other ethers, such as TAME and ETBE. However, these processes have slightly different characteristics and therefore are not as good candidates for reactive distillation in the MTBE process.

Reactors and separation devices are just some pieces that chemical engineers can use to build a process. There are many possible combinations, but only a few have real commercial importance. High isobutylene conversion can also be obtained by using a combination of reactors and distillations. This technology is not widely used in MTBE production. However, it has some unique advantages over reactive distillation when TAME, ETBE, and heavier ethers are produced.

B. TAME Processes

TAME units are located in conjunction with refineries. They get their feedstock from FCC units as shown in Figure 6. FCC gasoline contains less than 10% "reactive" isoamylenes (2-methyl-1-butene and 2-methyl-2-butene) [12–14]. The third isoamylene (3-methyl-1-butene) does not have tertiary carbon and therefore it does not react with alcohol in the presence of resin catalyst. Though exothermic reactions are analogous to MTBE synthesis, reaction kinetics are much slower. Also, reaction equilibrium does not favor product ether as much in the case of MTBE. 2-Methyl-1-butene is more reactive due to the double bond at the alpha site compared to 2-methyl-2-butene. Unfortunately, the beta version is dominant in FCC gasoline.

Since typical FCC gasoline contains a large range of hydrocarbons from C5 to C12, it must be fractionated to suitable fractions to be an economically feasible TAME unit. The split of FCC gasoline is done differently depending on the type of TAME unit. The conventional TAME process, as illustrated in Figure 11, is similar to the MTBE

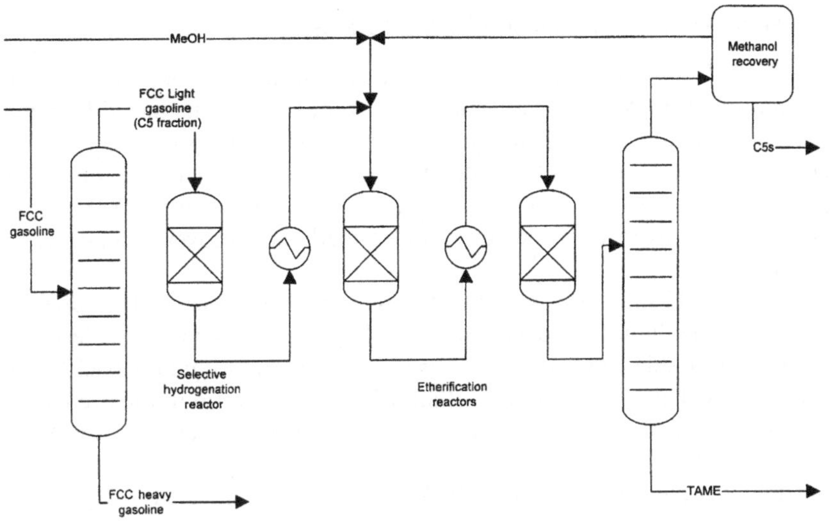

Figure 11 Conventional TAME process with FCC gasoline splitter and selective hydrogenation.

process. It has a set of prereactors and a product distillation, a methanol extraction unit, and a methanol/water separation unit. Oxygenate removal is not needed since unreacted hydrocarbons are typically routed to a gasoline pool, where small amounts of by-product oxygenates can be tolerated.

Only a few TAME units have been built. The main reason for this is that isoamylenes already have a high octane rating and are good gasoline components. However, by converting them to TAME, a refiner can further increase gasoline pool volumes (due to converting methanol to gasoline), octane numbers, and decrease Reid vapor pressure (RVP). The costs are much closer to benefit than in the MTBE process. Benefits of the TAME process are reduced due mainly to a lower isoamylene conversion rate, which in the conventional process is only about 65%. However, by advanced technology it is possible to achieve higher conversion.

By adding a catalyst section into a TAME product distillation tower, isoamylene conversion can be improved to 90% [15,16]. This is still not as high as in MTBE processes, but high enough to increase profitability. Unfortunately, catalyst contamination in TAME services is a much bigger problem than in MTBE units. Reduced catalyst life inside the distillation tower increases operating costs. However, it is possible to combine reactors and distillation to achieve high isoamylene conversion and long catalyst life. This technology is called reaction with distillation.

1. Reaction with Distillation—TAME Process

The TAME process based on reaction with distillation looks similar to the conventional TAME process. There are, though, two distinct differences. Methanol/isoamylene-rich stream is circulated from the product distillation column to prereactors and the column is operated in such a manner as to prevent excess methanol concentrating in the overhead product [17]. High isoamylene conversion is achieved and a methanol recovery section is not needed.

Figure 12 illustrates a high-conversion TAME process which does not require a methanol recovery section. High conversion is achieved by redirecting a side draw from the product distillation column back to the prereactors. This increases the methanol/isoamylene ratio in the prereactors and results in a 90+% isoamylene conversion. C4/methanol azeotrope ensures that only a small amount of methanol leaves with the overhead product. C4 concentration must be kept low (1% in the feed to the TAME unit). It is possible to route the C4 stream to the gasoline pool directly, or combine it with the bottom product. The distillation column is not used to separate ether from unreacted hydrocarbons and methanol. It is used as a tool to concentrate methanol in prereactors and to prevent methanol from leaving the process. Methanol conversion of 99% can be achieved.

The bottom product of the distillation column contains TAME and all unreacted hydrocarbons, other than C4s. Methanol is lighter and is boiled upward in the column. This process cannot produce pure TAME without an

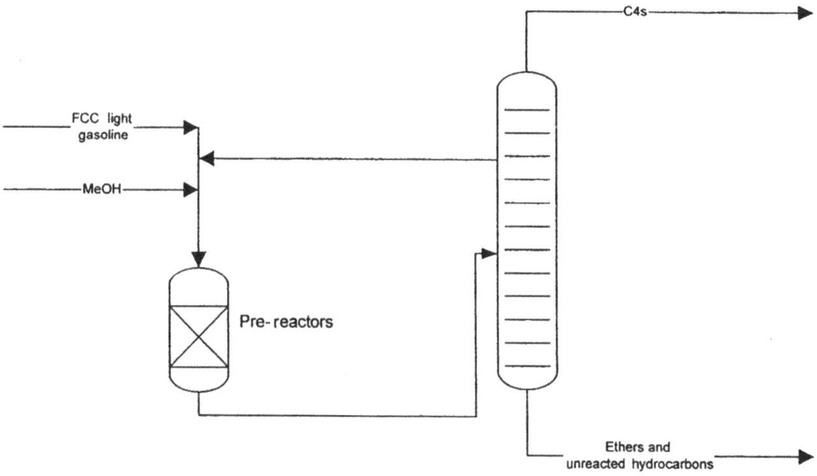

Figure 12 Reaction with distillation TAME process.

additional distillation step. However, TAME is always produced for captive use as a gasoline pool component. Chemical-grade TAME is not sold in the market.

Reaction with the distillation TAME process can easily produce other methyl ethers besides TAME [18]. By allowing C6 and C7 hydrocarbons in the TAME unit feed, a refiner can significantly increase ethers production. Conversion of tertiary C6 olefins can reach almost 70%. If C4 hydrocarbons are fed into a specially designed etherification unit, it is possible to produce MTBE, TAME, and heavier ethers in one unit.

Since there is no catalyst inside the distillation column, catalyst change-out is easy. The bottom product contains formed ethers and practically all unreacted components. It also contains propionitrile, the main poison of the resin catalyst. Propionitrile has relatively low once-through effect on the catalyst, but it can concentrate in the catalyst zone of the reactive distillation column. When unreacted hydrocarbons are left in the bottom product with TAME, propionitrile is also effectively removed from the process.

C. ETBE Process

ETBE can be produced by a similar type of process as MTBE and TAME [19]. Isobutylene conversion in ETBE processes is 10% lower compared to MTBE processes. ETBE processes, which have a water wash extraction section, also require a ethanol purification section, since water and ethanol cannot be separated by conventional distillation. High ethanol purity can be achieved by utilizing membrane or adsorption-based separation technology. Recirculating ethanol back to the prereactors without purification leads to increased amounts of side products, especially *tertiary*-butyl alcohol.

In addition to higher operating and investment costs, the economics of ETBE production are also heavily dependent on the price of ethanol, which is typically more expensive than methanol. Some countries offer tax benefits to ethanol producers, but even with these, ETBE is typically a less attractive product than MTBE. Since ethanol is used to produce ETBE, it can be considered a partly renewable gasoline component. This is an increasingly important factor in the world today. ETBE, as well as pure ethanol, allows an outlet of excess farm products produced in some countries.

V. PRETREATMENT AND CATALYST DEACTIVATION

Feedstock to world-scale MTBE units has to go through a dehydrogenation section and must be thoroughly clean of any kind of impurities. The catalyst life expectancy in these units is many years, if not decades.

Raffinate I stream has a high concentration of butadiene, which has to be removed prior to etherification. Butadiene will polymerize in the presence of resin catalyst. These polymers can block the pores of the resin or end up in the product. Polymers and oligomers can easily be detected in the product, due to their yellow color.

Since resin catalyst is acidic, many basic compounds can deactivate the catalyst. In the MTBE process, the main concern is acetonitrile, which is a weak basic compound but can generate ammonia in situ. Acetonitrile originates from the FCC unit. FCC light gasoline, a feedstock to a TAME plant, contains propionitrile, which is also a weak acid that can produce a stronger acid inside TAME reactors.

A. Nitriles Removal

Acetonitrile would not be harmful for etherification units if it were only processed once through the reactors. However, due to the use of excess methanol and the circulation of methanol back to the reactors, acetonitrile concentrates in the process. This increases catalyst deactivation significantly. Since ammonia, which is produced by decomposition of acetonitrile, is the real catalyst deactivator, decomposition of acetonitrile dictates the rate of deactivation. Higher temperatures enhance the decomposition rate [19].

Catalyst deactivation caused by acetonitrile can be avoided by having a water wash section prior to the MTBE unit. This operation is very effective in removing practically all acetonitrile.

TAME units are more problematic, since water wash is not very effective for propionitrile removal. The best way to protect a TAME unit is to use a process which does not circulate methanol and does not concentrate the propionitrile within the process.

Selective adsorbents have also been considered for nitriles removal. However, they have not gained commercial acceptance.

B. Selective Hydrogenation

Selective hydrogenation converts diolefins to olefins to avoid polymer formation in etherification reactors. This is especially important in TAME units, but captive refinery and raffinate I MTBE units also benefit from selective hydrogenation. A selective hydrogenation unit typically has one fixed-bed reactor, which has operating temperatures between 50 and 140°C. Some processes require a gas/liquid separation device after the reactor. However, it is possible to combine selective hydrogenation and etherification units in such a way that gas separation prior to the etherification unit is not required. Also, it is possible to combine FCC gasoline separation and selective hydrogenation. This reactive distillation combination can be cost-effective before the TAME unit [20].

Palladium is typically used as the catalyst for selective hydrogenation of di-olefins. Like other noble metal catalysts, palladium is somewhat sensitive to sulfur compounds, especially hydrogen sulfide. Also sodium, iron, and arsine compounds may be poisonous to the catalyst. Many compounds, especially sulfur, may temporarily deactivate the catalyst, but by increasing operating temperature it is possible to desorb temporary poisons. Sometimes it is possible to return to lower operating temperatures after desorbing, but quite often, by gradually increasing operating temperatures, it is possible to have the longest run length. A separate desulfurization step before selective dehydrogenation can also by justified. The desulfurization unit may also be placed before the FCC unit. These units are quite expensive, but they provide benefits not only to the selective hydrogenation and etherification units but also reduce the total sulfur content in the gasoline pool.

VI. PRODUCTS

All fuel ethers, namely, MTBE, TAME, and ETBE, have excellent gasoline properties, as seen in Table 2. They all have high octane numbers. TAME and ETBE also have favorable vapor pressure. Commercially available MTBE is low in sulfur [21].

The use of oxygenates is mandated in some countries, and some countries provide a small tax incentive to encourage use of oxygenates. In many parts of the world, fuel ethers are used, because they are fundamentally good gasoline components. Ethers, when used in gasoline, decrease harmful tailpipe emissions and ground-level ozone formation. Many areas have been able to improve air quality by using ethers in gasoline; however, many areas still need to do so.

Table 2 Properties of Fuel Ether

Ether	(RON + MON)/2	Boiling point (°C)	Vapor pressure (bar)	Oxygen content (w%)
MTBE	109	55	0.6	18.2
TAME	105	86	0.1	15.7
ETBE	110	72	0.3	15.7

REFERENCES

1. T. W. Evans, K. R. Edlund. Tertiary alkyl ethers preparation and properties. Ind Eng Chem, October 1936.
2. J. T. McNulty. The many faces of ion-exchange resins. Chem Eng, June 1997.
3. R. M. Singer. Review the basics of MTBE catalysis. Fuel Reformulation, November/December 1993.
4. C. R. Marston. Improve etherification plant efficiency and safety part 1. Fuel Reformulation, May/June 1994.
5. E. J. Chang, S. M. Leiby. Ethers help gasoline quality, Hydrocarbon Processing, February 1992.
6. F. Ancillotti, E. Pescarollo, E. Szatmari, L. Lazar. MTBE from butadiene-rich C4s. Hydrocarbon Processing, December 1987.
7. P. R. Sarathy, G. S. Suffridge. Etherify field butanes part 1. Hydrocarbon Processing, January 1993.
8. P. R. Sarathy, G. S. Suffridge. Etherify field butanes part 2. Hydrocarbon Processing, February 1993.
9. R. Trotta, E. Pescarollo, M. Hyland, S. Bertolli. Consider the advantges of water cooled reactors for ether production from refinery feedstocks. Fuel Reformulation, September/October 1994.
10. W. P. Stadig. Catalytic distillation—combining chemical reaction with product separation. Chem Processing, February 1987.
11. R. A. Becker, J. L. Degarmo, S. P. Davis, S. J. Frey. Advanced RWD technology for ethers provides gasoline flexibility. NPRA Annual Meeting, Washington, D.C., 1994.
12. A. O. I. Krause, L. G. Hammarström. Etherification of isoamylenes with methanol. Appl Catal, 30, 1987.
13. E. Pescarollo, R. Trotta, P. R. Sarathy. Etherify light gasolines. Hydrocarbon Processing, February 1993.
14. L. K. Rihko, A. O. I. Krause. Etherification of FCC light gasoline with methanol. Ind Eng Chem Res, 35 (8), 1996.
15. G. R. Patton, R. O. Dunn, B. Elridge. High conversion TAME processes: a technology comparison. Hydrocarbon Technol Int, Autumn 1995.
16. K. Rock. TAME: technology merits. Hydrocarbon Processing, May 1992.
17. J. Ignatius, H. Järvelin, P. Lindqvist. Use TAME and heavier ethers to improve gasoline properties. Hydrocarbon Processing, February 1995.
18. M. Koskinen, E. Tamminen, H. Järvelin. NExETHERS—a new technology for combined MTBE, TAME and heavier ethers production. NPRA Annual Meeting, San Antonio, TX, 1997.
19. C. R. Marston. Improve etherification plant efficiency and safety part 1. Fuel Reformulation, July/August 1994.
20. K. Rock, G. R. Gildert, T. McGuirk. Catalytic distillation extends its reach. Chem Eng, July 1997.
21. W. J. Piel. Diversify future fuel needs with ethers. Fuel Reformulation, March/April 1994.

10
Kinetics of *tertiary*-Alkyl Ether Synthesis

Faisal H. Syed
Chemical Market Resources, Inc., Houston, Texas, U.S.A.

Prakob Kitchaiya, Kyle L. Jensen, Tiejun Zhang, and Ravindra Datta
Worcester Polytechnic Institute, Worcester, Massachusetts, U.S.A.

Cory B. Phillips
University of Michigan, Ann Arbor, Michigan, U.S.A.

I. INTRODUCTION

tertiary-Alkyl ethers are industrially synthesized in the liquid phase over an acidic ion-exchange resin catalyst at temperatures ranging from 40 to 80°C and pressures of 100 to 300 psig [1,2]. The reaction is moderately exothermic and reversible, and the conversion is, consequently, thermodynamically limited, declining rapidly with temperature. Although any homogeneous or heterogeneous Brønsted acid may be used to catalyze the etherification reaction, such as sulfuric acid, zeolites, pillared silicates, and supported fluorocarbonsulfonic acid polymer, protonated cation-exchange resin catalysts are typically preferred in industry due to higher catalytic activity and the bound acid sites [3–7]. Amberlyst-15,^{TM} produced by Rohm and Haas, is currently the most widely used catalyst in industry for the liquid-phase synthesis of fuel ethers, providing high conversion and selectivity approaching 100%. It is a solid acid ion-exchange resin catalyst based on a styrene divinylbenzene backbone with bound sulfonic acid groups, as shown in Figure 1. Its macroreticular structure allows for high

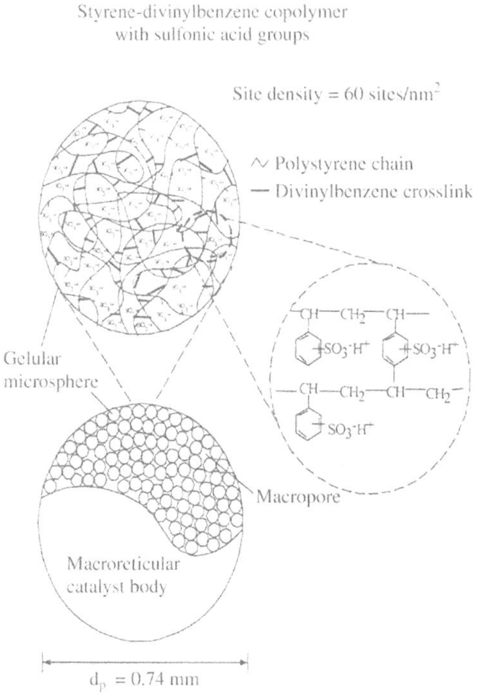

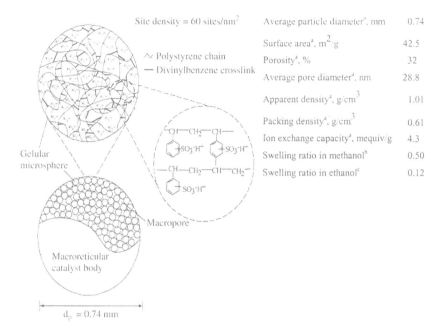

Average particle diameter[a], mm	0.74
Surface area[a], m^2/g	42.5
Porosity[a], %	32
Average pore diameter[a], nm	28.8
Apparent density[a], g/cm^3	1.01
Packing density[a], g/cm^3	0.61
Ion exchange capacity[a], mequiv/g	4.3
Swelling ratio in methanol[b]	0.50
Swelling ratio in ethanol[c]	0.12

Figure 1 Structure and physical properties of Amberlyst-15™ resin catalyst. ([a]From Ref. 56; [b]from Ref. 15; [c]from Ref. 29.)

surface area, and easier accessibility for reactants. Furthermore, the strength of the acid sites is optimal for *tertiary*-ether productions, suppressing side reactions such as olefin dimerization and hydration, and alcohol dehydration under typical operating conditions. However, thermal instability of the bound sulfonic acid groups above 120°C limits the range of operating temperature. Overheating causes the release of strongly acidic sulfonic and sulfuric acids [8]. If the temperature is allowed to increase beyond 140°C, the polymeric backbone of the catalyst also begins to decompose.

The reaction involves the addition of an alcohol to the double bond of a tertiary olefin. Only the tertiary olefins in the hydrocarbon streams are reactive under the above conditions, due to the stability of the tertiary carbonium ion intermediate [3,9]. Table 1 lists the reactive C4–C6 olefins, along with the corresponding ethyl ethers produced by reacting with ethanol. Methanol, on the other hand, would produce the corresponding methyl ethers. Thus, the C4 stream has only one reactive olefin, namely, isobutylene, but higher streams contain multiple reactive olefins. The C5

Table 1 Reactive Tertiary Olefins and Corresponding Ethers Formed with Ethanol

Reactive *tertiary* olefin	Ether formed with ethanol
Isobutylene (IB) CH_3 | $CH_2 = C$ | CH3	Ethyl *tert*-butyl ether (ETBE) CH_3 | $CH_3-C-O-CH_2-CH_3$ | CH_3
2-Methyl-1-butene (2M1B) CH_3 | $CH_3=C-CH_2-CH_3$ 2-Methyl-2-butene (2M2B) CH_3 | $CH_3-C=CH-CH_3$	*tert*-Amyl ethyl ether (TAEE) CH_3 | CH_2 | $CH_3-C-O-CH_2-CH_3$ | CH_3
2-Methyl-1-pentene (2M1P) CH_3 | $CH=C-CH_2-CH_2-CH_3$ 2-Methyl-2-pentene (2M2P) CH_3 | $CH_3-C=CH-CH_2-CH_3$	*tert*-Hexyl ethyl ether (THEE1) CH_3 | CH_2 | CH_2 | $CH_3-C-O-CH_2-CH_3$ | CH_3
2,3-Dimethyl-1-butene (2,3-DM1B) CH_3CH_3 | | $CH_2=C-CH-CH_3$ 2,3-Dimethyl-2-butene (2,3-DM2B) CH_3CH_3 | | $CH_3-C=C-CH_3$	*tert*-Hexyl ethyl ether (THEE2) CH_3 | CH_3-CH | $CH_3-C-O-CH_2-CH_3$ | CH_3

Table 1 Continued

Reactive *tertiary* olefin	Ether formed with ethanol
cis-3-Methyl-2-pentene (C3M2P) 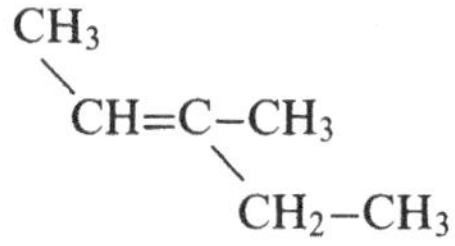	*tert*-Hexyl ethyl ether (THEE3)
trans-3-Methyl-2-pentene (T3M2P) 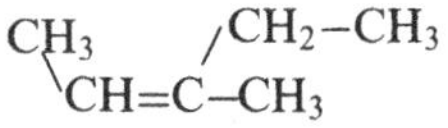	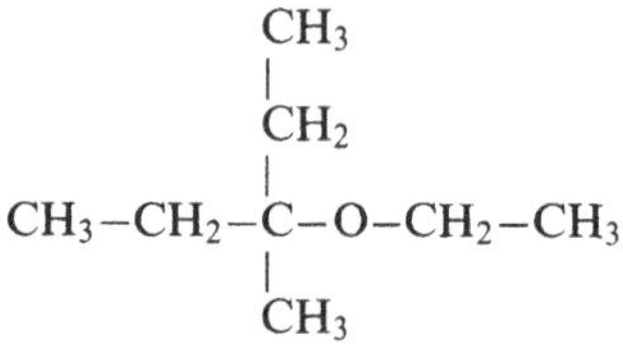
2-Ethyl-1-butene (2E1B) 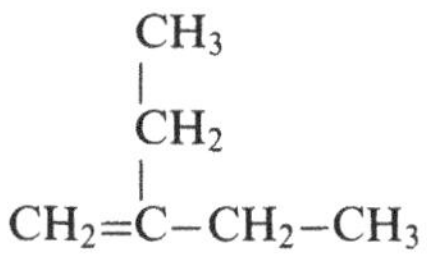	

stream contains two reactive olefins, 2-methyl-1-butene (2M1B) and 2-methyl-2-butene (2M2B), which can react with an alcohol to form one isomeric form of the ether. In addition to the etherification reaction, the olefins can simultaneously isomerize on the catalyst. The C6 stream has seven reactive olefins, and can react to form three different isomers of the respective ether. A general reaction network for simultaneous etherification and isomerization of tertiary olefins with alcohol can be depicted as

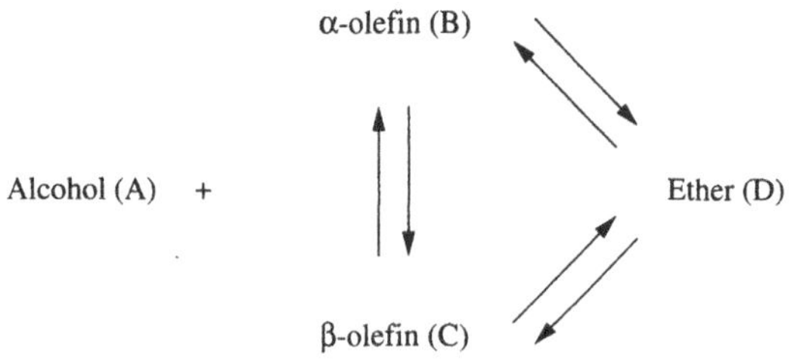

(1)

Initial work on liquid-phase kinetics of methyl *tertiary*-butyl ether (MTBE) over Amberlyst-15 was done by Ancillotti et al. [3] and later by Gicquel and Torck [10], Voloch et al. [11], Subramaniam and Bhatia [12],

Table 2 Proposed Rate Expressions in Terms of Concentrations, from the Literature

Ether	Rate expression	Author (Ref.)
MTBE	$r_{\mathrm{MTBE}} = kc_{\mathrm{IB}} - k'c_{\mathrm{MTBE}}$	Gicquel and Torck (10)
MTBE	$r_{\mathrm{MTBE}} = kc_{\mathrm{IB}}c_{\mathrm{MeOH}}$	Voloch et al. (11)
MTBE	$r_{\mathrm{MTBE}} = \dfrac{k(c_{\mathrm{IB}}c_{\mathrm{MeOH}} - c_{\mathrm{MTBE}}/K)}{1 + K_{\mathrm{MeOH}}c_{\mathrm{MeOH}} + K_{\mathrm{MTBE}}c_{\mathrm{MTBE}}}$	Subramaniam and Bhatia (12)
MTBE	$r_{\mathrm{MTBE}} = \dfrac{k(c_{\mathrm{IB}}^{0.5}c_{\mathrm{MeOH}} - c_{\mathrm{MTBE}}^{1.5}/K)}{1 + K_{\mathrm{MeOH}}c_{\mathrm{MeOH}} + K_{\mathrm{MTBE}}c_{\mathrm{MTBE}}}$	Al-Jarallah et al. (13)
ETBE	$r_{\mathrm{ETBE}} = k\left[c_{\mathrm{IB}} - \alpha\dfrac{c_{\mathrm{ETBE}}}{c_{\mathrm{EtOH}}} + \dfrac{\beta(c_{\mathrm{IB}}c_{\mathrm{EtO}} - \alpha c_{\mathrm{ETBE}})}{c_{\mathrm{IB}} + Fc_{\mathrm{EtOH}}^2 + c_{\mathrm{ETBE}}}\right]$	Francoisse and Thyrion (52)
TAME	$r_{\mathrm{TAME}} = kc_{iC5} - k'c_{\mathrm{TAME}}$	Gallo and Mulard (53)
TAME	$r_{\mathrm{TAME}} = k\left(c_{iC5} - \dfrac{c_{\mathrm{TAME}}}{K_{\mathrm{eq}}x_{\mathrm{MeOH}}^{0.4}}\right)$	Piccoli and Lovisi (54)

and Al-Jarallah et al. [13]. Some of the proposed rate expressions reported in the literature are shown in Table 2. These, as is the common practice in kinetics, are in terms of concentrations. However, for this highly nonideal reaction mixture it is more appropriate to develop kinetic expressions in terms of activities. Thus, rate expressions in terms of species activities were derived for MTBE by Rehfinger and Hoffmann [14,15] and Zhang and Datta [16]. Fité et al. [17] and Jensen [1] proposed activity-based rate expressions for ethyl *tertiary*-butyl ether (ETBE) synthesis. Rihko and Krause [18], Kitchaiya [19], and Linnekoski et al. [20] studied the kinetics of etherification of isoamylene with methanol and ethanol. Ethanol-based C6 ethers have been studied by Zhang and Datta [21]. These expressions, written in terms of activities, are summarized in Table 3. Arguments used to justify the use of activities in rate expressions include consistence with irreversible and equilibrium thermodynamics. Here, we discuss a more fundamental justification for this. It is provided here for ethanol-derived family of tertiary ethers, namely, ethyl *tertiary*-butyl ether (ETBE) from isobutylene, *tertiary*-amyl ethyl ether (TAEE) from the C5 tertiary olefins 2-methyl-1-butene (2M1B) and 2-methyl-2-butene (2M2B), and *tertiary*-hexyl ethyl ethers (THEE) from the C6 tertiary olefins, i.e., THEE1 from 2-methyl-1-pentene (2M1P) and 2-methyl-2-penene (2M2P), THEE2 from 2,3-dimethyl-1-butene (2,3-DM1B) and 2,3-dimethyl-2-butene (2,3-DM2B), and THEE3 from *cis*-3-methyl-2-pentene (C3M2P), *trans*-3-methyl-2-pentene (T3M2P), and 2-ethyl-1-butene (2E1B). It is based on the application of transition-state theory to the elementary steps involved in the overall catalytic reaction within the framework of the Langmuir-Hinshelwood-Hougen-Watson (LHHW) approach [22].

II. TRANSITION-STATE THEORY

Although the conventional transition-state theory (TST) of reaction rates has its shortcomings, it has frequently been utilized in the analysis of reactions in the gas phase as well as in solution [23,24]. It has also been applied to enzymatic reactions [23]. It is now being increasingly utilized in heterogeneous catalysis [24–26]. The conventional TST is based on the potential energy surface for a reaction and involves the following key postulates: (1) the reacting system in traversing from the initial to the final state passes through a region called the transition state that is the highest in its path on the potential energy surface; (2) the species corresponding to the transition state, namely, the transition state complex (TSC), is assumed to be in dynamic equilibrium with the reactants; and finally, (3) the rate of

Table 3 Proposed Rate Expressions in Terms of Species Activities, from the Literature

Ether	Rate expression	Author (Ref.)
MTBE	$r_{\mathrm{MTBE}} = k\left(\frac{a_{\mathrm{IB}}}{a_{\mathrm{MeOH}}} - \frac{1}{K_a}\frac{a_{MTBE}}{a_{\mathrm{MeOH}}^2}\right)$	Rehfinger and Hoffmann (14) Zhang and Datta (16)
ETBE	$r_{\mathrm{ETBE}} = k\left(\frac{a_{\mathrm{IB}}a_{\mathrm{EtOH}} - a_{\mathrm{ETBE}}/K}{a_{\mathrm{EtOH}}^3}\right)$	Fité et al. (17)
ETBE	$r_{\mathrm{ETBE}} = k\left(\frac{a_{\mathrm{IB}}}{a_{\mathrm{EtOH}}} - \frac{1}{K_a}\frac{a_{\mathrm{ETBE}}}{a_{\mathrm{EtOH}}^2}\right)$	Sundmacher and Hoffmann (55)
ETBE	$r_{\mathrm{ETBE}} = k\frac{a_{\mathrm{EtOH}}^2[a_{\mathrm{IB}} - a_{\mathrm{ETBE}}/(Ka_{\mathrm{EtOH}})]}{(1 + K_{\mathrm{EtOH}}a_{\mathrm{EtOH}})^3}$	Jensen and Datta (30)
TAME	$r_{\mathrm{TAME}} = \frac{-k_2\frac{K_{\mathrm{TAME}}}{K_{\mathrm{MeOH}}^2}a_{\mathrm{TAME}}\left(1 - K_1\frac{a_{\mathrm{MeOH}}a_{\mathrm{1B}}}{a_{\mathrm{TAME}}}\right) - k_4\frac{K_{\mathrm{TAME}}}{K_{\mathrm{MeOH}}^2}a_{\mathrm{TAME}}\left(1 - K_2\frac{a_{\mathrm{MeOH}}a_{\mathrm{2B}}}{a_{\mathrm{TAME}}}\right)}{\left(\frac{K_{\mathrm{TAME}}}{K_{\mathrm{MeOH}}}a_{\mathrm{TAME}} + a_{\mathrm{MeOH}} + \frac{K_{\mathrm{1B}}}{K_{\mathrm{MeOH}}}a_{\mathrm{1B}} + \frac{K_{\mathrm{2B}}}{K_{\mathrm{MeOH}}}a_{\mathrm{2B}}\right)^2}$	Rihko and Krause (18)
TAEE	$r_{\mathrm{TAEE}} = k\frac{a_{\mathrm{EtOH}}^2[a_{iC5} - a_{\mathrm{TAEE}}/(Ka_{\mathrm{EtOH}})]}{(1 + K_{\mathrm{EtOH}}a_{\mathrm{EtOH}})^3}$	Kitchaiya and Datta (32)
THEE	$r_{\mathrm{THEE}} = k\frac{[a_{\mathrm{EtOH}}a_{iC6} - a_{\mathrm{TAEE}}/K_{eq}]}{(1 + K_{\mathrm{EtOH}}a_{\mathrm{EtOH}})^2}$	Zhang and Datta (21)

product(s) formation is proportional to the product of a universal frequency and the *concentration* of the TSC.

A. Kinetics of Elementary Reactions

Consider an elementary reaction i possessing a single transition state:

$$\sum_{j=1}^{n} \nu_{ij} A_j = 0 \tag{2}$$

where the subscripts of species $A_j, j = 1, 2, \ldots, r$, are the reactants and $j = r+1, r+2, \ldots, n$ are the product species. Then, according to TST for the elementary reaction,

$$|\nu_{i1}|A_1 + |\nu_{i2}|A_2 + \cdots |\nu_{ir}|A_r \Leftrightarrow [X]^{\ddagger} \rightarrow \nu_{i,r+1}A_{r+1} + \cdots + \nu_{in}A_n \tag{3}$$

where the reactant species form a TSC in pseudo-equilibrium with reactants. The rate of formation of product,

$$r_i = \kappa \nu \vec{C}_{[X]^{\ddagger}} \tag{4}$$

is proportional to the universal frequency, ν, a transmission coefficient, κ, and the concentration of the TSC, $\vec{C}_{[X]^{\ddagger}}$, moving toward the products. Boudart [27] has shown that the universal frequency, ν, is

$$\nu = \frac{k_B T}{h} \tag{5}$$

Therefore, Eq. (4) becomes

$$r_i = \kappa \frac{k_B T}{h} \vec{C}_{[X]^{\ddagger}} \tag{6}$$

For the assumed pseudo-equilibrium between reactants and the TSC, the equilibrium constant is

$$\vec{K}_i^{\ddagger} = \frac{\vec{a}_{[X]^{\ddagger}}}{\prod_{j=1}^{r} a_j^{|\nu_{ij}|}} \tag{7}$$

where the activity of species j, a_j, is

$$a_j = \gamma_j \frac{C_j}{C^o} \equiv \gamma_j^C C_j \tag{8}$$

where C^o is the standard-state concentration to define the activity coefficient [28].
Thus,

$$\vec{C}_{[X]^\ddagger} = \frac{C^o \vec{a}_{[X]^\ddagger}}{\gamma_{[X]^\ddagger}} = \frac{C^o \vec{K}_i^\ddagger}{\gamma_{[X]^\ddagger}} \prod_{j=1}^{r} a_j^{|\nu_{ij}|} \tag{9}$$

Using Eq. (9) in Eq. (6) gives

$$\vec{r}_i = \vec{k}_i \prod_{j=1}^{r} a_j^{|\nu_{ij}|} \tag{10}$$

where

$$\vec{k}_i = \left(\frac{\kappa k_B T C^o \vec{K}_i^\ddagger}{h \gamma_{[X]^\ddagger}} \right) = \frac{\kappa k_B T \vec{K}_i^\ddagger}{h \gamma_{[X]^\ddagger}^C} \tag{11}$$

Similarly, for a reversible elementary reaction, the rate of reaction for the reverse reaction is

$$\overleftarrow{r}_i = \overleftarrow{k}_i \prod_{j=r+1}^{n} a_j^{|\nu_{ij}|} \tag{12}$$

where

$$\overleftarrow{k}_i = \left(\frac{\kappa k_B T C^o \overleftarrow{K}_i^\ddagger}{h \gamma_{[X]^\ddagger}} \right) = \frac{\kappa k_B T \overleftarrow{K}_i^\ddagger}{h \gamma_{[X]^\ddagger}^C}. \tag{13}$$

Finally, the *net* rate of reaction is

$$r_i = \vec{r}_i - \overleftarrow{r}_i \tag{14}$$

B. Thermodynamic Interpretation of the Rate Constant

The thermodynamic interpretation of the TST takes the leap from the statistical mechanics-based TST into the realm of thermodynamics by replacing the potential energy by free energy and, thus, allowing utilization of thermodynamic quantities in place of mechanical concepts. This makes extension of the thermodynamic transition-state theory (TTST) to nonideal systems relatively straightforward. $K_i^{\ddagger}$ is expressed in terms of the corresponding standard Gibbs free-energy change:

$$\Delta \vec{G}_{iT}^{\ddagger 0} = -RT \ln \vec{K}_i^{\ddagger}. \tag{15}$$

We will use

$$\Delta\vec{G}_{iT}^{\ddagger} = \Delta\vec{H}_{iT}^{\ddagger} - T\Delta\vec{S}_{iT}^{\ddagger} \tag{16}$$

along with a form of the van't Hoff equation,

$$\frac{d \ln \vec{K}_i^{\ddagger}}{d(1/T)} = -\frac{\Delta\vec{H}_i^{\ddagger 0}}{R} \tag{17}$$

Combining Eqs. (11), (16), and (17),

$$\vec{k}_i = \kappa\left(\frac{k_B T}{h\gamma_{[X]^{\ddagger}}^{C}}\right) \exp\left(\frac{\Delta\vec{S}_{iT}^{\ddagger 0}}{R}\right) \exp\left(-\frac{\Delta\vec{H}_{iT}^{\ddagger 0}}{RT}\right) \tag{18}$$

and with some rearrangement,

$$\frac{d \ln \vec{K}_i^{\ddagger}}{d(1/T)} = \frac{d \ln \vec{k}_i}{d(1/T)} + \frac{d \ln(1/T)}{d(1/T)} \tag{19}$$

where it has been assumed that the preexponential term on the right-hand side in Eq. (18) is substantially independent of temperature. Using Eq. (17),

$$-\frac{\Delta\vec{H}_{iT}^{\ddagger 0}}{R} = \frac{d \ln \vec{k}_i}{d(1/T)} + T \tag{20}$$

and comparing with the experimental Arrhenius equation,

$$\vec{k}_i = A_i \exp\left(\frac{-\vec{E}_i}{RT}\right) \tag{21}$$

for an elementary reaction $-\Delta\vec{H}_{iT}^{\ddagger 0} = \vec{E}_i - RT$, where $RT \approx$ 2–4 kJ/mol (0.6–1 kcal/mol) and is usually small compared to $\vec{E}_i$, and thus may be neglected. Using Eq. (21) in Eq. (18),

$$\vec{k}_i = \left[\left(\frac{\kappa k_B T}{h\gamma^C_{[X]^\ddagger}}\right)\exp\left(\frac{\Delta\vec{S}_{iT}^{\ddagger 0}}{R}\right)\right]\exp\left(-\frac{\vec{E}_i}{RT}\right) \tag{22}$$

and comparing with Eq. (21),

$$\vec{A}_i = \left(\frac{\kappa k_B T}{h\gamma^C_{[X]^\ddagger}}\right)\exp\left(\frac{\Delta\,\vec{S}_{iT}^{\ddagger 0}}{R}\right) \tag{23}$$

which is the preexponential factor for an elementary reaction and often weakly dependent on temperature. Thus, the Arrhenius equation adequately describes k_i for elementary reactions. Further, it allows for determination of Gibbs free energy of activation,

$$\frac{\Delta\vec{G}_{iT}^{\ddagger 0}}{RT} = \ln\left(\frac{T}{\vec{k}_i}\right) + \ln\left(\frac{\kappa k_B T}{h\gamma^C_{[X]^\ddagger}}\right) \tag{24}$$

Finally, $\Delta\vec{S}_{iT}^{\ddagger 0}$ may be determined from Eq. (16) based on knowledge of $\vec{k}_i$ for elementary reactions with reaction rates of the form of Eq. (10). Alternatively, it can be estimated from statistical mechanics [25].

III. TTST TREATMENT OF ETHERIFICATION AND ISOMERIZATION KINETICS

It is assumed here that a composite catalytic reaction involves several elementary steps, e.g., adsorption, surface reaction, and desorption, which may individually be treated according to TTST, i.e., each step is assumed to possess its own transition state, as shown in Figure 2. The TTST is applied to a series of assumed elementary steps of adsorption, surface reaction, and desorption, i.e.,

$$\mathrm{A} + \mathrm{S} \underset{k'_A}{\overset{k_A}{\leftrightarrow}} \mathrm{A}\cdot\mathrm{S} \qquad \text{(rapid)} \tag{25a}$$

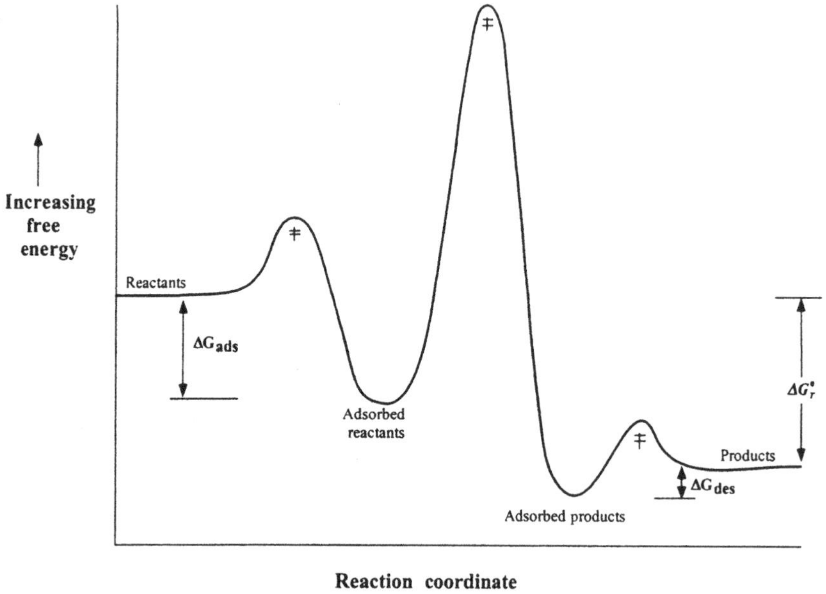

Figure 2 Reaction coordinate diagram for a composite catalytic reaction involving elementary steps of adsorption, surface reaction, and desorption, each involving a transition-state complex at a point of highest free energy.

$$\mathrm{B} + \mathrm{S} \underset{k'_B}{\overset{k_B}{\leftrightarrow}} \mathrm{B} \cdot \mathrm{S} \qquad \text{(rapid)} \qquad (25b)$$

$$\mathrm{C} + \mathrm{S} \underset{k'_C}{\overset{k_C}{\leftrightarrow}} \mathrm{C} \cdot \mathrm{S} \qquad \text{(rapid)} \qquad (25c)$$

$$m\mathrm{A} \cdot \mathrm{S} + \mathrm{B} \cdot \mathrm{S} + s\mathrm{S} \underset{k'_1}{\overset{k_1}{\leftrightarrow}} \mathrm{D} \cdot \mathrm{S} + (m-1)\mathrm{A} \cdot \mathrm{S} + (s+1)\mathrm{S} \qquad \text{(slow)} \qquad (25d)$$

$$m\mathrm{A} \cdot \mathrm{S} + \mathrm{C} \cdot \mathrm{S} + s\mathrm{S} \underset{k'_2}{\overset{k_2}{\leftrightarrow}} \mathrm{D} \cdot \mathrm{S} + (m-1)\mathrm{A} \cdot \mathrm{S} + (s+1)\mathrm{S} \qquad \text{(slow)} \qquad (25e)$$

$$\mathrm{B} \cdot \mathrm{S} + m\mathrm{A} \cdot \mathrm{S} + s\mathrm{S} \underset{k'_3}{\overset{k_3}{\leftrightarrow}} \mathrm{C} \cdot \mathrm{S} + m\mathrm{A} \cdot \mathrm{S} + s\mathrm{S} \qquad \text{(slow)} \qquad (25f)$$

$$\mathrm{D} \cdot \mathrm{S} \underset{k'_D}{\overset{k_D}{\leftrightarrow}} \mathrm{D} + \mathrm{S} \qquad \text{(rapid)} \qquad (25g)$$

in the overall ether (D) synthesis reaction, A+B or C↔D, from alcohol (A) and olefin (B) and (C), which may also simultaneously isomerize on the catalyst, B↔C, as shown in Eq. (1). Equations (25a)–(25f) are generally applicable, but in the case of C4 stream, with only one reactive olefin (isobutylene), Eqs. (25c) and (25f) may be neglected. To maintain generality, Eqs. (25d)–(25f) assume that m sites of adsorbed alcohol react with an adsorbed olefin site along with s vacant catalyst sites in the rate-determining step. Each elementary step is considered below in turn.

A. Adsorption

Consider the adsorption of alcohol, i.e., the forward of the step in Eq. (25a), which, according to TST, proceeds as follows through a TSC, $[X_A]^\ddagger$:

$$A + S \rightleftharpoons [X_A]^\ddagger \rightarrow A \cdot S \tag{26}$$

Applying TTST to this step, the forward rate of adsorption of alcohol is

$$\vec{r}_A = k_A a_A a_S \Theta_S \tag{27}$$

where the effective rate constant of adsorption now takes the form

$$k_A \equiv \frac{\kappa k_B T}{h}\left(\frac{\gamma_S^C}{\gamma_{[X_A]^\ddagger}^C}\right)\vec{K}_A^\ddagger = \frac{\kappa k_B T}{h}\left(\frac{\gamma_S^C}{\gamma_{[X_A]^\ddagger}^C}\right)\exp\left(-\frac{\Delta\vec{G}_{AT}^{\ddagger 0}}{RT}\right) \tag{28}$$

with the equilibrium constant in terms of the corresponding standard Gibbs free-energy change of activation of adsorption process, $\Delta\vec{G}_{AT}^{\ddagger 0}$, and involves the ratio of activity coefficients of vacant sites and the TSC.

B. Desorption

For the desorption process, i.e., the reverse of the step shown in Eq. (25a), the TST postulates

$$A \cdot S \rightleftharpoons [X_A]^\ddagger \rightarrow A + S \tag{29}$$

Based on the principle of microscopic reversibility [23], the TSC in Eq. (29) is the same as that involved in Eq. (26). Then, in a manner similar to that used for the adsorption step, the rate of desorption is

$$\overleftarrow{r}_A = k'_A C_t \Theta_{A \cdot S} \tag{30}$$

where the effective rate constant of desorption of compound A is

$$k'_A = \frac{\kappa k_B T}{h}\left(\frac{\gamma^C_{A\cdot S}}{\gamma^C_{[X_A]^\ddagger}}\right)\overleftarrow{K}^\ddagger_A = \frac{\kappa k_B T}{h}\left(\frac{\gamma^C_{A\cdot S}}{\gamma^C_{[X_A]^\ddagger}}\right)\exp\left(-\frac{\Delta \overleftarrow{G}^{\ddagger 0}_{AT}}{RT}\right) \tag{31}$$

and where $\Delta \overleftarrow{G}^{\ddagger 0}_{AT}$ is the Gibbs free energy of activation of the desorption process. Therefore, the *net* rate of adsorption of A is

$$r_A = \vec{r}_A - \overleftarrow{r}_A = k_A C_t\left(a_A \Theta_S - \frac{\Theta_{A\cdot S}}{K_A}\right) \tag{32}$$

with the adsorption equilibrium constant

$$K_A \equiv \frac{k_A}{k'_A} = \left(\frac{\gamma^C_S}{\gamma^C_{A\cdot S}}\right)\frac{\vec{K}^\ddagger_A}{\overleftarrow{K}^\ddagger_A} = \left(\frac{\gamma^C_S}{\gamma^C_{A\cdot S}}\right)\exp\left(-\frac{\Delta G^0_{AT}}{RT}\right) \tag{33}$$

which also involves ratio of the activity coefficients of the vacant and adsorbed sites. Thus, the adsorption equilibrium constant defined by Eq. (33) may be composition dependent, unless $\gamma^C_S \propto \gamma^C_{A\cdot S}$, or if the surface is ideal, as universally assumed in the LHHW formalism [22]. In Eq. (33), Gibbs free energy of adsorption $\Delta G^0_{AT} \equiv \Delta \vec{G}^{\ddagger 0}_{AT} - \overleftarrow{G}^{\ddagger 0}_{AT}$. Further, using $\Delta G^0_{AT} = \Delta H^0_{AT} - T\Delta S^0_{AT}$,

$$K_A = \left[\frac{\gamma^C_S}{\gamma^C_{A\cdot S}}\exp\left(\frac{\Delta S^0_{AT}}{R}\right)\right]\exp\left(-\frac{\Delta H^0_{AT}}{RT}\right) \tag{34}$$

Syed and Datta [29] and Kitchaiya [19], assuming a constant enthalpy, provided the following adsorption equilibrium constant for methanol and ethanol on Amberlyst-15:

$$K_A = 120\exp\left[\frac{15{,}000}{R}\left(\frac{1}{T} - \frac{1}{303}\right)\right] \tag{35}$$

and

$$K_A = 27\exp\left[\frac{11{,}000}{R}\left(\frac{1}{T} - \frac{1}{303}\right)\right] \tag{36}$$

respectively, which, owing to the exothermicity of adsorption, decreases with increasing temperature.

For pseudo-equilibrium, $r_A \to 0$ and, consequently, from Eq. (32), $\Theta_{A\cdot S} \to K_A a_A \Theta_S$. Similar expressions may be derived for the adsorption of *B*, *C*, and *D*.

C. Surface Reaction

The TTST formulation applied to the forward step of the surface etherification reaction, Eq. (25d),

$$m\mathrm{A}\cdot\mathrm{S} + \mathrm{B}\cdot\mathrm{S} + s\cdot\mathrm{S} \rightleftharpoons [\mathrm{X}_R]^{\ddagger} + (m-1)\mathrm{A}\cdot\mathrm{S} + (s+1)\mathrm{S} \tag{37}$$

gives the rate of reaction as

$$\vec{r}_R = k_1 C_t^{m+s+1} \Theta_{A\cdot S}^m \Theta_{B\cdot S} \Theta_S^s \tag{38}$$

where the rate constant

$$k_1 = \frac{\kappa k_B T}{h}\left[\frac{\left(\gamma_{A\cdot S}^C\right)^m \gamma_{B\cdot S}^C \left(\gamma_S^C\right)^s}{\gamma_{[X_1]^{\ddagger}}^C}\right]\vec{K}_1^{\ddagger} \tag{39}$$

Similarly, the backward surface reaction is

$$\overleftarrow{r}_1 = k_1' C_t^{m+s+1} \Theta_{D\cdot S} \Theta_{A\cdot S}^{m-1} \Theta_S^{s+1} \tag{40}$$

where

$$k_1' = \frac{\kappa k_B T}{h}\left[\frac{\left(\gamma_{A\cdot S}^C\right)^{m-1} \gamma_{D\cdot S}^C \left(\gamma_S^C\right)^{s+1}}{\gamma_{[X_1]^{\ddagger}}^C}\right]\overleftarrow{K}_1^{\ddagger} \tag{41}$$

The *net* reaction may be written as

$$r_1 = \vec{r}_1 - \overleftarrow{r}_1 = k_1 C_t^{m+s+1}\left(\Theta_{A\cdot S}^m \Theta_{B\cdot S} \Theta_S^s - \frac{\Theta_{A\cdot S}^{m-1} \Theta_{D\cdot S} \Theta_S^{s+1}}{k_1/k_1'}\right) \tag{42}$$

Similar equations can be derived for Eq. (25e) and (25f). Assuming pseudo-equilibrium of the adsorbed surface species, i.e.,

$$\Theta_{A\cdot S} = K_A a_A \Theta_S \tag{43a}$$

$$\Theta_{B \cdot S} = K_B a_B \Theta_S \tag{43b}$$

$$\Theta_{C \cdot S} = K_C a_C \Theta_S \tag{43c}$$

$$\Theta_{D \cdot S} = K_D a_D \Theta_S \tag{43d}$$

along with the site balance, the rate expressions for the etherification and isomerization reactions can be written as

$$r_1 = k_1 K_A^m K_B C_t^{m+s+1} \frac{a_A^m [a_B - a_D/(K_1 a_A)]}{(1 + K_A a_A + K_B a_B + K_C a_C + K_D a_D + K_I a_I)^{m+s+1}} \tag{44a}$$

$$r_2 = k_2 K_A^m K_B C_t^{m+s+1} \frac{a_A^m [a_C - a_D/(K_2 a_A)]}{(1 + K_A a_A + K_B a_B + K_C a_C + K_D a_D + K_I a_I)^{m+s+1}} \tag{44b}$$

$$r_3 = k_3 K_A^m K_B C_t^{m+s+1} \frac{a_A^m (a_B - a_C/K_3)}{(1 + K_A a_A + K_B a_B + K_C a_C + K_D a_D + K_I a_I)^{m+s+1}} \tag{44c}$$

where $K_1 = K_A K_B k_1/(K_D k_1')$, $K_2 = K_A K_C k_2/(K_D k_2')$, and $K_3 = K_B k_3/(K_C k_3')$ are the thermodynamic equilibrium constants of the three overall reactions, and $s + m + 1$ is the total number of catalyst sites involved in the surface reaction. Expressions for these thermodynamic equilibrium constants for MTBE, ETBE, *tertiary*-amyl methyl ether (TAME), TAEE, and THEE as functions of temperature are provided in Table 4 [16,30–34], based on equilibrium experiments.

This, thus, justifies the use of activities based on TTST treatment of the elementary reactions in the overall catalytic reaction [1,14,16,17,21,33]. Hougen and Watson [22] first suggested that the rate expressions of catalytic reactions should be written in terms of activities. The activity coefficients in the expressions above may be calculated from the UNIFAC method, which has been shown to be appropriate for the family of tertiary ethers.

D. Most Abundant Surface Species Assumption

The above rate expressions are further simplified by considering that the alcohol (methanol and ethanol) adsorbs highly preferentially on the acid sites of Amberlyst-15 resin as compared with olefins, paraffins, and ethers

Table 4 Coefficients of Liquid-Phase Etherification and Isomerization Reaction Equilibrium Constant Correlations, $\ln K_i = \lambda_{i1} + \lambda_{i2}/T + \lambda_{i3} \ln T + \lambda_{i4} T + \lambda_{i5} T^2 + \lambda_{i6} T^3$, T in K

Reaction I	λ_{i1}	λ_{i2}	λ_{i3}	λ_{i4}	$\lambda_{i5} \times 10^5$	$\lambda_{i6} \times 10^8$
MeOH + IB ⇆ MTBE[a]	−13.482	4388.68	1.2353	−0.01385	2.5923	−3.1881
EtOH + IB ⇆ ETBE[b]	10.387	4060.59	−2.8906	−0.01915	5.2859	−5.3298
MeOH + 2M1B ⇆ TAME[c]	−39.065	5018.61	4.6866	0.00773	−2.635	1.547
MeOH + 2M2B ⇆ TAME[c]	−34.798	3918.02	3.9168	0.01293	−3.121	1.805
2M1B ⇆ 2M2B[c]	−4.159	1100.69	0.7698	−0.00521	0.4865	−0.258
EtOH + 2M1B ⇆ TAEE[d]	22.809	3136.3	−5.8227	0.0179	−0.6395	−1.672
EtOH + 2M2B ⇆ TAEE[d]	26.779	2078.6	−6.5925	0.0231	−1.126	−1.414
2M1B ⇆ 2M2B[d]	−3.970	1057.7	0.7698	−0.0052	0.4865	−0.258
EtOH + 2M1P ⇆ THEE1[e]	−71.4519	7149.74	11.0547	−0.04006	3.8979	−3.6812
EtOH + 2M2P ⇆ THEE1[e]	−84.4308	6870.67	13.0318	−0.03783	3.3181	−3.2602
2M1P ⇆ 2M2P[e]	12.9789	279.07	−1.9771	−0.00223	0.5798	−0.0421
EtOH + 2,3-DM1B ⇆ THEE2[e]	−76.2082	6000.86	12.5825	−0.04762	4.7885	−4.0210
EtOH + 2,3-DM2B ⇆ THEE2[e]	−61.4243	4660.20	9.4324	−0.02728	2.7866	−2.9684
2,3-DM1B ⇆ 2,3-DM2B[e]	−14.7839	1340.66	3.1501	−0.02034	2.0019	−1.0526
EtOH + C3M2P ⇆ THEE3[e]	−72.1147	5433.52	11.4115	−0.03462	3.1287	−3.2044
EtOH + T3M2P ⇆ THEE3[e]	−85.5225	5730.14	13.6057	−0.03808	3.252	−3.1902
EtOH + 2E1B ⇆ THEE3[e]	−75.2419	6296.29	12.2972	−0.04153	3.7941	−3.5474
C3M2P ⇆ T3M2P[e]	13.4078	−296.62	−2.1942	0.00346	−0.1233	−0.0142
C3M2P ⇆ 2E1B[e]	−3.1272	862.77	0.8857	−0.00691	0.6654	−0.343
T3M2P ⇆ 2E1B[e]	10.2806	566.15	−1.3085	−0.00345	0.5421	−0.3572

[a] *Source*: Ref. 16.
[b] *Source*: Ref. 30.
[c] *Source*: Ref. 31.
[d] *Source*: Ref. 32.
[e] *Source*: Ref. 33.

[14,19,21]. Hence, applying the most abundant adsorbed species assumption (MASSA) for the alcohol species, the rate expressions can be simplified to

$$r_1 \approx k_{s1} \frac{a_A^m[a_B - a_D/(K_1 a_A)]}{(1 + K_A a_A)^{m+s+1}} \tag{45a}$$

$$r_2 \approx k_{s2} \frac{a_A^m[a_C - a_D/(K_2 a_A)]}{(1 + K_A a_A)^{m+s+1}} \tag{45b}$$

$$r_3 \approx k_{s3} \frac{a_A^m(a_B - a_C/K_3)}{(1 + K_A a_A)^{m+s+1}} \tag{45c}$$

where $k_{s1} \equiv k_1 K_A^m K_B C_t^{m+s+1}$, $k_{s2} \equiv k_2 K_A^m K_C C_t^{m+s+1}$, and $k_{s3} \equiv k_3 K_A^m K_B \times C_t^{m+s+1}$. Kitchaiya [19] determined that for concentrations of ethanol higher than 0.3 mol/L ($x_A \geq 0.04$) in a mixture of ethanol and inert paraffin, the amount of ethanol adsorbed on Amberlyst-15 resin has already reached saturation and the majority of the acid sites may be assumed to be covered with ethanol. In practice, since alcohol-to-olefin ratio greater than 1 (typically ~1.05) is likely to be utilized, for all practical purposes, the surface would be largely covered with alcohol. Under these conditions (for $x_A \geq 0.04$), the significance of vacant site in Eq. (45) may also be negligible, which agrees with the assumption of Rehfinger and Hoffmann [14], Fité et al. [17], and Rihko and Krause [18]. Then, except for $x_A < 0.04$, the rate expressions may be further simplified to

$$r_1 \approx k_{r1} \frac{[a_B - a_D/(K_1 a_A)]}{(a_A)^{s+1}} \tag{46a}$$

$$r_2 \approx k_{r2} \frac{[a_C - a_D/(K_2 a_A)]}{(a_A)^{s+1}} \tag{46b}$$

$$r_3 \approx k_{r3} \frac{(a_B - a_C/K_3)}{(a_A)^{s+1}} \tag{46c}$$

where $k_{r1} \equiv k_1 K_B C_t^{m+s+1}/K_A^{s+1}$, $k_{r2} \equiv k_2 K_C C_t^{m+s+1}/K_A^{s+1}$, $k_{r3} \equiv k_3 K_B \times C_t^{m+s+1}/K_A^{s+1}$. It may be noted these rate expressions are independent of m. Kitchaiya [19] and Jensen [1] next correlated their experimental data (for

$x_A \geq 0.04$) on the basis of Eq. (45) for various ethanol-derived ethers and found

$$r_1 \approx k_{r1}\left(\frac{a_B}{a_A} - \frac{a_D}{K_1 a_A^2}\right) \tag{47a}$$

and

$$r_2 \approx k_{r2}\left(\frac{a_C}{a_A} - \frac{a_c}{K_2 a_A^2}\right) \tag{47b}$$

as the expressions that best fitted their experimental results for the etherification reactions, while for the isomerization reaction they obtained

$$r_3 \approx k_{r3}\left(\frac{a_B}{a_A^2} - \frac{a_c}{K_3 a_A^2}\right) \tag{47c}$$

indicating $s = 0$ for etherification and $s = 1$ for isomerization. The Arrhenius parameters for k_{ri} in these expressions are listed in Table 5 and 6 [19].

E. Uniformity of Acid Sites

As noted above, m could not be established from these kinetic experiments [19]. In order to determine the total number of catalyst sites involved in the rate-determining step, i.e., $m + s + 1$, experiments were performed with partially deactivated catalysts so as to vary C_t. A question that needs to be addressed first, however, is the uniformity, or lack thereof, of the activity of the acid sites of Amberlyst-15 resin catalyst, since if the acid sites were nonuniform, the kinetics on deactivated catalyst could lead to erroneous conclusions regarding the number of active sites involved in the rate-determining step.

Rys and Steinegger [35] measured the acidity of Amberlyst-15 ion-exchange resin in terms of the Hammett acidity function (H_0) and found that several bases with different basicity provided almost the same H_0 for Amberlyst-15 resin in each solvent, e.g., $H_0 = -1.5$ in ethanol. This indicates substantial homogeneity of the acid sites in the Amberlyst-15 ion-exchange resin. This is also in accord with the conclusions of Raman spectra [36] and nuclear magnetic resonance (NMR) [37] studies. Gates and Johanson [38], who studied the kinetics of the dehydration of methanol catalyzed by ion-exchange resin, found an agreement between the Langmuir adsorption equilibrium constants determined from independent adsorption

Table 5 Arrhenius Kinetics Parameters for Ether Synthesis in Eqs. (47) and (48)

	k_{ri} [Eq. (47)]		k_{si} [Eq. (48)]	
Type of olefin	$A_{ri} \times 10^{-12}$ (mol/h-g)	E_{ri} (kJ/mol)	$A_{si} \times 10^{-11}$ (mol/h-g)	E_{si} (kJ/mol)
MeOH + IB $\leftrightarrows$ MTBE	6.30	85.4	1.50	39.4
EtOH + IB $\leftrightarrows$ ETBE	15.78	87.2	74.18	60.4
EtOH + 2M1B $\leftrightarrows$ TAEE	2.75	85.6	3.12	54.9
EtOH + 2M2B $\leftrightarrows$ TAEE	3.89	89.5	4.4	58.8
EtOH + 2M1P $\leftrightarrows$ THEE1	2.62	86.9	2.97	56.1
EtOH + 2M2P $\leftrightarrows$ THEE1	119.0	100.9	134.85	70.1
EtOH + 2,3-DM1B $\leftrightarrows$ THEE2	12.5	93.3	14.17	62.5
EtOH + 2,3-DM2B $\leftrightarrows$ THEE2	8.87	96.2	10.05	65.4
EtOH + 2E1B $\leftrightarrows$ THEE3	6.54	89.9	7.41	59.2
EtOH + C3M2P $\leftrightarrows$ THEE3	1.66	89.6	1.88	58.9
EtOH + T3M2P $\leftrightarrows$ THEE3	3.03	90.9	3.43	60.1

Source: Ref. 19.

Table 6 Arrhenius Kinetics Parameters for Olefin Isomerization in Eqs. (47) and (48)

Reaction	k_{ri} [Eqs. (47)]		k_{si} [Eqs. (48)]	
	$A_{ri} \times 10^{-12}$ (mol/h-g)	E_{ri} (kJ/mol)	$A_{si} \times 10^{-11}$ (mol/h-g)	E_{si} (kJ/mol)
2M1B → 2M2B	2.15	85.0	5.02	64.5
2M2B → 2M1B	0.8	89.1	1.86	68.6
2M1P → 2M2P	1.25	87.1	2.08	66.6
2,3-DM1B → 2,3-DM2B	3.63	91.3	6.06	70.8
2,3-DM2B → 2,3-DM1B	16.6	102.7	27.4	82.2
2E1B → C3M2P	3.05	91.9	7.15	71.4
2E1B → T3M2P	4.32	90.2	10.1	69.7
C3M2P → T3M2P	1.38	89.9	3.22	69.4

Source: Ref. 19.

experiments and those in the rate expression based on the LHHW formalism involving the usual ideal Langmuirian surface assumption. They, thus, surmised that the array of sulfonic acid groups on ion-exchange resin constitutes a nearly ideal, homogeneous "surface." Beránek [39] further confirmed the consistence of the adsorption equilibrium constants of given species obtained from LHHW rate expressions of different reactions catalyzed by the same ion-exchange resin. The acidity of Amberlyst-15 resin was also determined in water and found to be equivalent to an acidity of about 35 wt% aqueous sulfuric acid [35].

F. Kinetics on Deactivated Catalysts

The initial rate data of ETBE and TAEE synthesis, along with isoamylene isomerization, as a function of the total concentration of acid sites on catalyst, C_t, are shown in Figure 3 [1,19]. Initial rate of etherification shows a more pronounced drop with C_t as compared to that of isomerization. For TAEE synthesis, the rate of etherification dropped by 50% when the number of acid sites decreased by 24% from the full capacity (4.8 mEq/g),

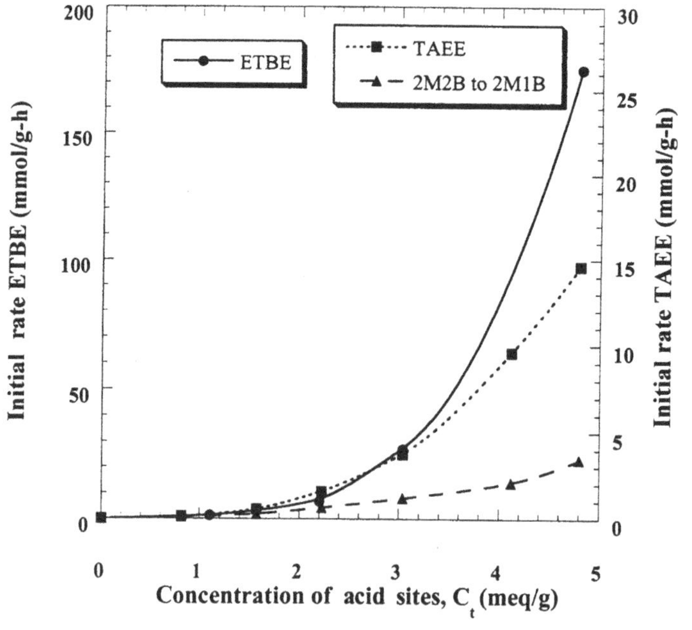

Figure 3 Initial rate of etherification for ETBE (1) and TAEE (19) synthesis, and isomerization of 2M2B to 2M1B as a function of catalyst acid site concentration.

while the rate of isomerization dropped by 43%. Further at an acid capacity of 0.8 mEq/g, or one-sixth of the full capacity of Amberlyst-15, the initial rate of etherification is only about 1% of that with the fully active catalyst. This also agrees with the results of Krause and Hammarstrom [40], who studied TAME synthesis in an integral reactor and found that no conversion was obtained for a resin with an acidic capacity between 0.7 and 1 mEq/g. The natural logarithm of the data shown in Figure 3 is plotted in Figure 4 to determine the slope of the resulting straight lines. A difference in the slopes and, hence, the order of the acidic groups involved in the etherification and isomerization reactions is quite evident. The order of acidic groups for etherification of ethanol with isobutylene and 2M2B, as obtained from linear regression of data in Figure 4, is 3.2 and 2.7, respectively. These numbers are close to that of about 3 reported by Ancillotti et al. [3] for MTBE synthesis, thus indicating $m \approx 2$ for etherification. The order of the acidic groups for isomerization reactions of both feeds, on the other hand, is almost exactly 2, thus yielding $m = 0$ for isomerization. This difference in the order of the acidic sites involved in etherification and isomerization reactions also implies that the rate-determining steps of the two reactions are different. From the mechanistic viewpoint, these high orders of acidic

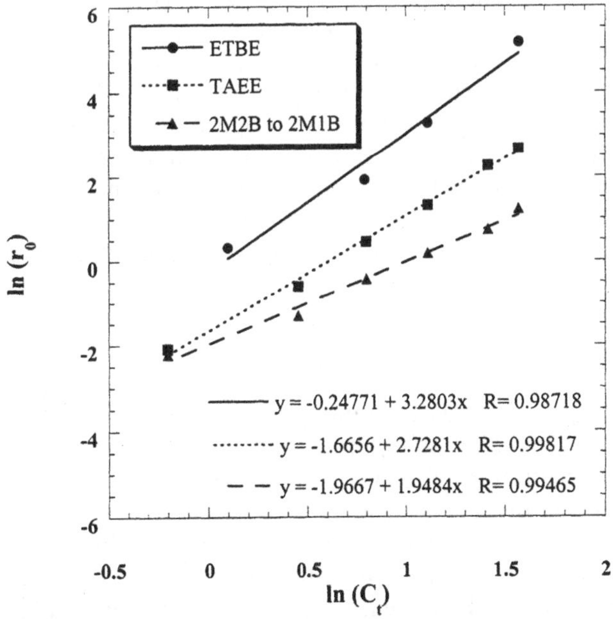

Figure 4 The order of the acidic groups involved in the etherification and isomerization reactions. (From Refs. 1, 19.)

groups could be explained perhaps by an associative structure of sulfonic acid groups, as proposed by Gates et al. [41] for the dehydration of t-butyl alcohol on Amberlyst-15 ion-exchange resin.

The following rate expressions for etherification and isomerization are thus proposed, as derived from Eqs. (44) with $m = 2$ and $s = 0$ for etherification [Eqs. (44a) and (44b)] and $m = 0$ and $s = 1$ for isomerization [Eq. (44c)], along with MASSA for the alcohol species:

$$r_1 = k_{s1}\frac{a_A^2[a_B - a_D/(K_1 a_A)]}{(1 + K_A a_A)^3} \tag{48a}$$

$$r_2 = k_{s2}\frac{a_A^2[a_C - a_D/(K_2 a_A)]}{(1 + K_A a_A)^3} \tag{48b}$$

and

$$r_3 = k_{s3}\frac{(a_B - a_C/K_3)}{(1 + K_A a_A)^2} \tag{48c}$$

where $k_{s1} \equiv k_1 K_A^2 K_B C_t^3$, $k_{s2} \equiv k_2 K_A^2 K_C C_t^3$, and $k_{s3} \equiv k_3 K_B C_t^2$. Equations (48) reduce to Eqs. (47) for $x_A \geq 0.04$, when the vacant sites can also be safely neglected. The Arrhenius parameters of k_{si} in Eqs. (48) are also listed in Tables 5 and 6. Since the effective rate constants k_{ri} [Eqs. (47)] and k_{si} [Eqs. (48)] of etherification reactions are proportional to $1/K_A$ and K_A^2, respectively, the activation energies of the etherification rate constants k_{ri} [Eqs. (47)] are larger than k_{si} [Eqs. (48)] by three times of the enthalpy of adsorption of ethanol. For $x_A \geq 0.04$, the simpler expressions, Eqs. (47), provide an excellent correlation and may be used under these conditions.

G. Etherification and Isomerization Selectivity

Etherification selectivity, defined as the ratio of the initial rate of etherification and the sum of the initial rates of etherification and isomerization, decreases with the concentration of acidic sites on the resin as shown in Figure 5, due to the higher order of acid sites of etherification as compared with the isomerization reaction. During the industrial production of MTBE, the resins become progressively deactivated due to the basic and cationic impurities in the feed. The presence of hot spots in the packed-bed reactor results in a further loss of the acidic groups (56). To offset this activity decline, temperature in an industrial reactor is concomitantly

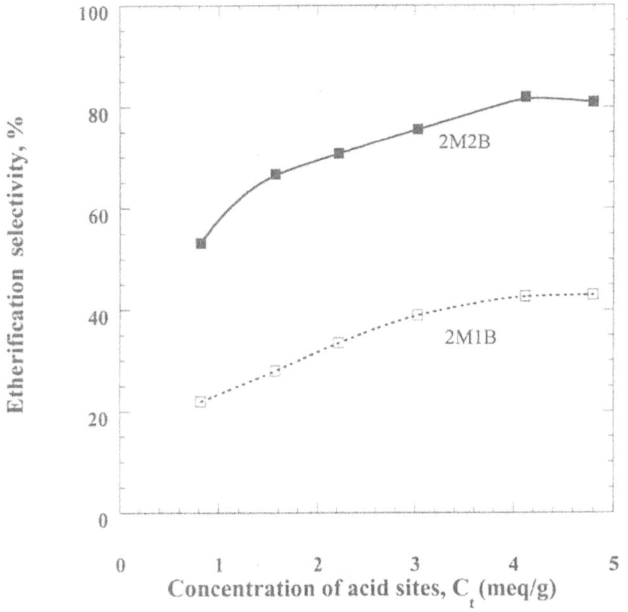

Figure 5 Etherification selectivity as a function of catalyst acid site concentration. (From Ref. 19.)

increased to maintain the desired high conversion, thus further exacerbating the loss of acidic groups. Eventually, thus, the deactivated catalyst must be discarded. However, due to the attendant isomerization reaction, the ether productivity may continue to decline despite raising the temperature because of the reduced selectivity obtained over partially deactivated resins.

H. Linear Free-Energy Relation

According to the extrathermodynamic linear free-energy relationship (LFER), the change in Gibbs free energy of activation of a reaction *i*, for instance, due to change in structure in a reaction series, is assumed to be proportional to the change in Gibbs free-energy change of reaction *i* [24,42], i.e.,

$$d(\Delta G_{iT}^{\ddagger 0}) = \alpha_i d(\Delta G_{iT}^0) \tag{49}$$

where α_i, the symmetry factor, may considered to be a measure of the similarity of the transition-state complex to reactants or products. Normally, α_i varies between 0 and 1, and it is close to 0 or 1 for reactant-like or

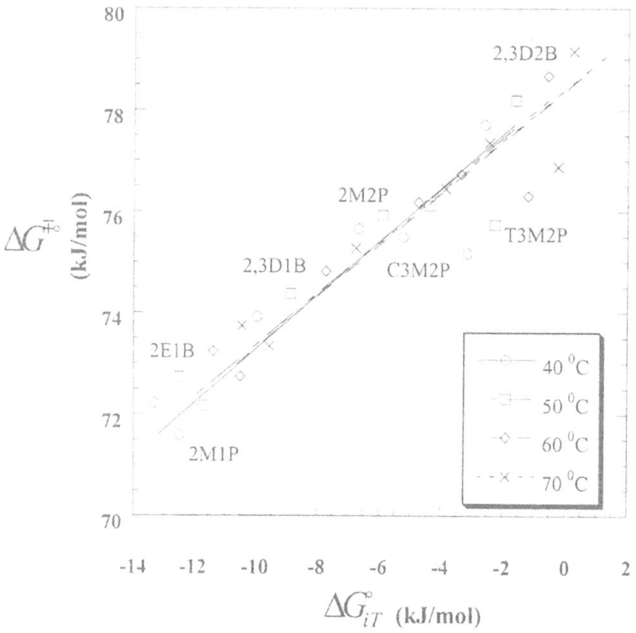

Figure 6 Linear free-energy relation for tertiary hexyl ethyl ethers. (From Ref. 51.)

product-like transition state complex, respectively. Frequently, $\alpha_i \approx 0.5$. Integrated form of Eq. (49) is

$$\Delta G_{iT}^{\ddagger 0} = \alpha_i \Delta G_{iT}^{0} + \Delta G_{iT0}^{\ddagger 0} \tag{50}$$

where $\Delta G_{iT^0}^{\ddagger 0}$ is the intrinsic barrier for the reaction series, which may be viewed as $\Delta G_{iT}^{\ddagger 0}$ of the isergonic reaction, i.e., for the case when $\Delta G_{iT}^{0} = 0$.

The kinetic and thermodynamic data of the six C6 tertiary olefins that form the *tertiary*-hexyl ethyl ethers (THEE) etherification reaction series is used to demonstrate the validity of the linear free-energy relation [Eq. (50)] in Figure 6 [21,33,34]. The constant α_i was found to be 0.45, implying that the transition-state structure is near the middle of the reactant-like and product-like structures.

I. Proposed Mechanism

The rather large number of acid catalyst sites apparently involved in the rate-determining step of these reactions suggests that an associative structure of sulfonic acid groups may be operative. This is consistent with the view of Zundel [43] that sulfonic acid groups on ion exchangers in the absence of water ($\varepsilon = 78.5$ at 25°C) and other highly polar compounds are associated and form a network involving hydrogen bonds. Extremely strong hydrogen bonds are formed between the hydrogen bond donor, OH acid group, and the acceptor, two double-bonded O atoms of the $-SO_2OH$ group [44]. Kampschulte-Scheuing and Zundel [44] demonstrated the presence of the associative structure of sulfonic acid groups in methanol ($\varepsilon = 32.6$ at 25°C) by using infrared spectroscopy. This associative structure increases with decreasing polarity of the medium. A concerted mechanism involving the associative structure of sulfonic acid groups has been suggested in the dehydration of alcohols on ion-exchange resin [41,45,46] and also in MTBE synthesis [47].

The enthalpy of adsorption of liquid methanol and ethanol on Amberlyst-15 resin are −15 kJ/mol and −11 kJ/mol, respectively [19,29], which is in the range of the hydrogen bond strength (≤ 30 kJ/mol) [48]. The enthalpy of formation of an ion pair, on the other hand, is more than 50 kJ/mol [48]. It thus seems unlikely that the adsorbed alcohol is protonated by the sulfonic acid groups of Amberlyst-15. On the other hand, it is well accepted that the reactions of tertiary olefins catalyzed by acid catalysts usually involve a tertiary carbenium ion, which will, hence, be assumed here.

A concerted mechanism is thus proposed in Figure 7, shown for the case of TAEE synthesis, to explain the above observations [19]. For isomerization (Fig. 7a), the reaction begins with the rapid adsorption of olefin on an acid site involving protonation to form a *tertiary*-carbenium ion. An additional acid site then participates in the formation of the TSC. The second acid site stabilizes the first acid site that loses a proton to the tertiary olefin. Both the acid sites also form bonds between H of the iso-olefin and O of the sulfonated groups (Fig. 7a), the dotted lines depicting a partial bond. The positions of partial positive and negative charges are distributed among atoms that have large difference in electronegativity like O and H. The TSC is converted to the adsorbed isomer olefin product by different rate-determining steps (not shown), which finally desorbs. The proposed mechanism also explains the isomerization of olefins in the absence of alcohol [49,50].

The formation of ether as proposed in Figure 7b involves a concerted mechanism between one adsorbed olefin and two adsorbed alcohol molecules. An olefin adsorbs on an acid site via protonation to form a *tertiary*-carbenium ion, while alcohol molecules adsorb on acid sites through hydrogen bonding. The TSC is formed when two adsorbed alcohol molecules and an adjacent adsorbed olefin molecule form a bond between the O of adsorbed alcohol and C of the carbenium ion. At the same time, the O of the sulfonated group with adsorbed olefin also forms a bond with H (in the circle) of an adsorbed alcohol. It is easier for this H (in the circle) to form a

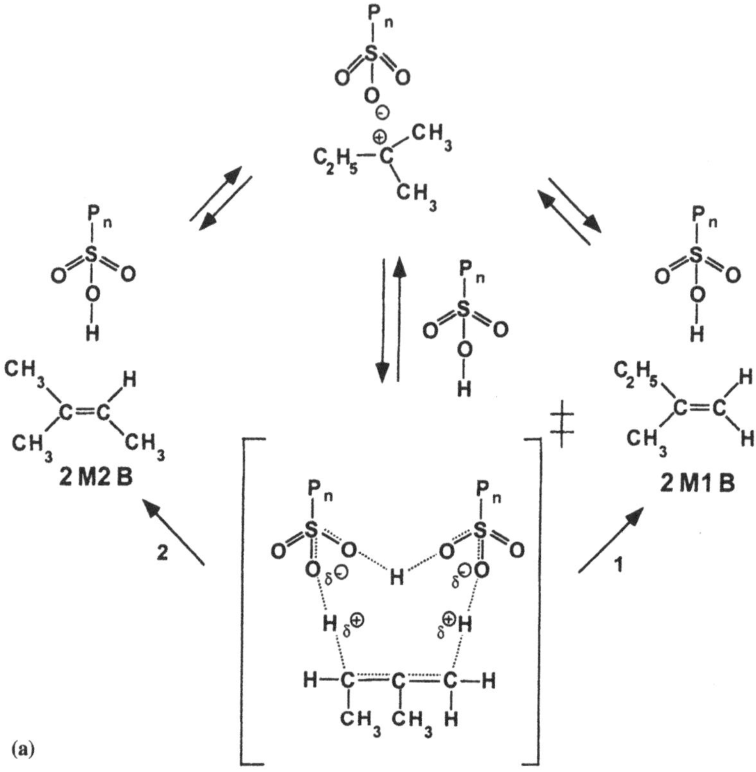

Figure 7a Proposed mechanism of isoamylene isomerization. (From Ref. 19.)

bond with the above O than the H (in the square) from the sulfonated group due to the steric effect of H (in the square) caused by adsorbed alcohol and the lower mobility of H (in the square) as compared to that of H (in the circle). Other partial bonds are shown as well in Figure 7b, along with the distribution of partial positive and negative charges. The TSC decomposes in the rate-determining step to form the adsorbed ether, which finally desorbs. The mechanism proposed here differs from that of Tejero et al. [47] for MTBE synthesis, who proposed that one isobutylene molecule adsorbs on two acid sites and then reacts with one adsorbed ethanol site.

IV. CONCLUSIONS

Rate expressions in terms of activities are provided for the liquid-phase synthesis of *tertiary*-alkyl ethers. The rate expressions are based on the application of the thermodynamic transition-state theory to the elementary

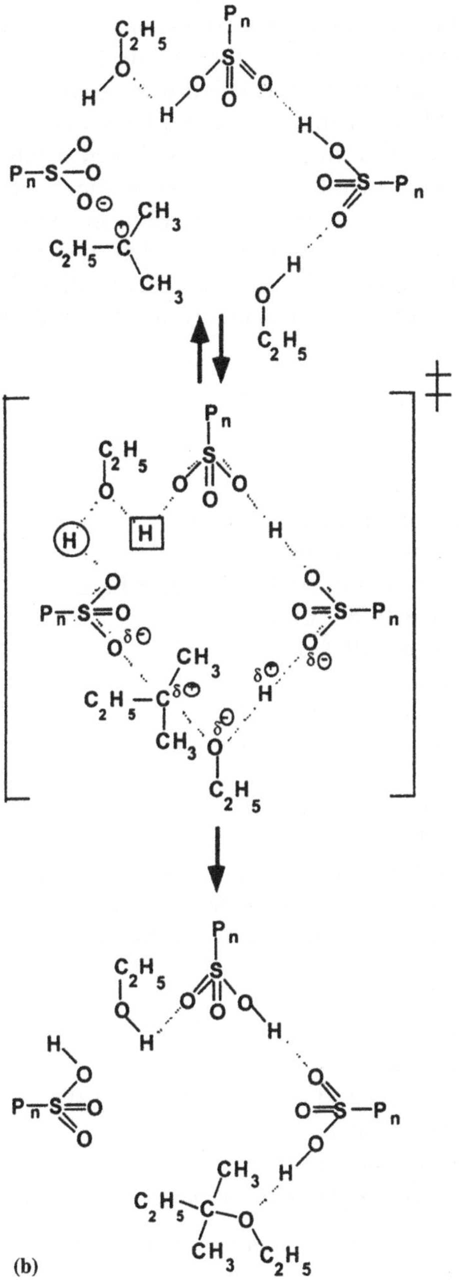

(b)

Figure 7b Continued.

steps within the LHHW formalism, resulting in rate expressions in terms of activities, which are more appropriate owing to the nonideality of the reaction system, as well as being consistent with reaction thermodynamics. The rate expressions involve equilibrium constants for the reversible reactions determined independently. Kinetics on partially deactivated catalyst confirms that the surface reactions involve three sites participating in the etherification reaction and two in the isomerization reaction, implying a concerted mechanism. Extrathermodynamic correlations that relate the kinetics to the reaction thermodynamics can also be rationalized within this framework and are experimentally observed for this family of tertiary ethers.

NOTATION

a_j	activity of species j, $\int \gamma_j x_j = \gamma_j^C C_j$
A_i	preexponential factor of reaction i, mol/h-g
C_j	concentration of species j, mol/dm^3
C_t	total acid site concentration of Amberlyst-15, mEq/g
E_i	activation energy of reaction i, kJ/mol
h	Planck's constant, 6.626×10^{-34} J-s
H_0	Hammett's acidity function
k_B	Boltzmann's constant, 1.38×10^{-23} J/K
k_i	rate constant of surface reaction (rate-determining step) of reaction i, mol/h-g
K_A	adsorption equilibrium constant of A $= 27 \exp[11{,}000(1/T - 1/303)/R]$
K_i	liquid-phase thermodynamic equilibrium constant of reaction i
K_j	adsorption equilibrium constant of species j on catalyst
m	number of ethanol adsorbed sites involved in the rate-determining step
r_i	rate of reaction i, mol/h-g
r_{i0}	initial rate of reaction i, mol/h-g
R	gas constant, 8.314 J/mol-K
s	number of vacant sites involved in rate-determining step
$[X]^\ddagger$	transition state complex (TSC)
α_i	sensitivity coefficient of reaction i
$\gamma^C_{[X]^\ddagger}$	activity coefficient of transition-state complex, $\equiv a_{[X]^\ddagger}/C_{[X]^\ddagger}$
$\Delta G'_{iT}$	standard Gibbs free energy of reaction i at temperature T, $= \sum_{j=1}^{n} \nu_{ij}\, \Delta G^{\circ}_{fjT}$, kJ/mol
$\Delta G^{\ddagger\circ}_{iT}$	effective standard Gibbs free energy of activation of reaction i at temperature T, $= \Delta G^{\circ}_{f[X]^\ddagger T} - \sum_{\text{Reactants}} \nu_{ij}\, \Delta G^{\circ}_{fjT}$, kJ/mol

ΔG°_{fjT}	standard Gibbs free energy of formation of species *j*, kJ/mol
$\Delta H^{\ddagger\circ}_{iT}$	effective standard enthalpy of activation of reaction *i* at temperature *T*, $=\Delta H^{\circ}_{f[X]^{\ddagger}T} - \sum_{\text{Reactants}} \nu_{ij}\, \Delta H^{\circ}_{fjT}$, kJ/mol
$\Delta S^{\ddagger\circ}_{iT}$	effective standard entropy of activation of reaction *i* at temperature *T*, $= \Delta S^{\circ}_{f[X]^{\ddagger}T} - \sum_{\text{Reactants}} \nu_{ij}\, \Delta S^{\circ}_{fjT}$, J/mol-K
ε	dielectric constant
κ	transmission coefficient
ν_{ij}	stoichiometric coefficient of species *j* in reaction *i*

Subscripts

A, B, C, D	ethanol, 2M1B, 2M2B, TAEE, respectively
i	of reaction *i*
j	of species *j*
1, 2, 3	reactions 1, 2, and 3, respectively

Superscripts

$'$	of reverse reaction
$\ddagger$	of transition state

Abbreviations

C3M2P	*cis*-3-methyl-2-pentene
ER	Eley-Rideal
ETBE	ethyl *tertiary*-butyl ether
LFER	linear free-energy relationships
LHHW	Langmuir-Hinshelwood-Hougen-Watson
MASSA	the most abundant surface species assumption
MTBE	methyl *tertiary*-butyl ether
TAEE	*tertiary*-amyl ethyl ether
TAME	*tertiary*-amyl methyl ether
THEE	*tertiary*-hexyl ethyl ether
THME	*tertiary*-hexyl methyl ether
TSC	transition-state complex
TST	transition-state theory
TTST	thermodynamic transition-state theory
T3M2P	*trans*-3-methyl-2-pentene
2M1B	2-methyl-1-butene
2M2B	2-methyl-2-butene
2M1P	2-methyl-1-pentene
2M2P	2-methyl-2-pentene
2,3-DM1B	2,3-dimethyl-1-butene
2,3-DM2B	2,3-dimethyl-2-butene
2E1B	2-ethyl-1-butene

REFERENCES

1. K. Jensen, Ethyl *tert*-butyl ether (ETBE) synthesis from ethanol and isobutylene over ion-exchange resin catalyst. PhD thesis, The University of Iowa, Iowa City, IA, 1996.
2. I. Miracca, L. Tagliabue, R. Trotta. Multitubular Reactors For Etherification. Chem Eng Sci, 51, 2349–2358, 1996.
3. F. Ancillotti, F. Mauri, E. Pescarollo. Ion exchange resin catalyzed addition of alcohols to alkenes. J Catal, 46, 49–57, 1977.
4. P. Chu, G. H. Kuhl. Preparation of, methyl *tert*-butyl ether (MTBE) over zeolite catalysts. Ind Eng Chem Res, 26, 365–369, 1987.
5. L. M. Tau, B. H. Davis. Acid catalyzed formation of ethyl tertiary butyl ether ETBE. Appl Catal, 53, 263–271, 1989.
6. S. I. Pien, W. J. Hatcher. Synthesis of methyl tertiary-butyl ether on HZSM-5 zeolite. Chem Eng Commun, 93, 257, 1990.
7. D. H. Fleitas, H. R. Macanio, C. F. Perez, O. A. Orio. Selective production of *t*-amyl methyl ether using sheet silicates. Reaction Kinet Catal Lett, 43, 183–188, 1991.
8. T. Takesono, Y. Fujiwara. Method for producing methyl *tert*-butyl ether and fuel composition containing the same. US Patent 4,182,913, 1980.
9. T. W. G. Solomon. Fundamentals of Organic Chemistry. Wiley, New York, 1982.
10. A. Gicquel, B. Torck. Synthesis of methyl tertiary butyl ether catalyzed by ion-exchange resin. Influence of methanol concentration and temperature. J Catal, 83, 9–18, 1983.
11. M. Voloch, M. R. Ladish, G. T. Tsao. Methyl t-butyl ether (MTBE) process catalyst parameters. Reactive Polymers, 4, 91–98, 1986.
12. C. Subramaniam, S. Bhatia. Liquid phase synthesis of methyl *tert*-butyl ether catalyzed by ion exchange resin. Can J Chem Eng, 65, 613–620, 1987.
13. A. M. Al-Jarallah, M. A. B. Siddiqui, A. K. K. Lee. Kinetics of methyl tertiary butyl ether synthesis catalyzed by ion exchange resin. Can J Chem Eng, 66, 802–807, 1988.
14. A. Rehfinger, U. Hoffmann. Kinetics of methyl tertiary butyl ether liquid phase synthesis catalyzed by ion exchange resin—II. Macropore diffusion of methanol as rate-controlling step. Chem Eng Sci, 45, 1619–1626, 1990.
15. A. Rehfinger, U. Hoffmann. Kinetics of methyl tertiary butyl ether liquid phase synthesis catalyzed by ion exchange resin—I. Intrinsic rate expression in liquid phase activities. Chem Eng Sci, 45, 1605–1617, 1990.
16. T. Zhang, R. Datta. Integral analysis of methyl *tert*-butyl ether synthesis kinetics. Ind Eng Chem Res, 34, 730–740, 1995.
17. C. Fité, M. Iborra, J. Tejero, J. Izquierdo, F. Cunill. Kinetics of liquid-phase synthesis of ETBE. Ind Eng Chem Res, 33, 583–589, 1994.
18. L. K. Rihko, A. O. I. Krause. Kinetics of heterogeneously catalyzed *tert*-amyl methyl ether reactions in the liquid phase. Ind Eng Chem Res, 34, 1172–1180, 1995.

19. P. Kitchaiya. Kinetics and thermodynamics of tertiary amyl ethyl ether (TAEE) synthesis on Amberlyst-15. PhD thesis, The University of Iowa, 1995.
20. J. Linnekoski, O. Krause, L. Rihko. Kinetics of the heterogeneously catalyzed formation of *tert*-amyl ethyl ether. Ind Eng Chem Res, 36, 310–316, 1997.
21. T. Zhang, R. Datta. Ethers from ethanol. 4. Kinetics of the liquid-phase synthesis of two *tert*-hexyl ethyl ethers. Ind Eng Chem Res, 34(7), 1995.
22. O. A. Hougen, K. M. Watson. Solid catalysts and reaction rates. General principles. Ind Eng Chem, 35, 529–541, 1943.
23. K. J. Laidler. Chemical Kinetics. 3rd ed. HarperCollins, New York, 1987.
24. K. A. Connors. Chemical Kinetics. The Study of Reaction Rates in Solution, VCH, 187–243, 1990.
25. J. Dumesic, D. Rudd, L. Aparicio, J. Rekoske, A. Trevino. The Microkinetics of Heterogeneous Catalysis. American Chemical Society, Washington, DC, 1993.
26. J. J. Rooney. Eyring transition-state theory and kinetics in catalysis. J Mol Catal A: Chem, 96, L1–L3, 1995.
27. M. Boudart. Kinetics of Chemical Processes. Prentice-Hall, Englewood Cliffs, NJ, 1968.
28. I. Levine. Physical Chemistry, 3rd ed. McGraw-Hill, New York, 1988.
29. F. H. Syed, R. Datta. Reactor multiplicity and temperature oscillations in the simultaneous synthesis of methyl *tert*-butyl ether (MTBE) and ethyl *tert*-butyl ether (ETBE) over Amberlyst-15. AIChE Annual Meeting, Los Angeles, CA, November 1997.
30. K. L. Jensen, R. Datta. Ethers from ethanol. 1. Equilibrium thermodynamic analysis of the liquid-phase ethyl *tert*-butyl ether reaction. Ind Eng Chem Res, 34, 392–399, 1995.
31. F. H. Syed, C. Egleston, R. Datta. *tert*-Amyl methyl ether (TAME): thermodynamic analysis of reaction equilibria in the liquid phase. J Chem Eng Data, 45(2), 319–323, 2000b.
32. P. Kitchaiya, R. Datta. Ethers from ethanol. 2. Reaction equilibria of simultaneous *tert*-amyl ethyl ether synthesis and isoamylene isomerization. Ind Eng Chem Res, 34, 1092–1101, 1995.
33. T. Zhang, R. Datta. Ethers from ethanol. 3. Equilibrium conversion and selectivity limitations in the liquid-phase synthesis of two *tert*-hexyl ethyl ethers. Ind Eng Chem Res, 34, 2237–2246, 1995.
34. T. Zhang, R. Datta. Ethers from ethanol. 5. Equilibria and kinetics of the coupled reaction network of liquid-phase 3-methyl-3-ethoxy-pentane synthesis. Chem Eng Sci, 51(4), 649–661, 1996.
35. P. Rys. W. J. Steinegger. Acidity function of solid-bound acids. J Am Chem Soc, 101, 4801–4806, 1979.
36. S. Lapanje, S. A. Rice. On the ionization of polystyrene sulfonic acid. J Am Chem Soc, 83, 496–497, 1961.
37. L. Kotin, M. Nagasawa. A study of the ionization of polystyrene sulfonic acid by proton magnetic resonance. J Am Chem Soc, 83, 1026–1028. 1961.

38. B. C. Gates, L. N. Johanson. Langmuir-Hinshelwood kinetics of the dehydration of methanol catalyzed by cation exchange resin. AIChE J, 17, 981–983, 1971.
39. L. Beránek. An examination of the Langmuir-Hinshelwood Model using ion exchange catalysts. Catal Rev Sci Eng, 16(1), 1–35, 1977.
40. A. O. I. Krause, L. G. Hammarstrom. Etherification of isoamylenes with methanol. Appl Catal, 30, 313–324, 1987.
41. B. C. Gates, J. S. Wisnouskas, H. W. Heath Jr. The dehydration of *t*-butyl alcohol catalyzed by sulfonic acid resin. J Catal, 24, 320–327, 1972.
42. J. E. Leffler, E. Grunwald. Rates and Equilibria of Organic Reactions. Wiley, New York, p 156, 1963.
43. G. Zundel. Hydration and Intermolecular Interaction, Infrared Investigations with Polyelectrolyte Membranes. Academic Press, pp 124–134, 1969.
44. I. Kampschulte-Scheuing, G. Zundel. Tunnel effect, infrared continuum, and solvate structure in aqueous and anhydrous acid solutions. J Phys Chem, 74, 2363–2368, 1970.
45. R. Thornton, B. C. Gates. Catalysis by matrix-bound sulfonic acid groups: olefin and paraffin formation from butyl alcohols. J Catal, 34, 275–287, 1974.
46. B. C. Gates. Catalytic Chemistry. Wiley, New York, p 207, 1992.
47. J. Tejero, F. Cunill, M. Iborra. Molecular mechanisms of MTBE synthesis on a sulphonic acid ion exchange resin. J Mol Catal, 42, 257–268, 1987.
48. E. Caldin, V. Gold. Proton-Transfer Reactions. Wiley, New York, 124, 1975.
49. A. Ozaki, S. Tsuchiya. The isomerization of *n*-butenes over a ion exchange resin. J Catal, 5, 537–539, 1966.
50. T. Uematsu, K. Tsukada, M. Fujishima, H. Hashimoto. The isomerization of 1-butene over cation-exchanged acidic resin. J Catal, 32, 369–375, 1974.
51. R. Datta, K. Jensen, P. Kitchaiya, T. Zhang. Thermodynamic transition-state theory and extrathermodynamic correlations for the liquid-phase kinetics of ethanol derived ethers. In: G. F. Froment, and K. C. Waugh, eds. Dynamics of Surfaces and Reaction Kinetics in Heterogeneous Catalysis. Elsevier, Amsterdam, pp 559–564, 1997.
52. O. Francoisse, F. C. Thyrion. Kinetics and mechanism of ethyl *tert*-butyl ether liquid-phase synthesis. Chem Eng Process, 30, 141–149, 1991.
53. S. Randriamahefa, R. Gallo, G. Raoult, P. Mulard. Synthesis of tertiary amyl methyl ether (TAME) by acid catalysis. J Mol Catal 49, 85–102, 1988.
54. R. Piccoli, H. Lovisi. Kinetics and thermodynamic study of the liquid-phase etherification of isoamylene with methanol. Ind Eng Chem Res, 34, 510–515, 1995.
55. K. Sundmacher, R. Zhang, U. Hoffmann. Mass transfer effects on kinetics of nonideal liquid phase ethyl *tert*-butyl ether formation. Chem Eng Technol, 18, 269–277, 1995.
56. F. Syed, K. Jensen, R. Datta. Thermodynamically consistent modeling of a liquid-phase nonisothermal packed-bed reactor. AICHE J, 46(2), 380–388, 2000.

11

Thermodynamics of Ether Production

Andrzej Wyczesany
University of Technology, Krakow, Poland

I. CHEMICAL EQUILIBRIA OF ETHER SYNTHESIS IN THE GAS PHASE

Commercially, ethers are synthesized from a combination of alcohols and olefins with a double bond on a tertiary carbon atom in the liquid phase. However, the literature presents works describing experimental equilibrium results of ether formation reactions in the gas phase. Thermodynamics provides the tools that allow prediction of equilibrium compositions of reacting mixtures, provided the reliable thermodynamic data are available. The data concerning olefins and alcohols are precise, but the properties of the tertiary ethers taken from various thermodynamic tables differ significantly. The experimental results of ether synthesis in the gas phase allow us to verify the data mentioned above. The verified data for the vapor phase can be used to predict the equilibrium compositions in the liquid phase.

Tejero et al. [1] determined experimentally the equilibrium constants for the reaction of methyl *tertiary*-butyl ether (MTBE) synthesis from isobutene and methanol in the vapor phase. The experiments were performed at atmospheric pressure in the temperature range 40–110°C. Knowing the equilibrium composition, the equilibrium constant K in the gas phase can be calculated from the following general equation:

$$K=\left(\prod_{i=1}^{N} y_i^{\nu_i}\right)\left(\prod_{i=1}^{N} \phi_i^{\nu_i}\right)\left(\frac{P}{P^o}\right)^{\sum_{i=1}^{N}\nu_i}=K_y K_\phi\left(\frac{P}{P^o}\right)^{\sum_{i=1}^{N}\nu_i} \qquad (1)$$

where y_i is the equilibrium mole fraction of species i in the gas phase, ϕ_i is its fugacity coefficient, P and P^o are the total and the standard pressure, respectively, N is the number of the reacting substances, and ν_i is the stoichiometric coefficient of species i—positive for products and negative for reactants.

The constant K can be also calculated from the thermodynamic properties through the equation

$$K = \exp\left(\frac{-\Delta G_T^o}{RT}\right) \tag{2}$$

where T is the absolute temperature, R is the gas constant, and ΔG_T^o is the standard Gibbs free energy of the reaction at T. Depending on the type of available data ΔG_T^o can be evaluated from one of two expressions:

$$\Delta G_T^o = \sum_{i=1}^{N} \nu_i \, \Delta G f_{i,T}^o \tag{3}$$

or

$$\Delta G_T^o = \Delta H_T^o - T\,\Delta S_T^o \tag{4}$$

where $\Delta Gf_{i,T}^o$ is the standard Gibbs free energy of formation for individual species i at T and ΔH_T^o and ΔS_T^o are the standard enthalpy and entropy of reaction at T, respectively.

In many cases, thermodynamic tables present the numerical values of $\Delta Gf_{i,T}^o$ at temperatures 298 K, 300 K, 400 K, etc. The method of least squares allows calculation of the coefficients (A–E) for function (5) using these values:

$$\Delta Gf_{i,T}^o = A + BT + CT^2 + DT^3 + ET \ln T \tag{5}$$

This temperature function $\Delta Gf_{i,T}^o$ is particularly convenient when the equilibrium calculations are performed with the Gibbs free-energy minimization method. However, more often the thermodynamic tables present numerical values of ΔGf_{298}^o, standard enthalpy of formation ΔHf_{298}^o, and the coefficients a–d of Eq. (6), allowing us to calculate the standard molar heat capacity C_p^o at T for the individual substances.

$$C_p^o = a + bT + cT^2 + dT^3 \tag{6}$$

Using these data, the Gibbs free energy of reaction at T can be obtained from the integrated form of the van't Hoff equation (7), the Kirchoff equation (8), and the relation (9) expressing the standard heat capacity of reaction ΔC_p^o.

$$\frac{\Delta G_T^o}{RT} = \frac{\Delta G_{298}^o}{298R} - \frac{1}{R}\int_{298}^{T} \frac{\Delta H_T^o}{T^2}\, dT \tag{7}$$

$$\Delta H_T^o = \Delta H_{298}^o + \int_{298}^{T} \Delta C_p^o\, dT \tag{8}$$

$$\Delta C_p^o = \sum_{i=1}^{N} \nu_i C_{p,i}^o \tag{9}$$

In accordance with the derivation presented by Jensen and Datta [2], if one assumes that Eq. (6) expresses the temperature function of molar heat capacity of individual species, after integration Eq. (8) gives the following form:

$$\Delta H_T^o = I_H + \Delta a T + \frac{\Delta b}{2} T^2 + \frac{\Delta c}{3} T^3 + \frac{\Delta d}{4} T^4 \tag{10}$$

where Δa–Δd are defined by Eq. (11) and I_H is the constant of integration including term ΔH_{298}^o and the last four constituents of Eq. (10) for $T = 298$.

$$\Delta a = \sum_{i=1}^{N} \nu_i a_i \qquad \Delta b = \sum_{i=1}^{N} \nu_i b_i \qquad \Delta c = \sum_{i=1}^{N} \nu_i c_i \qquad \Delta d = \sum_{i=1}^{N} \nu_i d_i \tag{11}$$

The use of Eq. (10) in Eq. (7) results in

$$\frac{\Delta G_T^o}{RT} = -I_K + \frac{I_H}{RT} - \frac{\Delta a}{R} \ln T - \frac{\Delta b}{2R} T - \frac{\Delta c}{6R} T^2 - \frac{\Delta d}{12R} T^3 \tag{12}$$

where I_K is the constant of integration including term $\Delta G_{298}^o/298\text{R}$ and the last five constituents of Eq. (12) for $T = 298$. The final form of the equation expressing the temperature function of the equilibrium constant can be obtained by inserting Eq. (12) into Eq. (2).

$$\ln K = I_K - \frac{I_H}{RT} + \frac{\Delta a}{R} \ln T + \frac{\Delta b}{2R} T + \frac{\Delta c}{6R} T^2 + \frac{\Delta d}{12R} T^3 \tag{13}$$

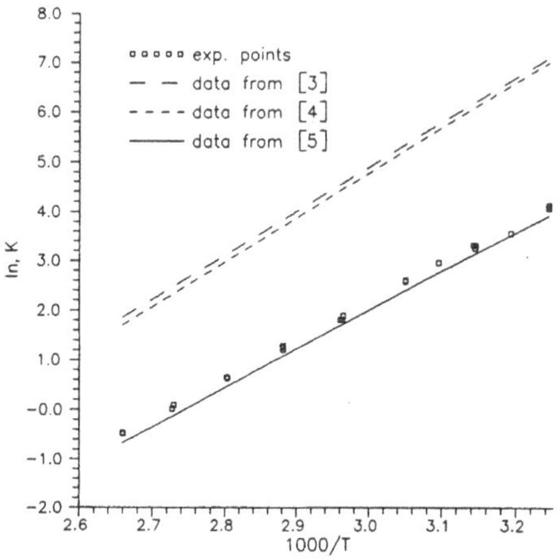

Figure 1 Plot of $\ln K$ against $1000/T$. Comparison between values obtained experimentally and those predicted from various thermal data.

The experimental data from [1] allow us to calculate the constant K from Eq. (1) for MTBE synthesis [Eq. (14)] with the assumption that the system is an ideal one ($\phi_i = 1$). The values of $\ln K$ versus $1000/T$ for the experimental results are shown in Figure 1 as individual points. In the same figure the $\ln K$ values obtained from Eq. (5) and thermodynamic data taken from Stull et al. [3] or from Eq. (13) and thermodynamic data taken from Reid et al. [4] are drawn as two different dashed lines, respectively. The calculated values differ significantly from the experimental results. Good agreement with experiment can be obtained when the thermodynamic data ($\Delta Gf^o_{298.15}$ and $\Delta Hf^o_{298.15}$) for MTBE are taken from Fenwick et al. [5] (the solid line in Fig. 1).

$$CH_3OH + CH_2{=}C(CH_3){-}CH_3 \rightarrow CH_3{-}O{-}C(CH_3)_3 \tag{14}$$

Table 1 presents the coefficients A–E of Eq. (5) for the reacting substances of Eq. (14). They were calculated from tables [3] for C_4H_8 and CH_3OH, whereas in the case of MTBE they were determined using $\Delta Gf^o_{298.15}$ and $\Delta Hf^o_{298.15}$ from [5] and the temperature function of C^o_p from tables [3].

Table 2 presents the equilibrium compositions calculated for the initial feed: 1 mole of isobutene and 1 mole of methanol at $P = 1$ bar and $T = 343.15$ K. For this temperature the constant K determined from Eqs. (5), (3), and (2) is equal to 5.66. The calculated results were obtained with the assumption that the reacting mixture is an ideal gas (IDEAL) and for two nonideal models. The first one (SRK) used the Soave-Redlich-Kwong

Table 1 Numerical Values of the Coefficients *A*–*E* of Eq. (5) (kJ/mol-K) for C_4H_8, CH_3OH, and MTBE

Compound	A	B	$C \times 10^{-4}$	$D \times 10^{-9}$	E
C_4H_8	3.6753	−0.26632	−0.37186	0.67328	0.080697
CH_3OH	−185.4550	−0.28008	−0.47233	7.1978	0.065036
MTBE	−243.6344	−0.49021	−1.1693	13.916	0.166304

equation of state (EOS) and the second one (VIR) applied the virial EOS truncated after the second term (the method of Hayden and O'Conell [6]). The equilibrium conversion of isobutene was equal to 61% regardless of the model applied.

Iborra et al. [7] investigated the synthesis of ethyl *tertiary*-butyl ether (ETBE) from isobutene and ethanol. They found that the equilibrium constant of this reaction determined from experimental data differed significantly from the value K calculated by Eq. (13) using published thermodynamic data. Therefore, they fitted the coefficients of Eq. (13) to the experimental data, obtaining the following expression:

$$\ln K = -4.10 + 7223T^{-1} - 3.102 \ln T + 8.432 \times 10^{-3}T - 3.155 \times 10^{-6}T^2 + 2.332 \times 10^{-9}T^3 \tag{15}$$

Heese et al. [8] studied the chemical equilibria of the synthesis of di-isopropyl ether (DIPE) from water and propylene over the catalyst Amberlyst 15. They published also the data referring to the chemical equilibria reached in decomposition of isopropanol (IPA) to water, propylene, and DIPE. These data enable verification of the accuracy of the published thermodynamic properties. The parameters of the investigated reaction were

Table 2 Equilibrium Compositions (mol) of Reaction (14) Obtained for the Methods Considered for Describing Nonideality

	IDEAL	SRK		VIR	
Compound	n_i	n_i	ϕ_i	n_i	ϕ_i
C_4H_8	0.3896	0.3897	0.9850	0.3897	0.9871
CH_3OH	0.3896	0.3897	0.9842	0.3897	0.9811
MTBE	0.6104	0.6103	0.9696	0.6103	0.9690
Total	1.3896	1.3897	—	1.3897	—

$T = 343.15$ K, $P = 1$ bar; initial feed composition isobutene = 1 mol, methanol = 1 mol.

Table 3 Comparison of Calculated Equilibrium Results (mol%) with Experimental Data

Compound	Exp.	IDEAL	SRK	PR
C_3H_6	46.9	48.07	48.06	48.06
H_2O	46.9	48.28	48.28	48.28
IPA	5.8	3.44	3.45	3.45
DIPE	0.4	0.21	0.21	0.21

$T = 120°C$, $P = 1.01$ bar; initial feed consists of pure isopropanol.

$T = 120°C$, $P = 1.01$ or 50 bar. At atmospheric pressure the equilibrium mixture was entirely in the gas phase. The experimental results in Ref. 8 were presented in the form of the plots showing the compositions versus the contact time. When the time was long enough the compositions were constant. Table 3 presents the comparison of the experimental data taken from the plot with the calculated equilibrium results. The last values were obtained for two nonideal models using the SRK or Peng-Robinson (PR) EOS and assuming that the reacting mixture was an ideal gas (IDEAL). The calculations were performed with the Gibbs free-energy minimization method described in detail in [9]. The values of $\Delta Gf^o_{i,T}$ for all components were taken from Ref. 3, and the critical parameters T_c and P_c as well as the acentric factors ω from Ref. 4. Analysis of Table 3 proves that the final result of the calculations does not depend on the model applied. The differences between the experimental and calculated results do not exceed 2.4 mol%. Considering the fact that the experimental data were taken from the plot, they agree with the values calculated within the limit of experimental error. Thus, the thermodynamic values used for the calculations can be treated as sufficiently precise data and will be applied in the next section for the equilibrium calculations in the liquid phase.

The results presented in Tables 2 and 3 point out that the reacting mixtures formed in ether syntheses in the gas phase can be treated as ideal gases at atmospheric pressure.

II. CHEMICAL EQUILIBRIA OF ETHER SYNTHESIS IN THE LIQUID PHASE

A. Tertiary Ethers

Commercially, MTBE is synthesized in the liquid phase at temperatures of 40–80°C and pressures of 7–20 bar in the presence of acid cation-exchange catalysts such as Amberlyst-15. Since the reacting mixture is highly non-ideal, the liquid-phase equilibrium constant of Eq. (14) is given by

$$K=\left(\frac{a_{\mathrm{MTBE}}}{a_{\mathrm{C_4H_8}}a_{\mathrm{CH_3OH}}}\right)=\left(\frac{x_{\mathrm{MTBE}}}{x_{\mathrm{C_4H_8}}x_{\mathrm{CH_3OH}}}\right)\left(\frac{\gamma_{\mathrm{MTBE}}}{\gamma_{\mathrm{C_4H_8}}\gamma_{\mathrm{CH_3OH}}}\right)=K_xK_\gamma \tag{16}$$

where a, x, and γ are activity, mole fraction, and activity coefficient of the liquid phase, respectively.

The constant K can also be calculated by Eqs. (6)–(13) when the values ΔHf^o_{298}, ΔGf^o_{298} as well as the coefficients a–d of Eq. (6) are known for all components in the liquid state. Unfortunately, these values presented in the thermodynamic tables are not precise enough, especially for the ethers.

The standard Gibbs free energies of formation in the liquid phase can be calculated as well from the following equation:

$$\Delta Gf^o_{i,T}(l)=\Delta Gf^o_{i,T}(g)+RT\ \ln\frac{P^s_i\phi^s_i}{P^o}+V_i(l)\left(P^o-P^s_i\right) \tag{17}$$

where P^s_i is the saturated vapor pressure of pure component i at T, ϕ^s_i is the fugacity coefficient of saturated vapor pressure of pure component i (calculated for T and P^s_i), and V_i is the liquid molar volume of pure component i at T. Letters (l) and (g) distinguish liquid and gas phases, respectively.

Equation (17) is very useful when the values $\Delta Gf^o_{i,T}(g)$ are known and vapor pressures of individual components do not exceed a few bar. In the last case, the coefficients ϕ^s_i are usually close to 1 and the last term in Eq. (17) has a very small value that can be neglected. In addition, the saturated vapor pressure can be calculated with high accuracy from the Antoine equation [(18a) or (18b)] for which coefficients AP, BP, and CP are presented in the literature for many substances. If the saturated vapor pressure of the given component exceeds a few bar, this can be calculated from the Wagner equation (19), which requires knowledge of four constants, AP–DP, and the critical parameters. These values are published in Ref. 4 for many substances.

$$\ln P^s=\mathrm{AP}-\frac{\mathrm{BP}}{T-\mathrm{CP}} \tag{18a}$$

or

$$\log P^s=\mathrm{AP}-\frac{\mathrm{BP}}{T-\mathrm{CP}} \tag{18b}$$

$$\ln \frac{P^s}{P_c} = \frac{\mathrm{AP} \times x + \mathrm{BP} \times x^{1.5} + \mathrm{CP} \times x^3 + \mathrm{DP} \times x^6}{1 - x}$$
$$\text{where} \quad x = 1 - \frac{T}{T_c} \tag{19}$$

In some published articles the authors have calculated the constant K for MTBE synthesis from the experimentally determined K_x, and the corresponding K_γ obtained through the UNIFAC method. The correlations expressing the temperature functions of constant K given by Colombo et al. [10], Rehfinger and Hoffmann [11], Izquierdo et al. [12], and Zhang and Datta [13] are presented below by Eqs. (20)–(23), respectively.

$$\ln K = -10.0982 + 4254.05T^{-1} + 0.2667 \ln T \tag{20}$$

$$\ln K = 357.094 - 1492.77T^{-1} - 77.4002 \ln T + 0.507563T - 9.12739 \times 10^{-4}T^2 + 1.10649 \times 10^{-6}T^3 - 6.27996 \times 10^{-10}T^4 \tag{21}$$

$$\ln K = 1145.0257 - 14714.411T^{-1} - 232.7593 \ln T + 1.065597T - 1.0775 \times 10^{-3}T^2 + 5.30525 \times 10^{-7}T^3 \tag{22}$$

$$\ln K = -13.482 + 4388.7T^{-1} + 1.2353 \ln T - 0.013849T + 2.5923 \times 10^{-5}T^2 - 3.1881 \times 10^{-8}T^3 \tag{23}$$

Table 4 presents the numerical values of constants K calculated in the temperature range 25–90°C from the four expressions given above and from Eq. (2), where the standard Gibbs free energy of reaction (14) was calculated from the standard Gibbs free energies of formation expressed by Eq. (17). In the last equation, $\Delta Gf^o_{i,T}(g)$ values were calculated with coefficients A–E taken from Table 1, P^s_i and $V_i(l)$ [by Eq. (24)] values with data from Table 5, whereas the coefficients ϕ^s_i were evaluated using the SRK EOS.

The values K calculated in this work were determined for data taken from the thermodynamic tables only. In spite of this, these constants never exceed experimental values in the range of temperatures investigated.

$$V(l) = \mathrm{VA} + \mathrm{VB} \times T + \mathrm{VC} \times T^2 \tag{24}$$

Table 4 Equilibrium Constants Calculated with Some Literature Expressions and the Method Used in This Chapter

Temperature, °C	25	30	40	50	60	70	80	90
Temperature, K	298.15	303.15	313.15	323.15	333.15	343.15	353.15	363.15
K (Colombo et al. [10])	295.7	234.8	151.3	100.2	68.0	47.3	33.5	24.2
K (Rehfinger and Hoffmann [11])	284.0	220.7	135.6	85.0	54.3	35.3	23.3	15.6
K (Izquierdo et al. [12])	242.4	188.9	117.2	74.6	48.6	32.3	21.9	15.1
K (Zhang and Datta [13])	272.2	210.5	128.7	80.9	52.0	34.2	22.9	15.6
K calculated in this chapter	246.5	194.8	123.8	80.4	53.2	35.8	24.5	17.1

Table 5 Coefficients Necessary for Calculation of P_i^s, ϕ_i^s, and $V_i(l)$ (L/mol)

	AP	BP	CP	DP	T_c, K [4]	P_c, bar [4]	ω [4]	VA	VB × 10^{-4}	VC × 10^{-7}
C_4H_8	−6.95542	1.35673	−2.45222	−1.46110	417.9	40.0	0.194	0.21514	−9.7304	19.173
CH_3OH	16.16967	3394.194	44.02	—	512.6	80.9	0.556	0.041846	−0.55903	1.7337
MTBE	6.14854	1211.690	35.719	—	496.4	33.7	0.269	0.12968	−2.3047	6.6301

$P^s_{C_4H_8}$ was calculated from Eq. (19), data from Ref. 4; $P^s_{CH_3OH}$ from Eq. (18a) (P^s in kPa, T in K), data from Ref. 14; and P^s_{MTBE} from Eq. (18b) (P^s in kPa, T in K), data from Ref. 15.

The constants VA–VC for the individual components were calculated from the numerical values of volumes taken from Ref. 16 through the method of least squares.

To determine the influence of the method describing nonideality on the equilibrium composition, the calculations were performed for six different models using the Gibbs free-energy minimization method [9]. The results of calculation at $T = 343.15$ K, $P = 10$ bar for initial feed of 1 mole of isobutene and 1 mole of methanol are presented in Table 6. The models used are defined as UNI = the UNIFAC method [4], UL = the UNIFAC method modified by Larsen et al. [17], UD = the UNIFAC method modified by Gmehling et al. [18], ASOG = the ASOG method [19,20], WIL = the Wilson equation [for which the numerical values of the coefficients $(\lambda_{ij} - \lambda_{ii})$ were taken from Ref. 21], and NRTL = the NRTL equation. In the last model the coefficients presented in Table 7 were estimated from the experimental values of vapor–liquid equilibria (VLE) through the minimization of the objective function defined by Eq. (25). The VLE data for the binaries C_4H_8–CH_3OH, C_4H_8–MTBE, and CH_3OH–MTBE were taken from Refs. 22, 23, and 24, respectively. The parameters PT in Eq. (25) mean P for isothermal and T for isobaric data.

$$\mathrm{OF} = \sum_i \left[\frac{\mathrm{PT}_i^{\exp} - \mathrm{PT}_i^{\mathrm{calc}}}{\mathrm{PT}_i^{\exp}}\right]^2 + \sum_i \left[\frac{y_{1,i}^{\exp} - y_{1,i}^{\mathrm{calc}}}{y_{1,i}^{\exp}}\right]^2 + \sum_i \left[\frac{y_{2,i}^{\exp} - y_{2,i}^{\mathrm{calc}}}{y_{2,i}^{\exp}}\right]^2 \tag{25}$$

As can be observed in Table 6, the model of nonideality description has a small but noticeable effect on the calculated equilibrium results. It is related to the fact that reaction (14) is shifted significantly to the right. Even for the Wilson equation and especially for the ASOG method, the considerable differences in the quotient $1/K\gamma$ do not cause great changes in the calculated compositions. The equilibrium conversions of isobutene at 343.15 K reach about 91% for all the versions of the UNIFAC method and the NRTL equation. This value estimated for the reaction in the liquid phase is about 30% higher than in the gas phase (cf. Table 2).

The same phenomenon can be observed in the reaction of forming higher ethers from the reactive olefins and methanol (ethanol). This fact can be explained as follows. The values of the saturated vapor pressures of the reactants (alkene and alcohol) are greater than for the products (ether). According to this, the Gibbs free energies of the individual etherification reactions in the liquid phase are more negative than in the gas phase at the same temperature. Therefore, the tendency of olefins to react to ethers is

Table 6 Equilibrium Compositions (mol) of MTBE Synthesis and Activity Coefficients Obtained for the Methods Considered for Describing Nonideality

	UNI		UL		UD		ASOG		WIL		NRTL	
	n_i	γ_i	n_i	γ_i	n_i	γ_i	n_i	γ_i	n_i	γ_i	n_i	γ_i
C_4H_8	0.092	1.232	0.090	1.130	0.095	1.028	0.071	1.497	0.081	1.253	0.095	1.101
CH_3OH	0.092	2.691	0.090	3.025	0.095	2.952	0.071	3.696	0.081	3.414	0.095	2.787
MTBE	0.908	1.003	0.910	1.003	0.905	1.004	0.929	1.007	0.919	1.008	0.905	1.002
$1/K\gamma$	—	3.305	—	3.408	—	3.023	—	5.494	—	4.244	—	3.062

$T = 343.15\,K$, $P = 10\,bar$; initial feed composition isobutene = 1 mol, methanol = 1 mol.

Table 7 NRTL Parameters $(g_{ij}-g_{jj})/R$, $(g_{ji}-g_{ii})/R$ in K^{-1} and $a_{ij}=a_{ji}$

	$(g_{ij}-g_{jj})$			a_{ij}		
	C_4H_8	CH_3OH	MTBE	C_4H_8	CH_3OH	MTBE
C_4H_8	0.0	755.26	258.94	0.0	0.46260	0.41408
CH_3OH	477.07	0.0	286.21	0.46260	0.0	0.69438
MTBE	− 190.20	242.39	0.0	0.41408	0.69438	0.0

much greater in the liquid phase than in the gas phase. Additionally, the numerical values of the coefficients γ_i are often greater than 3 for alcohols and close to 1 for olefins and ethers. Owing to this, the quotient $1/K\gamma$ exceeds usually the value 3, increasing the constant K_x in relation to the value K.

Izquierdo et al. [12] and Jensen and Datta [2] investigated the synthesis of ethyl *tertiary*-butyl ether (ETBE) from isobutene and ethanol. Using experimental equilibrium data and the activity coefficients calculated by the UNIFAC method, they proposed the following expressions that give constant K versus temperature:

$$\ln K=-9.5114+4262.21T^{-1} \tag{26}$$

$$\ln K=10.387+4060.59T^{-1}-2.89055\ln T-0.0191544T + 5.28586\times 10^{-5}T^2-5.32977\times 10^{-8}T^3 \tag{27}$$

The equilibrium conversion of isobutene with ethanol to ETBE is equal to 86.22% for the same conditions as in Table 6 when the value K was calculated from Eq. (27) and the coefficients γ_i through the UL method. So, this conversion is about 4% lower than in the case of MTBE.

The industry also produces methyl *tertiary*-amyl ether (TAME) from methanol and C_5 olefins. Since only olefins with a double bond on a tertiary carbon atom exhibit a tendency to react with methanol, only 2-methyl-1-butene (2MB1) and 2-methyl-2-butene (2MB2) are reactive.

$$CH_3OH+CH_2{=}C(CH_3){-}CH_2{-}CH_3 \rightarrow CH_3{-}O{-}C(CH_3)_2{-}CH_2{-}CH_3 \tag{28}$$

$$CH_3OH+CH_3{-}C(CH_3){=}CH{-}CH_3 \rightarrow CH_3{-}O{-}C(CH_3)_2{-}CH_2{-}CH_3 \tag{29}$$

When both of the above reactions attain the equilibrium state, the isomerization reaction of 2 MB1 to 2 MB2, as a dependent one, must reach the equilibrium as well. Several thermodynamic studies have been reported on TAME (Randriamahefa et al. [25], Safronov et al. [26], Rihko et al. [27], Hwang and Wu [28]; Piccoli and Lovisi [29]). However, none of these have been able to provide satisfactory agreement between experiments and theory. More recently, Syed et al. [30] stated that the value of the standard Gibbs free energy of TAME formation reported in the literature is inaccurate. Using experimental results at 298.15 K coupled with the linear regression of data at other temperatures, they estimated the value of $\Delta Gf^{o}_{\mathrm{TAME},298.15}(l)$ equal to -113.8 kJ/mol and defined the temperature functions of constants K1, K2, and K3 for reactions (28), (29), and isomerization of 2MB1 to 2MB2, respectively.

$$\ln K_1 = -39.065 + 5018.61T^{-1} + 4.6866 \ln T + 0.00773T - 2.635 \times 10^{-5}T^2 + 1.547 \times 10^{-8}T^3 \tag{30}$$

$$\ln K_2 = -34.798 + 3918.02T^{-1} + 3.9168 \ln T + 0.01293T - 3.121 \times 10^{-5}T^2 + 1.805 \times 10^{-8}T^3 \tag{31}$$

$$\ln K_3 = -4.159 + 1100.69T^{-1} + 0.7698 \ln T - 0.00521T + 4.865 \times 10^{-6}T^2 - 2.580 \times 10^{-9}T^3 \tag{32}$$

In the near future the tertiary ethers will be produced from ethanol and C_{5+} olefins. The chemical equilibrium of these reactions was investigated by Kitchaiya and Datta [31], Zhang and Datta [32] and Zhang et al. [33]. Using experimental equilibrium compositions and activity coefficients computed by the UNIFAC method, they estimated the correlations for the equilibrium constants in the liquid phase in a form similar to Eqs. (30)–(32). The numerical values of the coefficients of these correlations for the syntheses of 2-methyl-2-ethoxy butane (TAEE); 2-methyl-2-ethoxy pentane; 2,3-dimethyl-2-ethoxy butane, and 3-methyl-3-ethoxy pentane from ethanol and tertiary olefins are given in Ref. 33.

B. Side Reactions in *tertiary*-Ethers Synthesis

Although the reactions of alcohols with the reactive olefins are highly selective, side products can be formed. The dimerization of olefins (isobutene in MTBE or ETBE synthesis or 2MB1 and 2MB2 in

TAME synthesis) is the main side reaction. The hydrocarbon feed is the refinery C_4 or C_5 fraction, which should not contain di-olefins (especially butadiene) because of their strong tendency to polymerization. Other possible side reactions can form dimethyl ether (DME) from methanol or diethyl ether from ethanol. These processes consume the basic reactant alcohol and also produce water. The last one, together with H_2O contained in the feed as an impurity (especially in the case of ethanol), reacts with olefins to form alcohols—*t*-butanol (TBA) from isobutene or *t*-pentanol from C_5 olefins.

The next calculations were performed with the assumption that all the side reactions possible in MTBE synthesis reach the equilibrium state. According to this assumption the equilibrium product can also contain water, DME, TBA, and isobutene dimers: 2,4,4-trimethyl-1-pentene (244TMP1) and 2,4,4-trimethyl-2-pentene (244TMP2). The thermodynamic data referring to $\Delta Gf^o_T(g)$ for DME, H_2O, and TBA were taken from Ref. 3, and for di-olefins from Ref. 34 (values calculated by the Benson method). The coefficients of the Wagner equation (19) to calculate P^s_i for DME, H_2O, and TBA were taken from Ref. 4, and the coefficients of the Antoine equation (18b) for di-olefins from Ref. 35. The results of the calculations performed by the Gibbs free-energy minimization method [9] are presented in Table 8. The liquid-phase nonideality was described by the UL method.

These test calculations reveal that all side reactions are thermodynamically highly privileged. If all of them attained the equilibrium state, methanol would convert almost entirely to DME and water formed in this

Table 8 Calculated Compositions (mol) of Liquid Product Obtained with the Assumption That All Side Reactions Possible in MTBE Synthesis Reached Equilibrium State

	T (K)				
	313.15	323.15	333.15	343.15	353.15
C_4H_8	0.00096	0.00140	0.00199	0.00277	0.00377
CH_3OH	0.01567	0.01816	0.02084	0.02370	0.02672
MTBE	0.00434	0.00484	0.00534	0.00581	0.00628
H_2O	0.02400	0.02909	0.03488	0.04143	0.04878
DME	0.49000	0.48850	0.48691	0.48524	0.48350
TBA	0.46600	0.45941	0.45203	0.44381	0.43472
2,4,4-TMP1	0.05920	0.06302	0.06686	0.07072	0.07462
2,4,4-TMP2	0.20515	0.20415	0.20346	0.20308	0.20300

$P = 10$ bar; initial feed composition isobutene = 1 mol, methanol = 1 mol.

reaction would react nearly completely to TBA. The rest of the olefins, being in excess to water, would dimerize almost entirely, and in consequence the amount of the main product MTBE would be less than 1 mol%. Fortunately, in the presence of suitable catalyst the reactions of dimerization and methanol dehydration are very far from the equilibrium conversion, especially when the temperature is not too high. However, water contained in the feed reacts nearly completely to TBA.

C. Calculation of Equilibrium Composition by Excel Worksheet

Computation of the equilibrium composition of MTBE synthesis in the gas phase with the assumption of gas-phase ideality is simple. Such an operation for the liquid phase creates many more problems, since the activity coefficients are complicated functions of the composition to be calculated. The following example shows that the calculations can be performed with ease on the Excel worksheet using the tool "Solver."

Let us assume that 1 mole of C_4H_8 ($n^oC_4H_8$) reacts with 1 mole of CH_3OH (n^oCH_3OH) at 343.15 K ($K = 35.8$ from Table 4). The nonideality of the mixture is described by the NRTL equation with coefficients taken from Table 7. If one assumes that ξ is the equilibrium extent of reaction, the numbers of moles for the individual components in the equilibrium state can be calculated by Eq. (33). This allows defining the mole fractions x_i. Subsequently, Eq. (16) can be rearranged to Eq. (34), which is an objective function of only one unknown, ξ. The absolute value of this function is minimized by "Solver" and the numerical value in the cell containing the variable ξ is changed by "Solver." Excel's algorithm for searching of the minimum value is relatively sensitive to the initial guess. Therefore, the value ξ should be reasonably estimated before the calculations. The complete worksheet with the solution found is presented in Table 9. Definition of any constant or value calculated from an expression is placed above that cell in the preceding row. The values τ_{ij} and G_{ij} represent the auxiliary parameters of the NRTL equation. The auxiliary sums used in the expressions calculating the coefficients γ_i are placed in row 20.

$$n_i = n_i^o + \nu_i \xi \tag{33}$$

where ν_i is the stoichiometric coefficient of compound i—negative for reactants and positive for products.

$$\mathrm{OF} = \left| \left(\frac{x_{\mathrm{MTBE}}}{x_{\mathrm{C_4H_8}} x_{\mathrm{CH_3OH}}} \right) \left(\frac{\gamma_{\mathrm{MTBE}}}{\gamma_{\mathrm{C_4H_8}} \gamma_{\mathrm{CH_3OH}}} \right) \frac{1}{K} - 1 \right| = \mathrm{MIN} \tag{34}$$

Table 9 Results of Equilibrium Calculation Obtained by Excel Worksheet and Tool "Solver"

1	ξ	K	T, K			
2	0.905	35.8	343.15		OF	
3	n°_{C4H8}	n°_{CH3OH}	n°_{MTBE}		3.4E − 06	
4	1	1	0			
5	n_1	n_2	n_3	x_1	x_2	x_3
6	0.095	0.095	0.905	0.087	0.087	0.826
7	$(g_{11}-gg_{11})/R$	$(g_{12}-g_{22})/R$	$(g_{13}-g_{33})/R$	α_{11}	α_{12}	α_{13}
8	0	755.26	258.94	0	0.4626	0.41408
9	$(g_{21}-g_{11})R$	$(g_{22}-g_{22})/R$	$(g_{23}-g_{33})/R$	α_{21}	α_{22}	α_{23}
10	477.07	0	286.21	0.4626	0	0.69438
11	$(g_{31}-g_{11})/R$	$(g_{32}-g_{22})/R$	$(g_{33}-g_{33})/R$	α_{31}	α_{32}	α_{33}
12	− 190.2	242.39	0	0.41408	0.69438	0
13	τ_{11}	τ_{12}	τ_{13}	G_{11}	G_{12}	G_{13}
14	0.00000	2.20096	0.75460	1.00000	0.36126	0.73164
15	τ_{21}	τ_{22}	τ_{23}	G_{21}	G_{22}	G_{23}
16	1.39027	0.00000	0.83407	0.52564	1.00000	0.56037
17	τ_{31}	τ_{32}	τ_{33}	G_{31}	G_{32}	G_{33}
18	− 0.55428	0.70637	0.00000	1.25799	0.61233	1.00000
19	$\Sigma\tau_{i1}G_{i1}x_i$	$\Sigma\tau_{i2}G_{i2}x_i$	$\Sigma\tau_{i3}G_{i3}x_i$	$\Sigma G_{i1}x_i$	$\Sigma G_{i2}x_i$	$\Sigma G_{i3}x_i$
20	− 0.51270	0.42645	0.08854	1.17198	0.62419	0.93851
21	γ_1	γ_2	γ_3			
22	1.101	2.785	1.002			

Subscripts 1, 2, and 3 mean isobutene, methanol, and MTBE, respectively.

D. Thermodynamics of DIPE Synthesis—Equation-of-State Methods

Besides the alkyl *tertiary*-alkyl ethers, which are currently produced and investigated for future applications, DIPE is beginning to attract attention. This ether can be produced from water and propylene via the intermediate IPA according to the following reactions:

$$CH_2{=}CH{-}CH_3 + H_2O \rightarrow CH_3{-}CH(OH){-}CH_3 \qquad (35)$$

$$CH_3{-}CH(OH){-}CH_3 + CH_2{=}CH{-}CH_3 \rightarrow (CH_3)_2CH{-}O{-}CH(CH_3)_2 \qquad (36)$$

$$CH_3-CH(OH)-CH_3+CH_3-CH(OH)-CH_3 \rightarrow (CH_3)_2CH-O-CH(CH_3)_2+H_2O \quad (37)$$

To increase conversion and the reaction rate, the hydration of propylene is carried out at elevated pressures (10–100 bar) and in the temperature range 100–140°C. Heese et al. [8] presented experimental equilibrium results of this process performed in the presence of the catalyst Amberlyst 15 at 120°C and 50 bar. In the first experiment the initial feed consisted of the stoichiometric (2:1) mixture of propylene and water, whereas in the second one, pure IPA. (The authors' intention was to define the primary and secondary reactions). In both cases the equilibrium product obtained was in the liquid phase far from the bubble point.

The mixture considered is highly nonideal and contains one supercritical component (propylene). Additionally, the chemical equilibrium is reached at elevated pressure; therefore the EOS methods should be applied to describe the system nonideality. To establish which of these methods is most accurate for the system regarded, eight different models were investigated. The first four apply the classical mixing rules with or without the binary interaction coefficient k_{ij}. The last four use the mixing rules resulting from combining EOS and excess free-energy models.

1. Soave-Redlich-Kwong EOS [Eqs. (38)–(39)] with the coefficients k_{ij} in Eq. (40) equal to 0 (denoted below as SRK)
2. Soave-Redlich-Kwong EOS with k_{ij} fitted from experimental VLE data (SRK-KIJ)
3. Peng-Robinson EOS with k_{ij} equal to 0 (PR)
4. Peng-Robinson-Stryjek-Vera EOS with k_{ij} being a function of temperature as well as composition [Eqs. (41)–(42)] [36] (PRSV)
5. Predictive SRK EOS [37] (PSRK)
6. MHV-2 model [38] (MHV2)
7. LCVM method [39], being the linear combination of Vidal and Michelsen models (LCVM)
8. The method based on the SRK EOS with the mixing rule worked out by Two and Coon [40] (TC)

All the above methods were incorporated into the algorithm calculating the chemical and phase equilibria presented in [9]. The models SRK, PR, PSRK, MHV2, and LCVM are entirely predictive. The rest of the methods require the fitting of parameters from the VLE data for each binary mixture. The SRK-KIJ model needs one parameter k_{ij}, whereas the method PRSV needs four parameters, ($k_{ij}^{(1)}, k_{ij}^{(2)}, k_{ji}^{(1)}$, and $k_{ji}^{(2)}$), necessary in Eqs. (42) and (41).

$$P=\frac{RT}{V-b}-\frac{a}{V(V+b)} \tag{38}$$

$$a=\frac{0.42478R^2T_c^2}{P_c}\left[1+\left(0.48+1.574\omega-0.176\omega^2\right)\left(1-\sqrt{\frac{T}{T_c}}\right)\right]^2$$
$$b=\frac{0.08664RT_c}{P_c} \tag{39}$$

$$a_{\text{mix}}=\sum_i\sum_j x_ix_j\sqrt{a_ia_j}\left(1-k_{ij}\right) \tag{40}$$

$$a_{\text{mix}}=\sum_i\sum_j x_ix_j\sqrt{a_ia_j}\left(1-x_ik_{ij}-x_jk_{ji}\right) \tag{41}$$

where

$$k_{ij}=k_{ij}^{(1)}+\frac{k_{ij}^{(2)}}{T}\qquad k_{ji}=k_{ji}^{(1)}+\frac{k_{ji}^{(2)}}{T} \tag{42}$$

The TC model uses the SRK EOS and the mixing rules defined by Eqs. (43) and (44), in which the reduced parameters marked with (*) are given by expression (45). The subscript vdW means the parameter is calculated from the mixing rules applied for a van der Waals fluid through Eqs. (46) and (47). The last equation also contains the interaction coefficient l_{ij} for the parameter b. The excess molar Helmholtz energy A^E is calculated in the TC model from the NRTL equation containing three adjustable parameters: $(g_{ij}-g_{jj})/R$, $(g_{ji}-g_{ii})/R$, and α_{ij}. The mixing rules worked out by Two and Coon are more flexible than in the remaining models. When the system contains polar substances, five adjustable parameters can be applied. Three of them describe the excess molar Helmholtz energy and the remaining two represent the van der Waals interaction coefficients (k_{ij} and l_{ij}). However, when A^E is set to 0, the mixing rules expressed by Eqs. (43) and (44) reduce to the van der Waals mixing rules with one or two adjustable parameters. The last case is often applied for systems of nonpolar compounds or for mixtures of these substances with light gases.

$$b^*=\frac{b_{\text{vdW}}^*-a_{\text{vdW}}^*}{1-\left[\left(a_{\text{vdW}}^*/b_{\text{vdW}}^*\right)-A^E/(\ln(2)RT)\right]} \tag{43}$$

$$a^* = b^*\left[\frac{a^*_{vdW}}{b^*_{vdW}} - \frac{A^E}{\ln(2)RT}\right] \tag{44}$$

$$a^* = \frac{Pa_{mix}}{R^2T^2} \qquad b^* = \frac{Pb_{mix}}{RT} \tag{45}$$

$$a_{vdW} = \sum_i \sum_j \sqrt{a_i a_j} x_i x_j (1 - k_{ij}) \tag{46}$$

$$b_{vdW} = \sum_j \sum_i x_i x_j \left(\frac{b_i + b_j}{2}\right)(1 - l_{ij}) \tag{47}$$

The reacting system of DIPE synthesis consists of four components (C_3H_6, H_2O, IPA, and DIPE), forming six binary mixtures. The required parameters of the three models mentioned above were fitted to VLE data using minimization of the objective function defined by Eq. (25). For the binary mixtures, C_3H_6–H_2O and H_2O–IPA, the high-pressure data VLE have been used ([41] and [42], respectively), whereas for the binaries IPA–DIPE and C_3H_6–IPA these data concerned atmospheric pressures ([43] and [44], respectively). No reliable data have been found for the remaining two binaries, C_3H_6–DIPE and H_2O–DIPE.

Tables 10 and 11 present the experimental results and the equilibrium compositions calculated as well for the ideal model (IDE) as for eight methods of nonideality description. Since pressure is elevated and the reactions proceed in the liquid phase, the influence of a method modeling nonideality of system on the calculated equilibrium composition is significant. The ideal model fails completely. In both cases the differences between the calculated and the measured compositions exceed 20 mol%.

Table 10 Comparison of Calculated Equilibrium Results (mol%) with Experimental Data

	Exp.	IDE[a]	SRK	SRK-KIJ	PR	PRSV	PSRK	MHV2	LCVM	TC
C_3H_6	31.5	18.75	30.37	31.29	30.46	29.81	29.27	24.66	29.57	30.89
H_2O	4.9	3.37	3.53	4.76	3.49	3.15	5.66	3.27	6.14	4.64
IPA	21.8	12.00	23.32	21.77	23.47	23.51	17.96	18.11	17.28	21.60
DIPE	41.8	65.88	42.78	42.18	42.57	43.52	47.11	53.96	47.01	42.87

$T = 120°C$, $P = 50$ bar; propylene:water molar ratio in feed = 2:1.

[a]This model predicts that in the equilibrium state the system consists of 20.34 mol% of the gas phase and 79.66 mol% of the liquid phase. The total composition of both phases is presented.

Table 11 Comparison of Calculated Equilibrium Results (mol%) with Experimental Data

	Exp.	IDE	SRK	SRK-KIJ	PR	PRSV	PSRK	MHV2	LCVM	TC
C_3H_6	10.1	0.00	7.94	11.52	8.07	8.41	8.60	4.26	8.71	11.68
H_2O	25.3	35.69	26.26	26.85	26.14	26.01	29.29	26.73	30.05	28.35
IPA	50.1	28.62	47.48	46.31	47.72	47.97	41.42	46.55	39.90	43.30
DIPE	14.5	35.69	18.32	15.33	18.07	17.61	20.69	22.46	21.34	16.67

$T = 120°C$, $P = 50$ bar; initial feed consists of pure isopropanol.

Additionally, for the parameters from Table 10, the ideal model predicts that the system in equilibrium is a vapor–liquid mixture. Three of the used methods combining EOS with the excess functions are also inaccurate. The MHV2 model gives errors in the range of 8–12 mol%. The values of errors for the methods PSRK and LCVM are lower (6–9%). Since both of them use the same version of the UNIFAC method, they give the similar results. Better accuracy is revealed for the TC model, especially for the case shown in Table 10. The simplest models, SRK and PR, work surprisingly well, giving similar errors of order 1.5–3.5%. Using the PRSV method increases the accuracy in comparison with the PR model in the case shown in Table 11 but lowers the precision for the parameters in Table 10. The best description presents the SRK-KIJ model, particularly in the case shown in Table 10, where the calculated results are nearly the same as the compositions obtained in the experiment.

As has been pointed out in Ref. 45, the SRK-KIJ model correlates and predicts VLE better than the MHV2, PSRK, and LCVM methods for systems containing at least one supercritical component. The last three models predominate over a simple cubic EOS with one adjustable parameter for mixtures containing polar substances, provided all components are subcritical ones. The same phenomenon can be observed for systems in which reactions take place. Since in the mixture considered, propylene is the supercritical component, the MHV2, PSRK, and LCVM methods are much less precise than the SRK-KIJ model.

The TC and PRSV methods use adjustable parameters estimated from VLE data. Only some of the required data have been found in the literature for the system considered. Perhaps the accuracy of these models would have increased if the parameters calculated from the experimental VLE data for all six binaries of this quaternary mixture had been used. The precision of the TC and PRSV methods also depends on the range of temperatures and pressures covered by VLE data. These values should be close to the ones at which the models have to do with the multicomponent mixture. This requirement is much less critical for the simple cubic EOS, with one adjustable parameter k_{ij}. In consequence of that, the SRK-KIJ model has appeared to be the most accurate.

III. SIMULTANEOUS CHEMICAL AND PHASE EQUILIBRIA IN MTBE SYNTHESIS

The problem of chemical and phase equilibria (CPE) calculation is important in reactive distillation, where both reaction and separation are taking place in one vessel. Barbosa and Doherty [46] have shown that chemical reactions alone can induce the formation of azeotropes that are not present in the uncatalyzed nonreacting systems. Such reactive azeotropes limit the range of feasible product compositions from a distillation column in the same way that ordinary azeotropes do in conventional distillation. The chemical reaction can also eliminate ordinary azeotropes that are present in the uncatalyzed nonreacting systems. To predict these phenomena, a thermodynamic model is necessary that describes the phase properties as precisely as possible.

The influence of the method of modeling nonideality on the calculated equilibrium composition in MTBE synthesis is insignificant in the case of gas phase and limited in the case of liquid phase (cf. Tables 2 and 6). However, this influence is much more serious for calculation CPE. Table 12 shows the results of calculations for the same parameters when the six various methods describing nonideality were applied. The value of pressure was set to form a two-phase system at equilibrium. The differences between the compositions of the individual phases are not significant for all the

Table 12 Equilibrium Composition (mole fractions) of Reaction (14) Obtained for Methods Considered for Describing Nonideality

		UNI	UL	UD	ASOG	WIL	NRTL
Vapor phase	C_4H_8	0.2971	0.2868	0.2919	0.2157	0.2466	0.2941
	CH_3OH	0.1366	0.1427	0.1423	0.1991	0.1728	0.1410
	MTBE	0.5663	0.5705	0.5658	0.5852	0.5806	0.5649
	Fraction	0.2053	0.1667	0.0824	0.8378	0.4397	0.1423
Liquid phase	C_4H_8	0.0677	0.0709	0.0816	0.0398	0.0546	0.0762
	CH_3OH	0.1091	0.0997	0.0950	0.1257	0.1124	0.1016
	MTBE	0.8232	0.8294	0.8234	0.8345	0.8330	0.8222
	Fraction	0.7947	0.8333	0.9176	0.1622	0.5603	0.8577

$T = 343.15$ K, $P = 2.43$ bar; initial feed consists of 1 mol of isobutene and 1 mol of methanol.

versions of the UNIFAC method and the NRTL equation. Greater discrepancies can be observed in the case of the Wilson equation and especially for the ASOG method. Serious differences appear in values concerning the vapor fractions, particularly for the last two methods.

Verification of the reliability of the individual methods should be done by comparison of the results calculated with experimental data. However, there is a lack of reliable experimental data for CPE of MTBE synthesis. Therefore, the following chain of reasoning and the resulting method of verification are proposed.

The chemical equilibrium constants can be calculated in both phases from data known from measurements ($\Delta Gf^o_{i,T}$, P^s_i, V_i). Constants K in the gas phase have the same numerical value for each of the methods considered. Small differences appear in these values for the liquid phase because the coefficients ϕ^s_i in Eq. (17) are determined from the SRK or the virial EOS. Since in the equilibrium state the minimum value of the Gibbs free energy of the whole mixture is reached (or for a given reaction the equilibrium constants have the same value for each model), the reliability of the calculated equilibrium composition depends on the correctness of the system nonideality description. The coefficients determining the nonideality are calculated the same way, regardless of whether the chemical reactions proceed or the system reaches vapor–liquid equilibrium only. Therefore, to confirm which of the methods gives the most reliable equilibrium composition for CPE, it is sufficient to state, on the basis of the experimental data, which of the methods gives the best prediction of VLE for the mixture considered. The VLE data should cover the same range of composition for the same species (including any inerts) encountered in the CPE composition.

The comparison of predictive capabilities of individual methods was carried out for all binary mixtures and for ternary mixture of the system considered. The results of calculations and references to the experimental data used are presented in Table 13. The value Δy represents absolute mean deviation between the experimental and calculated equilibrium compositions in the vapor phase (in mol%). ΔT is absolute mean deviation between the experimental and equilibrium temperatures (in K) for isobaric data, and ΔP represents absolute mean deviation between the experimental and equilibrium pressures, relative to the experimental P (in%) for isothermal data. The last two rows of Table 13 contain the average deviations for all systems investigated, separately for isothermal and isobaric data.

It is clear from Table 13 that the NRTL equation is the most accurate model. All three versions of the UNIFAC method are slightly less precise, and the accuracy decreases for the remaining two models, especially

Table 13 Comparison of VLE Predictions for Methods Considered for Describing Nonideality for Binary and Ternary Mixtures of the System C_4H_8–CH_3OH–MTBE

		UNI		UL		UD		ASOG		WIL		NRTL		
System	No. exp. p.	y (mol%)	$\Delta T/\Delta P$ (K/%)	Δy (mol%)	$\Delta T/\Delta P$ (K/%)	Δy (mol%)	$\Delta T/\Delta P$ (K/%)	Δy (mol%)	$\Delta T/\Delta P$ (K/%)	Δy (mol%)	$\Delta T/\Delta P$ (K/%)	Δy (mol%)	$\Delta T/\Delta P$ (K/%)	Ref.
C_4H_8–CH_3OH	10	0.61	2.92%	0.60	1.38%	1.00	1.68%	0.94	2.13%	0.85	1.46%	1.08	0.75%	22
C_4H_8–MTBE	8	0.69	3.71%	0.24	1.67%	0.27	1.30%	1.68	9.47%	0.79	4.03%	0.10	0.40%	23
	8	0.54	2.95%	0.11	1.18%	0.65	2.37%	2.37	11.9%	0.70	3.41%	0.18	0.17%	23
CH_3OH–MTBE	11	1.13	0.74%	1.70	4.70%	1.49	2.96%	3.24	10.2%	2.27	4.72%	1.11	2.57%	47
	17	0.58	1.16%	1.61	2.86%	1.05	1.32%	3.60	7.92%	1.77	2.79%	0.86	0.99%	48
	19	0.50	1.10%	1.28	2.37%	0.85	1.10%	3.07	7.30%	1.59	2.54%	0.55	0.60%	48
	28	0.78	0.56 K	1.41	0.43 K	1.18	0.20 K	4.27	1.93 K	1.52	0.48 K	1.11	0.20 K	49
	13	—	2.57 K	—	1.85 K	—	1.99 K	—	1.92 K	—	2.06 K	—	2.07 K	50
	13	0.98	0.68 K	1.29	0.36 K	1.32	0.35 K	3.34	1.44 K	1.64	0.79 K	1.08	0.30 K	51
	24	0.86	0.19 K	1.67	0.90 K	1.17	0.48 K	3.92	2.20 K	2.02	0.93 K	0.77	0.33 K	24
	22	0.89	0.47 K	0.94	0.60 K	0.67	0.34 K	2.91	1.83 K	1.61	0.83 K	0.39	0.13 K	24
	22	1.83	0.61 K	0.73	0.48 K	1.37	0.38 K	1.97	1.78 K	1.74	0.79 K	1.16	0.24 K	24
	48	0.83	0.42 K	0.74	0.56 K	0.31	0.20 K	2.99	1.97 K	0.84	0.61 K	0.65	0.17 K	52
C_4H_8–CH_3OH	7	—	2.76%	—	2.13%	—	3.68%	—	9.39%	—	5.46%	—	2.81%	53
–MTBE	5	—	1.91%	—	1.97%	—	4.09%	—	10.1%	—	5.21%	—	3.00%	53
	7	—	1.61%	—	1.75%	—	4.51%	—	9.44%	—	5.79%	—	2.51%	53
avg. T=const	92	0.66	1.86%	1.09	2.38%	0.93	2.17%	2.70	8.27%	1.45	3.55%	0.76	1.31%	
avg. T=const	170	0.99	0.63 K	1.07	0.66 K	0.88	0.43 K	3.24	1.91 K	1.44	0.81 K	0.82	0.36 K	

in the case of the ASOG method. Similarly, the compositions of the individual phases presented in Table 12 are nearly alike for the NRTL, UNI, UL, and UD models. The differences increase for the Wilson equation and the ASOG method. Furthermore, the last two methods also predict much greater vapor fractions for CPE. Therefore, the predictive capabilities of the individual methods defined on the base of VLE calculations are in agreement with the results from Table 12 referring to CPE.

IV. THERMODYNAMICS OF ETHERIFICATION OF LIGHT FCC GASOLINE WITH METHANOL

Refinery ether production can be increased using the tertiary C_5–C_7 olefins from light FCC gasolines. Pescarollo et al. [54] stated that the amount of reactive olefins in these fractions ranges from 25% to 40%. It should be emphasized that processing of C_5 and C_6 olefins is more advantageous than C_7 ones, because the former give ethers of higher octane numbers and lower Reid vapor pressures (RVPs). C_7 olefins have a proper RVP and produce ethers of similar octane numbers. So, the etherification of C_7 alkenes (reducing the amount of unsaturated hydrocarbons capable of forming polymeric compounds) improves only the stability of gasoline.

Reference 54 presents experimental results of the etherification of light FCC gasoline with methanol in the liquid phase in a pilot plant. The analysis of these data confirms the fact observed for C_4 and C_5 olefins, that only alkenes with a double bond on a tertiary carbon atom exhibit a tendency to react with methanol. The amounts of other hydrocarbons contained in gasoline (paraffins, naphthenes, aromatics, and the remaining olefins) are nearly the same in the feed and in the reaction products, which implies that the remaining olefins do not isomerize to reactive ones. Table 14 shows the list of all (33) possible reactions forming C_5–C_8 ethers from methanol and the tertiary C_4–C_7 olefins.

Owing to the large number of components, the thermodynamic analysis of the process considered is complicated. A model based on the equilibrium constants of 33 reactions would lead to a system of 33 equations with 33 unknowns, practically impossible to solve. In such cases the Gibbs free-energy minimization method is much more efficient. In this approach the function g is minimized subject to side conditions of elemental abundances [Eq. (49)] and non-negativity of the mole numbers [Eq. (50)].

$$g = \sum_{i=1}^{N} n_i \, \mu_i = \min \tag{48}$$

Table 14 List of the Reactions

Isobutene	C_4 class + methanol = MTBE	
2-Methyl-2-butene 2-Methyl-1-butene	C_5 class + methanol = 2-methyl-2-methoxybutane	
C_6 class		
2-Methyl-1-pentene 2-Methyl-2-pentene	+ methanol	= 2-methyl-2-methoxypentane
cis-3-Methyl-2-pentene *trans*-3-Methyl-2-pentene 2-Ethyl-1-butene	+ methanol	= 3-methyl-3-methoxypentane
2,3-Dimethyl-1-butene 2,3-Dimethyl-2-butene	+ methanol	= 2,3-dimethyl-2-methoxybutane
1-Methylcyclopentene	+ methanol	= 1-methyl-1-methoxycyclopentane
C_7 class		
2-Methyl-1-hexene 2-Methyl-2-hexene	+ methanol	= 2-methyl-2-methoxyhexane
cis-3-Methyl-2-hexene *trans*-3-Methyl-2-hexene *cis*-3-Methyl-3-hexene *trans*-3-Methyl-3-hexene 2-Ethyl-1-pentene	+ methanol	= 3-methyl-3-methoxyhexane
2,3-Dimethyl-1-pentene 2,3-Dimethyl-2-pentene	+ methanol	= 2,3-dimethyl-2-methoxypentane
cis-3,4-Dimethyl-2-pentene *trans*-3,4-Dimethyl-2-pentene 2-Ethyl-3-methyl-1-butene	+ methanol	= 2,3-dimethyl-3-methoxypentane
2,4-Dimethyl-1-pentene 2,4-Dimethyl-2-pentene	+ methanol	= 2,4-dimethyl-2-methoxypentane
3-Ethyl-2-pentene	+ methanol	= 3-ethyl-3-methoxypentane
2,3,3-Trimethyl-1-butene	+ methanol	= 2,3,3-trimethyl-2-methoxybutane
1-Ethylcyclopentene	+ methanol	= 1-ethyl-1-methoxycyclopentane
1,2-Dimethylcyclopentene 1,5-Dimethylcyclopentene	+ methanol	= 1,2-dimethyl-1-methoxycyclopentane
1,3-Dimethylcyclopentene 1,4-Dimethylcyclopentene	+ methanol	= 1,3-dimethyl-1-methoxycyclopentane
1-methylcyclohexene	+ methanol	= 1-methyl-1-methoxycyclohexane

$$\sum_{i=1}^{N} A_{ji} n_i = B_j \qquad j = 1, 2, \ldots, M \tag{49}$$

$$n_i \geq 0 \qquad j = 1, 2, \ldots, N \tag{50}$$

where g is the Gibbs free energy of the entire mixture, n_i is the number of moles of species i, μ_i is the chemical potential of species i, N is the number of species, A_{ji} is the number of atoms of element j in species i, B_j is the moles of element j in feed, and M is the number of elements.

In nonideal systems the chemical potential μ_i of component i in the gas and in the liquid phase is defined by Eqs. (51) and (52), respectively.

$$\mu_i = \mu_i^o + RT \ln P + RT \ln n_i - RT \ln n_T + RT \ln \phi_i \tag{51}$$

$$\mu_i = \mu_i^o + RT \ln P_i^s + RT \ln \phi_i^s - V_i^L (P - P_i^s) + RT \ln n_i - RT \ln n_T + RT \ln \gamma_i \tag{52}$$

where μ_i^o is the standard chemical potential of species i referred to the ideal gas state and identified with the standard Gibbs free energy of formation and n_T is the total number of moles including inerts.

The method described is very efficient in the case of large reacting systems and calculates the equilibrium compositions, being the result of all thermodynamically privileged reactions—this is the so-called full equilibrium case. However, only the reactions listed in Table 14 proceed in the process considered. We do not observe the other thermodynamically favored reactions, such as isomerization of ethers and all kinds of hydrocarbons or the reactions between the olefins of different classes. The reactions of isomerization of olefins attain equilibrium only in the case of alkenes forming the same ethers. This results from the fact that these reactions are dependent ones. The above remarks point out that the thermodynamic model of the process regarded is an example of so-called restricted equilibrium. According to the assumption about the absence of the isomerization reactions, the individual ethers are formed from the corresponding olefins only (for example, 2-methyl-2-methoxypentane from 2-methyl-1-pentene and 2-methyl-2-pentene). Therefore, it was necessary to distinguish individual groups of components by adding one fictitious element X_i to the corresponding reactants and products. Thus, 2-methyl-1-butene, 2-methyl-2-butene, and 2-methyl-2-methoxybutane can be described by formulas such as $C_5H_{10}(X_1)_1$, $C_5H_{10}(X_1)_1$, and $C_6H_{14}O_1(X_1)_1$, and 2-methyl-1-pentene, 2-methyl-2-pentene, and 2-methyl-2-methoxypentane as $C_6H_{12}(X_2)_1$,

$C_6H_{12}(X_2)_1$, and $C_7H_{16}O_1(X_2)_1$, and so on. The number of fictitious elements introduced this way was equal to 16. This procedure precluded formally, for example, reactions between C_5 and C_7 olefins giving 2 molecules of alkenes C_6 or between 2-methyl-2-pentene and *cis*-3-methyl-2-pentene but permitted the reactions of methanol with every reactive olefin. The B_j values [Eq. (49)] of the individual fictitious elements X_i were equal to the total mole numbers of the olefins contained in the feed and marked by the X_i element.

The next problem to be taken into consideration was the presence of inert components in the feed. These compounds do not react, but they affect the total equilibrium composition by the activity coefficients and/or the fugacity coefficients. It was assumed that the inerts contained in the light FCC gasoline were represented by selected hydrocarbons: paraffins by *n*-butane, *n*-pentane, 2-methylbutane, *n*-hexane, 2,2-dimethylbutane, *n*-heptane, and 2,2,3-trimethylbutane; inert olefins by 1-butene, 1-pentene, 3-methyl-1-butene, 1-hexene, 3,3-dimethyl-1-butene, and 1-heptene; cycloparaffins by cyclopentane, cyclohexane, and methylocyclohexane; and aromatics by benzene. In the case of the two-phase reacting systems, the equilibrium mole numbers of the individual inert species in each phase are the result of the phase equilibrium. Therefore, the inert species must be treated as the reacting components. This is not necessary for the one-phase systems.

According to Eq. (52), the following values have an influence on the calculated equilibrium composition in the liquid phase: μ_j^0, P_i^s, V_i^L, and the coefficients ϕ_i^s and γ_i. In etherification processes the accuracy of the computed results is particularly sensitive to the standard chemical potentials and to the vapor pressures. The coefficients ϕ_i^s and the fourth term of Eq. (52) are of minor importance. As has been pointed out in Section II.A (Table 6), the coefficients γ_i can be calculated by any version of the UNIFAC method.

The thermodynamic tables publish reliable values of $\Delta Gf_{i,T}^o(g)$ for methanol, C_4–C_6 olefins [3], and MTBE [5]. There is a lack of such values for the C_7 reactive olefins and the C_6–C_8 ethers. Data referring to the saturated vapor pressures of methanol, MTBE, TAME, C_4–C_6 tertiary olefins, C_7 chain olefins, and methylcyclohexene can be found in Refs. 4, 15, and 35. For five C_7 reactive cycloolefins and C_7–C_8 ethers, p_i^s values were calculated by the Riedel method [4].

The results of the equilibrium calculations performed for the process of etherification of light FCC gasoline with methanol have already been presented in Ref. 55. The missing values of μ_i^o were calculated there using two essentially different group contribution methods worked by Benson and Yoneda [4]. However, it appears that while these methods give quite good estimates for hydrocarbons, they inaccurately predict μ_i^o values in the case of tertiary ethers.

More recently, Syed et al. [30], set on the basis of the experiments, that $\Delta Gf^o_{\text{TAME},298.15}(l) = -113.8$ kJ/mol. This value differs significantly and is equal to −120.69 kJ/mol when we use the Benson method, P^s_{TAME}, and Eq. (17). Rihko and Krause [56] presented experimental equilibrium constants of the reactions of C_6 tertiary olefins with methanol to C_7 ethers at 343 and 353 K. Using these data, one can calculate $\Delta Gf^o_{i,343}(l)$ for the following C_7 ethers: 2-methyl-2-methoxypentane, 3-methyl-3-methoxypentane, 2,3-dimethyl-2-methoxybutane, and 1-methyl-1-methoxycyclopentane. The evaluated numbers are equal to −66.98, −65.10, −59.62, and −37.18 kJ/mol, respectively. The following different values are calculated using the Benson method and the appropriate saturated vapor pressures: −76.85, −71.31, −71.58, and −45.97 kJ/mol. From the examples given above, it is evident that in the case of tertiary ethers the Benson method significantly lowers the values of the standard Gibbs free energies of formation. The Yoneda method exhibits a similar effect.

Because of these facts, the values of ΔGf^o_T of C_6–C_8 ethers applied in the present model were determined as follows. The temperature function of $\Delta Gf^o_T(l)$ for TAME was established from the values $\Delta Gf^o_{298.15}(l)$ and $\Delta Hf^o_{298.15}(l)$ reported in Ref. 30, and the temperature function of Cp was taken from Ref. 16. It was impossible to determine the temperature function of $\Delta Gf^o_{i,T}(l)$ for C_7 ethers. Therefore, the equilibrium calculations were performed at $T = 343$ K using the values of $\Delta Gf^o_{i,343}(l)$ calculated above from Ref. 56. Since there are no experimental data allowing us to evaluate the standard Gibbs free energies of formations of C_8 ethers, they were estimated arbitrarily. In the case of C_7 ethers the Benson method predicts numerical values $\Delta Gf^0_{343}(l)$ which are more negative than the experimental ones. An average error of estimation is equal to about 14%. Therefore, it was assumed that the same error of estimation could refer to the C_8 ethers and the missing data were corrected by this value.

Table 15 presents the feed composition and the equilibrium results calculated at $T = 343$ K and $P = 10$ bar for 100 mol of gasoline and 30 mol of methanol. The weight ratio of gasoline to methanol is close to 100:12. The feed composition with respect to individual classes and groups of hydrocarbons was established on the basis of the figures and Table 3 of Ref. 54. The more detailed gasoline composition required for the equilibrium calculations was estimated arbitrarily. The results of calculations shown in Table 15 were re-counted to 100 kg of gasoline feed and the values obtained are presented in Table 16 together with the pilot-plant results from Ref. 54. Although the detailed experimental feed composition and temperature of the process were not specified, the agreement between theoretical and experimental results is quite good.

Table 15 Compositions (mol) of Feed and Equilibrium Product Calculated at $T = 343$ K and $P = 10$ bar

Compound	Feed	Prod.	Compound	Feed	Prod.
Methanol	30.000	15.888	2,4-Dimethyl-1-pentene	0.180	0.055
Isobutene	0.810	0.026	2,4-Dimethyl-2-pentene	0.230	0.216
MTBE	—	0.784	2,4-Dimethyl-2-methoxypentane	—	0.139
2-Methyl-2-butene	8.140	3.892	3-Ethyl-2-pentene	0.230	0.224
2-Methyl-1-butene	3.740	0.408	3-Ethyl-3-methoxypentane	—	0.006
2-Methyl-2-methoxybutane	—	7.580	2,3,3-Trimethyl-1-butene	0.010	0.000
2-Methyl-1-pentene	1.190	0.108	2,3,3-Trimethyl-2-methoxybutane	—	0.010
2-Methyl-2-pentene	2.150	1.090	1-Ethylcyclopentene	0.080	0.012
2-Methyl-2-methoxypentane	—	2.143	1-Ethyl-1-methoxycyclopentane	—	0.068
cis-3-Methyl-2-pentene	1.520	0.884	1,2-Dimethylcyclopentene	0.130	0.039
trans-3-Methyl-2-pentene	2.690	1.997	1,5-Dimethylcyclopentene	0.030	0.002
2-Ethyl-1-butene	0.530	0.074	1,2-Dimethyl-1-methoxycyclopentane	—	0.119
3-Methyl-3-methoxypentane	—	1.784	1,3-Dimethylcyclopentene	0.070	0.003
2,3-Dimethyl-1-butene	0.380	0.071	1,4-Dimethylcyclopentene	0.080	0.004
2,3-Dimethyl-2-butene	0.310	0.346	1,3-Dimethyl-1-methoxycyclopentane	—	0.144
2,3-Dimethyl-2-methoxybutane	—	0.274	1-Methylcyclohexene	0.010	0.000
1-Methylcyclopentene	0.970	0.601	1-Methyl-1-methoxycyclohexane	—	0.010

1-Methyl-1-methoxycyclopentane	—	0.369	*n*-Butane	0.690	0.690
2-Methyl-1-hexene	0.140	0.012	1-Butene	0.670	0.670
2-Methyl-2-hexene	0.310	0.153	*n*-Pentane	10.560	10.560
2-Methyl-2-methoxyhexane	—	0.284	2-Methylbutane	10.560	10.560
cis-3-Methyl-2-hexene	0.310	0.272	1-Pentene	7.320	7.320
trans-3-Methyl-2-hexene	0.310	0.256	3-Methyl-1-butene	2.980	2.980
cis-3-Methyl-3-hexene	0.170	0.087	Cyclopentane	1.140	1.140
trans-3-Methyl-3-hexene	0.400	0.349	*n*-Hexane	7.910	7.910
2-Ethyl-1-pentene	0.100	0.010	2,2-Dimethylbutane	7.910	7.910
3-Methyl-3-methoxyhexane	—	0.315	1-Hexene	6.000	6.000
2,3-Dimethyl-1-pentene	0.210	0.023	3,3-Dimethyl-1-butene	2.380	2.380
2,3-Dimethyl-2-pentene	0.470	0.605	Cyclohexane	0.960	0.960
2,3-Dimethyl-2-methoxypentane	—	0.052	Benzene	3.080	3.080
cis-3,4-Dimethyl-2-pentene	0.230	0.241	*n*-Heptane	3.600	3.600
trans-3,4-Dimethyl-2-pentene	0.230	0.259	2,2,3-Trimethylbutane	3.600	3.600
2-Ethyl-3-methyl-1-butene	0.080	0.009	1-Heptene	3.790	3.790
2,3-Dimethyl-3-methoxypentane	—	0.031	Methylcyclohexane	0.410	0.410
Total				130.00	115.89

Table 16 Comparison of Calculated Equilibrium Composition (wt%) from Table 15 with Results of Pilot Plant

	Equil. comp. from Table 15	Pilot plant results	Test calc. results
Isobutene	0.018	0.09	0.047
C_5 reactive olefins	3.761	3.24	8.618
C_6 reactive olefins	5.411	5.84	10.076
C_7 reactive olefins	3.465	3.77	6.829
Conjugated diolefins	—	0.04	—
Inert hydrocarbons	73.914	74.09	54.205
Dimethyl ether	—	0.05	—
MTBE	0.862	0.51	1.528
TAME	9.657	12.71	15.219
C_7 ethers	6.612	5.19	9.728
C_8 ethers	1.906		2.795
Other formed	—	0.15	—
Total	105.605	105.70	109.045
Methanol	6.348	6.32	3.407
Isobutene conversion, %	96.80	84.2	170.63
C_5 reactive olefins conversion, %	63.80	68.8	100.54
C_6 reactive olefins conversion, %	47.05	42.9	68.94
C_7 reactive olefins conversion, %	29.43	23.2	42.92

The results of the test calculation were obtained with the assumption that equilibrium was also reached for the reactions of isomerization of olefins.

The equilibrium conversions of olefins to appropriate ethers lower significantly as the number of carbon atoms in the olefin increases. Increase of these conversions would be possible if the catalyst was able to speed up not only the reactions of tertiary olefins with methanol but also the isomerization of all the olefins of a given class to the reactive alkenes. The fourth column of Table 16 contains the results of a test calculation obtained with the assumption that chemical equilibrium was reached for both kinds of reactions mentioned above. With such an assumption the model used 5 fictitious elements X_i. These were added to the chain olefins C_5, C_6, C_7, cycloolefins C_6, C_7, and the corresponding ethers formed from them.

In Table 16 it can be seen that if the isomerization reactions of olefins also reached equilibrium, a significant amount of the primary and the secondary alkenes would convert to the tertiary olefins, which would subsequently react with methanol to the corresponding ethers. Thus, the conversions of the tertiary C_4–C_5 olefins would even exceed the value of

100% and the considerable increase of conversions of the C_6 and C_7 olefins could be observed as well.

Such a catalyst, which would speed up the isomerization reactions together with etherification ones without simultaneous promoting of disadvantageous dimerization of olefins, does not exist. Nevertheless, the test results presented in Table 16 reveal the potential possibilities of improving the process.

REFERENCES

1. J. Tejero, F. Cunill, J. F. Izquierdo. Ind Eng Chem Res, 27: 338–343, 1988.
2. K. Jensen, R. Datta. Ind Eng Chem Res, 34:392–399,1995.
3. R. D. Stull, E. W. Prophet, G. C. Sinke. The Chemical Thermodynamics of Organic Compounds. New York: Wiley, 1969.
4. R. C. Reid, J. M. Prausnitz, B. E. Poling. The Properties of Gases and Liquids. 4th ed. New York: McGraw-Hill, 1987.
5. J. O. Fenwick, D. Harrop, A. J. Head. J Chem Thermodynam, 7: 943–954, 1975.
6. J. G. Hayden, J. P. O'Connel. Ind Eng Chem Process Des Dev, 14: 209–216, 1975.
7. M. Iborra, J. F. Izquierdo, J. Tejero, F. Cunill. J Chem Eng Data, 34: 1–5, 1989.
8. F. P. Heese, M. E. Dry, K. P. Möller. Catal Today, 49: 327–335, 1999.
9. A. Wyczesany. Ind Eng Chem Res, 32: 3072–3080, 1993.
10. F. Colombo, L. Corl, L. Dalloro, P. Delogu. Ind Eng Chem Fund, 22: 219–223, 1983.
11. A. Rehfinger, U. Hoffmann. Chem Eng Sci, 45: 1605–1617, 1990.
12. J. F. Izquierdo, F. Cunill, M. Vila, M. Iborra, J. Tejero. Ind Eng Chem Res, 33: 2830–2835, 1994.
13. T. Zhang, R. Datta. Ind Eng Chem Res, 34: 730–740, 1995.
14. A. Mączyński, A. Biliński, Z. Mączyńska. Verified Vapor–Liquid Equilibrium Data. Binary Systems of Alcohols and Oxygen Compounds. Warszawa: PWN, 1984; vol. 8, pp 20.
15. A. Heine, K. Fischer, J. Gmehling. J Chem Eng Data, 44: 373–378, 1999.
16. Chemcad-V, Process Flowsheet Simulator, Data Bank. Chemstations, Huston, TX, 1999.
17. B. L. Larsen, P. Rasmussen, A. Fredenslund. Ind Eng Chem Res, 26: 2274–2286, 1987.
18. J. Gmehling, J. Li, M. Schiller. Ind Eng Chem Res, 32: 178–193, 1993.
19. K. Kojima, K. Tochigi. Prediction of Vapor–Liquid Equilibria by ASOG Method. Tokyo: Kodansa, 1979.
20. A. Correa, J. Tojo, J. M. Correa, A. Blanco. Ind Eng Chem Res, 28: 609–611, 1989.
21. D. Barbosa, M. F. Doherty. Chem Eng Sci, 43: 529–540, 1988.

22. V. N. Churkin, V. A. Gorshkov, S. Yu Pavlov, E. N. Levicheva, L. L. Karpacheva. Zh Fiz Khim, 52(2): 488, 1978.
23. L. Ah-Dong, D. B. Robinson. J Chem Eng Data, 44: 398–400, 1999.
24. S. Loras, A. Aucejo, R. Muñoz, J. Wisniak. J Chem Eng Data, 44: 203–208, 1999.
25. S. Randriamahefa, R. Gallo, G. Raoult, P. Mulard. J Mol Catal, 49: 85–102, 1988.
26. V. V. Safronov, K. G. Sharonov, A. M. Rozhnov, B. I. Alenin, S. A. Siborov. Zh Prik Khim, 62: 824–828, 1989.
27. L. Rihko, J. A. Linnekoski, O. I. Krause. J Chem Eng Data, 39: 700–704, 1994.
28. W. S. Hwang, J. C. Wu. J Chin Chem Soc, 41: 181–186, 1994.
29. R. L. Piccoli, H. R. Lovisi. Ind Eng Chem Res, 34: 510–515, 1995.
30. F. H. Syed, C. Egleston, R. Datta. J Chem Eng Data, 45: 319–323, 2000.
31. P. Kitchaiya, R. Datta. Ind Eng Chem Res, 34: 1092–1101, 1995.
32. T. Zhang, R. Datta. Ind Eng Chem Res, 34: 2237–2246, 1995.
33. T. Zhang, K. Jensen, P. Kitchaiya, C. Philips, R. Datta. Ind Eng Chem Res, 36: 4586–4594, 1997.
34. R. A. Alberty, C. A. Gehrig. J Chem Phys Ref Data, 14: 803–820, 1985.
35. B. J. Zwoliński, R. C. Wilhoit. Handbook of Vapor Pressures and Heats of Vaporization of Hydrocarbons and Related Compounds. Thermodynamic Research Center, Department of Chemistry, Texas A&M University, College Station, TX, 1971.
36. R. Stryjek, J. H. Vera . Can J Chem Eng, 64: 334–340, 1986.
37. T. Holderbaum, J. Gmehling. Fluid Phase Equil, 70: 251–265, 1991.
38. S. Dahl, Aa Fredenslund, P. Rasmussen. Ind Eng Chem Res, 30: 1936–1945, 1991.
39. C. Boukouvalas, N. Spiliotis, P. Coutsikis, N. Tzouvaras, D. Tassios. Fluid Phase Equil, 92: 75–106, 1994.
40. C. H. Twu, J. E. Coon. AIChE J, 42: 3212–3222, 1996.
41. C. C. Li, J. J. McKetta. J Chem Eng Data, 8: 271–275, 1963.
42. F. Barr-David, B. F. Dodge. J Chem Eng Data, 4: 107–121, 1959.
43. L. A. J. Verhoeye, J Chem Eng Data, 15: 222–226, 1970.
44. S. Horstmann, H. Gardeler. R. Bölts, J Rarey, J Gmehling. J Chem Eng Data, 44: 383–387, 1999.
45. A. Wyczesany. Inż Chem i Proc, 22: 153–173, 2001 (in Polish).
46. D. Barbosa, M. F. Doherty. Chem Eng Sci, 43: 541–550, 1988.
47. E. Valasco, M. J. Cocero, F. Mato. J Chem Eng Data, 35: 21–23, 1990.
48. J. Fárková, J. Linek, I. Wichterle. Fluid Phase Equil, 109: 53–65, 1995.
49. K. Aim, M. J. Ciprian. J Chem Eng Data, 25: 100–103, 1980.
50. V. N. Churkin, V. A. Gorshkov, S. Yu. Pavlov , M. F. Basner. Prom Sint Kauch, 4: 2–4, 1979.
51. G. R. Ascota, R. E. Rodriguez, M. P. De La Guardia. Rev Inst Mex Petrol, 12: 40–46, 1980.
52. A. Arce, J. Martinez-Ageitos, A. Soto. J Chem Eng Data, 41: 718–723, 1996.

53. A. Vetere, I. Miracca, F. Cianci. Fluid Phase Equil, 90: 189–203, 1993.
54. E. Pescarollo, R. Trotta, P. R. Saranty. Hydrocarbon Processing, 73(2): 53–60, 1993.
55. A. Wyczesany. Ind Eng Chem Res, 34: 1320–1326, 1995.
56. L. K. Rihko, A. O. I. Krause. Ind Eng Chem Res, 35: 2500–2507, 1996.

12

Air Control Technologies for MTBE: Bioremediation Aspects of MTBE

Daniel P. Y. Chang
University of California at Davis, Davis, California, U.S.A.

I. AIR EMISSION CONTROL TECHNOLOGIES FOR MTBE

In the design of any emission control system, the first consideration should be whether it is possible to prevent the emission in the first place. As an example, for methyl *tertiary*-butyl ether (MTBE), source control can be achieved for sensitive water bodies by fuel substitution, as has been practiced for recreational watercraft activities at Lake Tahoe and at other California drinking water reservoirs, or by improving storage (underground tanks and piping leak detection) and filling systems (spill control).

When MTBE is initially in the aqueous phase, its treatment there is sensible since it partitions favorably to water, and can be readily biodegraded in the presence of sufficient oxygen. Furthermore, if a co-contaminant such as *tertiary*-butyl alcohol (TBA) is present, air stripping becomes impractical because of TBA's unfavorable Henry's coefficient. Demonstrated ex-situ treatment [1], and technical advances for in-situ biological treatment, are promising for both MTBE and TBA [2–6].

The probable reasons for emission of MTBE vapors, other than from direct fuel transfer operations or emission of unburned fuel, result from contaminated water or soil clean-up activities. Transfer of MTBE to the gas phase is commonly practiced, especially in soil vapor extraction (SVE) applications used in leak and spill clean-up, or from air stripping practiced by water utilities at the point of withdrawal [7]. In the case of a distributed well system feeding into a community's water supply, it is often impractical to install a treatment system requiring significant area. Also, for reasons of reliability and historical reluctance to employ biological treatment for drinking water supplies, adoption of microbial treatment of drinking water supplies has been slow, even though biofilm growth on granular activated carbon (GAC) used for water treatment occurs naturally.

The treatment of air streams contaminated with MTBE is similar to that of other volatile organic compounds (VOCs) associated with gasoline vapors, and a range of options exists. Control of these air emissions, even at quite low levels, has been required because of air toxics concerns [8,9]. Technologies that have been attempted for gas-phase MTBE include:

Granular activated carbon (GAC)
Thermal and catalytic oxidation
Biological treatment
Advanced oxidation

In this chapter, conventional techniques such as GAC and thermal or catalytic oxidation are only briefly mentioned. An extensive review of such control measures has recently been conducted, and the interested reader is referred to that work [10]. Costs of treatment are case-specific, though some generally useful trends in cost considerations for carbon adsorption or catalytic oxidation can be discerned in Table 1. Standard references for organic vapor treatment by conventional means such as adsorption or combustion are abundant in texts and handbooks [11–14]. In that regard,

Table 1 Comparison of Treatment Costs of MTBE from Air-Stripping Operation for GAC and Recuperative Thermal Oxidation

System flow		Vapor-phase GAC ($/1000 gal)		Thermal oxidation[b] ($/1000 gal)	
Water (gal/min)	Air[a] (cfm)	0.5 ppm_v influent MTBE	5 ppm_v influent MTBE	0.5 ppm_v influent MTBE	5 ppm_v influent MTBE
60	1,200	$0.54	$1.86	$1.18	$1.18
600	12,000	$0.24	$1.56	$0.54	$0.54
6000	120,000	$0.23	$1.55	$0.44	$0.44

Costs are in 1999 dollars.
[a]Based on an air:water ratio of 150.
[b]Recuperative thermal oxidation at 60 and 600 gal/min and recuperative flameless thermal oxidation at 6000 gal/min.
Source: Adapted from Ref. 10.

once in the gas phase, MTBE is similar to other VOCs and does not require special consideration. Attention is given in this chapter to emerging gas-phase treatment approaches for MTBE by biofiltration.

For purposes of comparison, cost estimates for liquid-phase biological treatment of MTBE are provided in Tables 2 and 3, from a study by Converse and Schroeder [15]. In that study it was assumed that

> ...because fixed film biological MTBE degradation has not been practiced in the industry the cost of liquid phase biological MTBE removal was estimated by two methods to obtain low cost and high cost boundaries. The low cost boundary was determined by estimating the construction and operating cost of a trickling filter with a rotary distribution system similar to those used for secondary wastewater treatment. A material cost estimating method was used for the various trickling filters required for the flow scenarios. The high cost boundary was determined by estimating the construction and operating cost of a carbon adsorption system similar to those used to treat drinking water. The carbon adsorption system would have biologically active carbon and would not function as an adsorption unit.

In the above analyses, it is seen that if the estimates are reasonably accurate, the crossover point for a liquid- or gas-phase treatment system occurs in the range from 100 to 1000 gal/min, i.e., for larger installations, liquid-phase treatment is more favorable.

Table 2 Low-Cost Estimate Boundary for Liquid-Phase MTBE Biodegradation for 5- and 20-Year Return Periods

		MTBE concentration ppb_w			
		Less that 1000 ppb_w		5000 ppb_w	
Flow gal/min (L/min)		Cost in cents per 1000 gal ($\$/m^3$)			
		20-year return period			
10	(37.85)	340	(0.90)	396	(1.05)
100	(378.5)	53	(0.14)	84	(0.22)
1000	(3785)	13	(0.034)		
5000	(18,930)	8	(0.021)		
		5-year return period			
10	(37.85)	377	1.00	547	(1.44)
100	(378.5)	70	0.185	165	(0.436)
1000	(3785)	25	0.066		
5000	(18,930)	20	0.053		

Source: Ref. 15.

Table 3 High-Cost Estimate Boundary for Liquid-Phase MTBE Biodegradation for 5- and 20-Year Return Periods

		MTBE concentration ppb_w			
		Less that 1000 ppb_w		5000 ppb_w	
Flow gal/min (L/min)		Cost in cents per 1000 gal ($/m^3$)			
		20-year return period			
10	(37.85)	746	(1.97)	1068	(2.82)
100	(378.5)	167	(0.44)	429	(1.13)
1000	(3785)	67	(0.177)		
5000		44	(0.116)		
		5-year return period			
10	(37.85)	1373	(3.63)	2338	(6.18)
100	(378.5)	368	(0.972)	1078	(2.85)
1000	(3785)	168	(0.444)		
5000	(18,930)	110	(0.291)		

Source: Ref. 15.

II. CONSIDERATIONS SPECIFIC TO GAS-PHASE TREATMENT OF MTBE

Two considerations in the treatment of MTBE-contaminated air are (1) concentration in the air and (2) presence of other contaminants. In water treatment applications where air stripping is involved, air:water ratios as large as 200:1, are used because of the unfavorable Henry's law coefficient. (See Table 4.) The resulting air concentrations can range from tens of parts per million by volume (ppm) to less than 1 ppm. If off-gas treatment is required and MTBE aqueous influent concentration to the air stripper is low (<200 μg/L), "vapor phase GAC is generally the most cost-effective off-gas technology because carbon usage rates are low [as a result of the very dilute MTBE in the air stream] and, thus, O&M costs remain low" [10]. If aqueous-phase MTBE influent concentrations are larger (e.g., >2000-μg/L scenario), thermal oxidation is the recommended technology for an air stream from an air-stripping system. In the case of a concentrated wastewater such as may be present at MTBE production or gasoline-blending operations, application of recycled heat and air from a catalytic oxidation unit can be used to improve stripping efficiency, thereby reducing column size and total air:water ratio. Such an approach may be economically justified [9,16]. An interesting integrated application of hollow-fiber membranes used to air-strip MTBE under vacuum, from

Table 4 Properties of the Most Soluble Gasoline Components (Benzene, Toluene, Ethyl Benzene, and Xylene) and MTBE (CAS No. 1634-04-4)

Compound	Molecular formula	Molecular weight	Solubility (mg/L)	V_p (mmHg)	H^a (@25°C)	Log K_{ow} (@25°C)
Benzene	C_6H_6	78	1780	75	0.244	2.13
Toluene	C_7H_8	92	515	22	0.28	2.69
Ethyl benzene	C_8H_{10}	106	150	7.1	0.325	3.15
Xylene[b]	C_8H_{10}	106	580	6.8	0.21	3.20
MTBE	$C_4H_9OCH_3$	88	48,000	240	0.033	0.94–1.16

[a]Dimensionless Henry's coefficient.
[b]Average of *ortho-* and *para*-xylene.
Source: Ref. 18.

an SVE operation and using an internal combustion engine (ICE) to supply the vacuum for the SVE and air-stripping operations as well as destroying the VOC, has been described by Keller et al. [17]. While costs cannot be directly compared, because the cost of the SVE cannot be separated from the rest of the system costs in the study, the pilot-scale total unit costs were in the range of $\$4/m^3$ to $\$4.50/m^3$ treating water with 2000 to 8000 μg/L MTBE.

At many sites, the extracted water includes other contaminants including TBA, ethyl *tertiary*-butyl alcohol (ETBE), *tertiary*-amyl methyl ether (TAME), diisopropyl ether (DIPE), benzene, toluene, ethylbenzene, xylene (BTEX), and possibly other solvents such as tetrachloroethene or perchloroethylene (PCE), trichloroethene (TCE), trichloroethane (TCA), dichloroethene (DCE), dichloroethane (DCA), carbon tetrachloride, and chloroform. The presence of chlorinated solvents in the same wastewater streams should signal caution with regard to catalytic oxidation systems, since some data have been obtained recently indicating that low levels of chlorinated dioxins may be formed in catalytic oxidation units treating SVE effluent with organics and chlorinated solvents [19].

III. BIOLOGICAL AIR TREATMENT

The statement above, that "vapor phase GAC is generally the most cost-effective off-gas technology," was made at a time when few biofilter installations were available to treat MTBE in air streams. Nevertheless, costs of MTBE treatment do not differ from conventional biofilter costs. At a minimum, the replacement carbon costs can be reduced. Biological treatment of MTBE-contaminated air has been achieved in laboratory and pilot-scale applications [18,20–29]. The technology generally applied is biofiltration, most commonly with compost-bed media. The interested reader is directed to the monograph by Devinny, Deshusses, and Webster [30] for general information on biofiltration. In the study by Stocking et al. [10], described above, few biofilter installations for gas-phase treatment were in operation. Nevertheless, it is worthwhile noting that one of the most successful microbial consortia for MTBE biodegradation actually originated from a pilot-scale biofilter located at the Joint Water Pollution Control Plant (JWPCP) in Carson, California, operated by the County Sanitation Districts of Los Angeles County. That biofilter continued to successfully degrade MTBE vapor for over 2 years (Rob Morton, JWPCP, personal communication, 1998). The biofilter was not specifically designed for MTBE control, but developed a consortia that established itself in the biofilter in about 1 month, after a year of operation with no detectable MTBE biodegradation. (See Fig. 1.) The primary organism strain, PM-1, isolated from that consortia has undergone numerous evaluations by other researchers, and has been found to be robust and competitive in various environments [1,2,31,32].

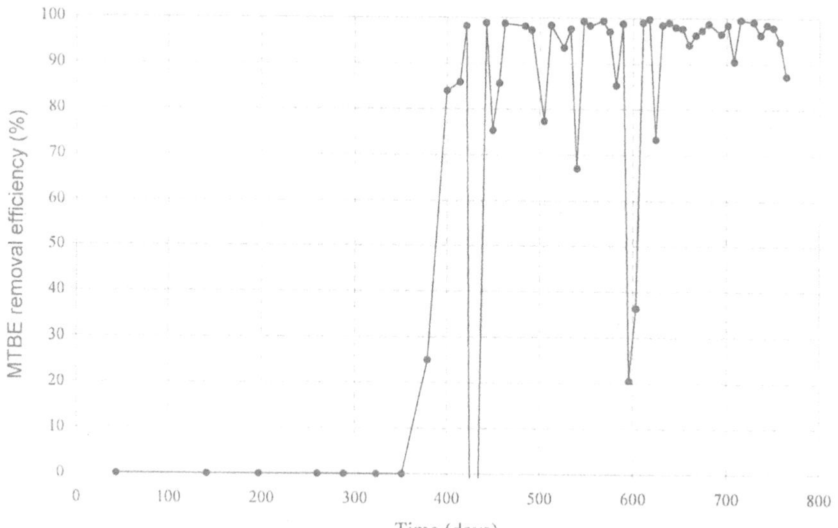

Figure 1 MTBE removal by Los Angeles County Sanitation Districts' biofilter treating mixed liquor channel air. (From Ref. 21.)

Considered nonbiodegradable as late as the mid-1990s, or at least highly recalcitrant, MTBE has recently been shown to be biodegraded aerobically [2,3,21,26,27,33], co-metabolically [29], and anaerobically [6]. The most rapid biodegradation rates have been observed with naturally occurring aerobic microorganisms. Although conceptually simple, successful application of biofiltration requires particular care and attention to certain process variables—moisture content, pH, and adequate nutrient supply—especially for high concentrations of MTBE and/or total organics.

Successful biodegradation requires that an organism capable of degrading the compound be present. Although several strains of aerobic degraders appear to be capable of degrading MTBE, the PM-1 strain seems to be particularly robust (Deshusses, personal communication, 1999). Using 16S-rRNA techniques, it was determined not to be a member of *Sphingomonas* as originally thought, but instead PM-1 is a member of the β-1 subgroup of *Proteobacteria* and related to the subgroup *Rubrivivax* [2]. A second strain present in the original consortia was denoted as, YM-1 and is in the genus *Pseudomonas*. However, because growth of that strain was not as vigorous, it was not pursued.

The original consortia from which PM-1 was isolated was characterized with regard to pH and temperature environmental requirements. Like many bacterial strains, the PM-1 organisms prefer a neutral pH range, but one that is relatively narrow, from about 6.5 to 8.0, as illustrated in Fig. 2. They are active at temperatures as low as 10°C, but biodegradation rates decline sharply, as shown in Fig. 3. These liquid-phase flask studies provide insights that are also applicable to conditions that should be maintained in a biofilter.

Laboratory-scale biofilter work with a compost biofilter has indicated that even with inoculation of the MTBE-degrading consortia, acclimation requires a longer period than for more rapidly growing organisms. Figure 4 illustrates the time course that might be expected, which is of the order of 1 or 2 months for significant removal when seeded from a liquid inoculum. While much shorter in length than the 1-year initial period of no removal observed in the JWPCP biofilter, it is still longer than for more widely degradable compounds such as toluene, which may require only a week or two for significant removal. The particular laboratory-scale biofilter used in the studies was 15 cm in diameter and consisted of four stainless steel sections, each 30 cm high. Each section was packed to a height of 25 cm with Celite^{TM} R-635 (Janus Scientific Inc., Fairfield, CA), an extruded diatomaceous earth pellet. The material is a highly porous, diatomaceous earth pellet, which is commonly used in laboratory-scale biofilters. The pellets have average dimensions of 0.64 cm in diameter and 1.27 cm long,

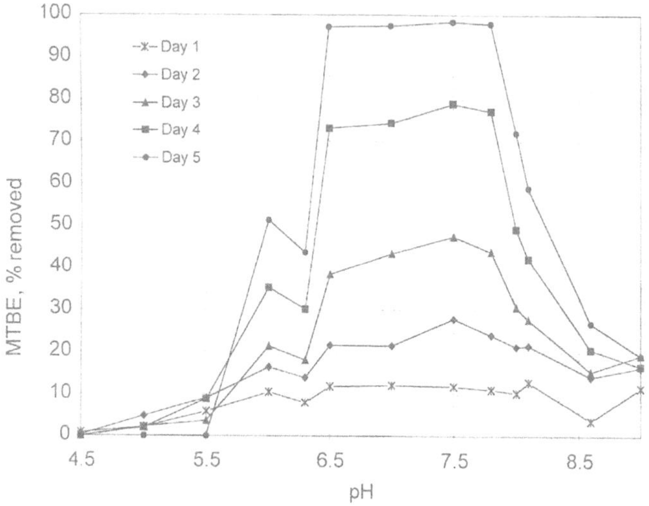

Figure 2 Effect of pH on MTBE biodegradation. Removal was monitored over 5 days, based on initial MTBE liquid-phase concentration of 35 mg/L in a series of flasks with different initial pH. (From Ref. 34.)

with a 20-μm mean pore diameter, a 0.27-m^2/g B.E.T. surface area, and a compacted bed density of 0.51 g/cm^3. Its cost would normally preclude use in full-scale biofilter operations, but as indicated earlier, lower-cost compost-based media will also support the consortia. The air flow rate in the lab-scale biofilter was 0.3 L/sec, resulting in an empty-bed residence time (EBRT) of approximately 1 min. As seen in Figure 5, MTBE was readily biodegraded, even in the presence of a more readily biodegradable compound such as toluene. Nevertheless, at too high a loading rate, nitrogen nutrient limitation was observed. Figure 6, based on the elimination capacity of individual sections of a similar four-section lab-scale biofilter as described above, illustrates the range of elimination capacity that might be expected. Approximately 300 g/m^3/day could be removed with greater than 99% removal efficiency in 25 cm of bed depth with a 15-sec EBRT. Mass transfer rate of the MTBE to the biomass does not appear to be a design issue; rather it is the distribution and growth of biomass within the biofilter that appears to be controlling, and that requires the greatest attention.

The rather slow spread of the consortia, observed in both laboratory and field, suggest a strategy such as utilization of "breeder" reactors. In such a situation, media are removed from one actively degrading bed and blended with fresh media so as to distribute acclimated colonies

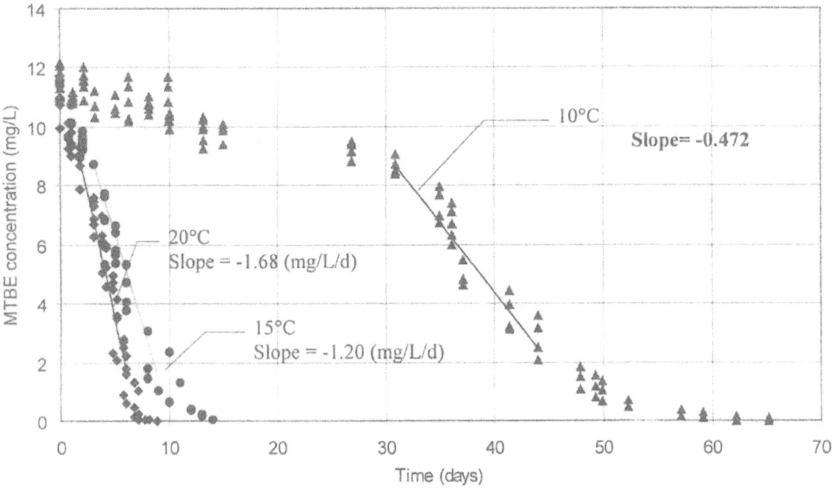

Figure 3 Results of headspace MTBE measurements of sealed 250-mL flasks during microcosm temperature experiment. The amount of MTBE degradation as a function of time for three different temperatures, 20, 15 and 10°C, is shown. The slopes were calculated to fit data points from approximately 80% to 20% of the initial MTBE concentration. Overall degradation rates were 1.24, 0.794, and 0.171 mg/L/day for 20, 15, and 10°C, respectively. The overall degradation rate at 10°C was more than 7 times slower than at 20°C, increasing to a factor of about 3 to 4 times slower after an acclimation period. (From Ref. 23.)

throughout the bed. The experience to date suggests that this approach may be required in order to obtain more rapid spread of the consortia throughout the biofilter. The liquid-phase inoculation in the lab-scale biofilter required almost 2 months for the organisms to develop well-distributed biomass.

Provision of a steady supply of MTBE during periods of shutdown or initial acclimation should be integral to the system design. The study of PM-1 substrate interaction by Deeb et al. [4] suggests that alternative substrates might be used to maintain the enzyme system required for MTBE biodegradation, but directly supplying MTBE would be a conservative approach. The same study indicated that inhibition of PM-1 by ethylbenzene and xylenes was observed. In a mixed-culture system, there may be a spatial separation of degradation of ethylbenzene and xylenes within the biofilter bed that would still allow for biodegradation of MTBE. Thus it would be wise to ensure that ethyl benzene and xylenes are not in the initial mixture fed to the biofilter during acclimation, but are introduced only after the MTBE-degrading strain(s) are established.

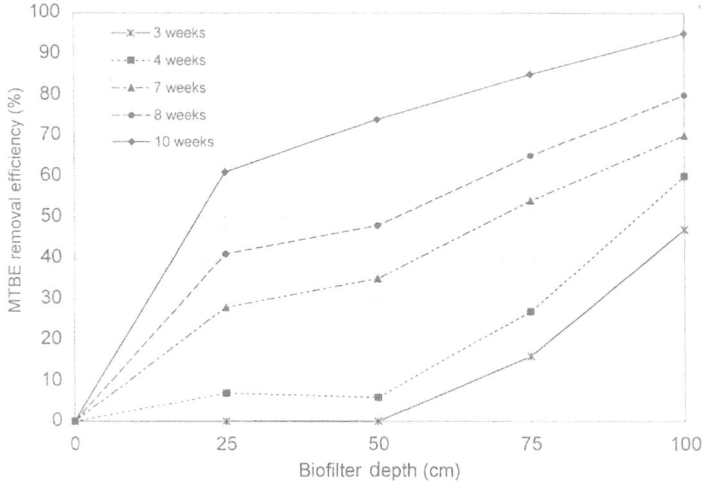

Figure 4 Acclimation of the UC Davis biofilter to MTBE degradation. The artificial media used in the study was Celite® R-635, Janus Scientific Inc., Fairfield, CA. The material is a highly porous, diatomaceous earth pellet, which is commonly used in laboratory-scale biofilters. This medium has average dimensions of 0.64 cm in diameter and 1.27 cm long, with a 20-μm mean pore diameter, a 0.27 m^2/g B.E.T. surface area, and a compacted bed density of 0.51-g/cm^3. The data points indicate percent removal at the end of each week listed. (From Ref. 34.)

Subsequent presence of low concentrations of ethyl benzene and xylenes may be able to be tolerated, since these compounds were likely present in a gasoline site SVE study where MTBE biodegradation was observed [24].

The sensitivity of the particular consortium to acidic conditions should also be taken into account. Oxidation of high concentrations of organic compounds accompanying the MTBE can result in acidic intermediates that would inhibit MTBE biodegradation. In the lab-scale biofilters, buffer was provided through a mineral-nutrient solution. In compost beds, it is typical to amend the compost with a slow-release buffer such as carbonate-containing shells. At the MTBE concentration levels anticipated from air-stripping operations, EBRT will be the determining design factor in the specific loading rate, and it is unlikely that acidification would be a problem.

Cost estimates for operation of a gas-phase biofilter control system for MTBE are presented in Table 5. In the analysis by Converse and Schroeder, an air:liquid ratio of 200:1 was assumed, with a 4% interest rate and two different return periods.

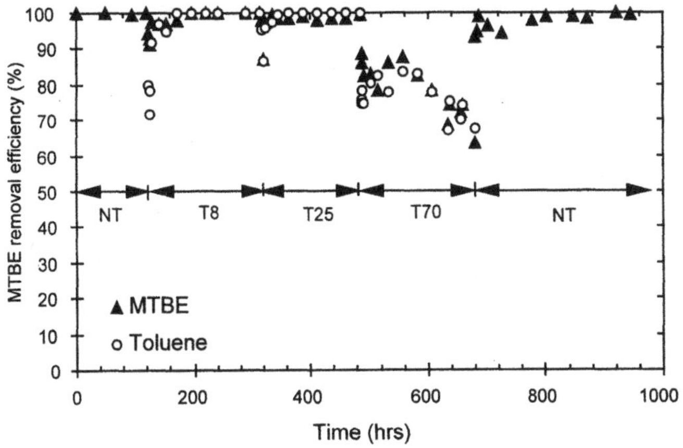

Figure 5 MTBE removal in successive sections of a four-section artificial media biofilter (Celite was selected as the packing material.) The biofilter was inoculated as illustrated in Figure 4 and acclimated over a period of 10 weeks. It was then operated with MTBE at 35 ppm inlet concentration as the only introduced carbon source for an additional 180 days. Toluene was then introduced in a stepwise increasing manner at 8 ppm, then 25 ppm, then 70 ppm, while maintaining MTBE at 35 ppm, before the biofilter performance declined. Ultimately, loss of performance was traced to nitrogen limitation. (From Ref. 22.)

> Low air phase MTBE concentrations cause the controlling design parameter to be the empty bed contact time (EBCT).... Experienced companies were contacted and provided with design parameters specific to MTBE and asked to estimate construction and operating cost for biofilters.

It can be observed from Tables 1 and 5 that if concentrations are sufficiently low (<5 ppm in the liquid phase), a GAC system may still be less costly for a lower liquid flow range (<10–100 gal/min). For higher concentrations, the replacement carbon costs result in the biofilter being more economical.

IV. CONCLUSIONS

The presence of MTBE in the vapor phase can be controlled by several means, including GAC adsorption, thermal oxidation, and biofiltration. Additional VOC control techniques may also be suitable, e.g., ultraviolet peroxone, but supporting data for these treatment technologies are not available at present. The choice of which technology to apply must be determined on a case-by-case basis, determined largely by economics.

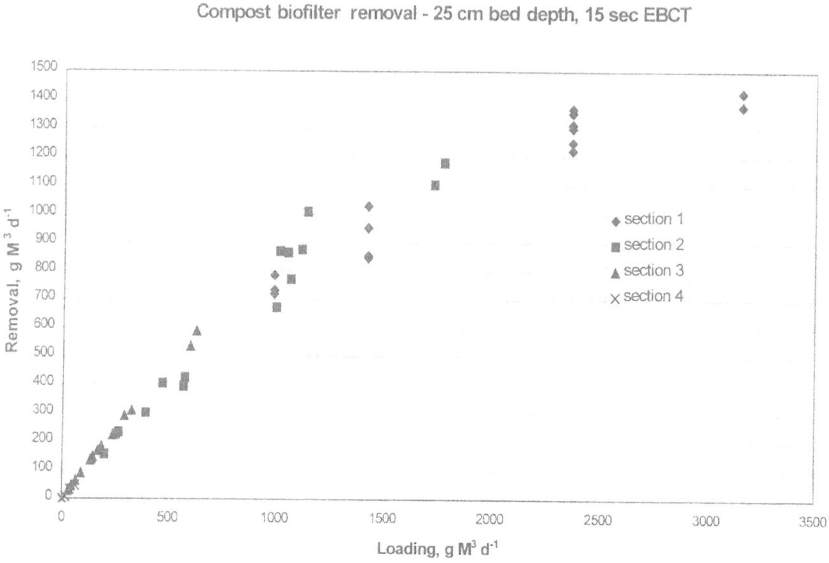

Figure 6 The elimination capacity of the pilot-scale Celite-based biofilter was additionally studied. The empty-bed contact time in each section is 15 sec. Removal efficiency approaches 100% for loading rates less than about 300 g/m^3/day. (From Ref. 18.)

For low concentrations and low air flow rates, GAC beds appear to be favored. As concentration and flow rate increase, biofiltration, either alone or in combination with GAC, becomes favorable. Only for the highest concentration streams associated with industrial wastewaters, i.e., greater than about 100 ppm$_v$ gas-phase MTBE and total petroleum hydrocarbon

Table 5 Gas-Phase Treatment Biofiltration Costs of MTBE from Air-Stripping Operation for an Assumed 200:1 Air:Liquid Ratio Independent of Liquid Concentration

		Cost in cents per 1000 gal treated ($/m^3)			
Water flow gal/min (L/min)		5-year return period		20-year return period	
10	(37.85)	449	(1.19)	147	(0.388)
100	(378.5)	73	(0.193)	24	(0.063)
1000	(3785)	24	(0.063)	8	(0.021)
5000		18	(0.048)	5.9	(0.016)

Source: Ref. 10.

concentration, would thermal oxidation be economically viable. (Note that a concentrated air stream can be diluted with a correspondingly increased size of biofilter.) Cost of GAC adsorption beds would be higher for the equivalent BTEX compound, whereas after a longer initial acclimation period, costs of biofiltration would be comparable to other VOCs.

ACKNOWLEDGMENTS

This chapter could not have been prepared without the efforts of colleagues E. D. Schroeder and K. M. Scow at the University of California, Davis, and a number of former students, B. Converse, J. B. Eweis, J. Hanson, M. Fan, J. Scarano, and N. Watanabe, whose research contributed to the material presented. The efforts on MTBE biodegradation had their roots in research carried out for the California Air Resources Board (ARB) and the National Institute of Environmental Health Sciences (NIEHS) as well as later work funded by the American Petroleum Institute, and gifts from Arco, Chevron Research, and a research collaboration with Environmental Resolutions, Inc. J. B. Eweis was partially supported by a traineeship, Grant IEHS 5 P42 ES04699 from the National Institute of Environmental Health Sciences, National Institutes of Health (NIH), with funding provided by the U.S. Environmental Protection Agency (EPA). The contents of this manuscript are solely the responsibility of the author and do not necessarily represent the official views of the ARB, NIEHS, NIH, or EPA. The author wishes to acknowledge the Department of Chemical and Process Engineering, University of Canterbury, Christchurch, New Zealand, for the hospitality and support afforded as a Visiting Erskine Fellow during the preparation of the chapter.

REFERENCES

1. J. E. O'Connell, D. E. Weaver. Fluidized bed bioreactor for treatment of MTBE and TBE in ground water or scrubber water. Poster presentation. 2000 USC-TRG Conference on Biofiltration. University of Southern California, Los Angeles, CA, Oct 19–20, 2000, pp 297–303.
2. J. R. Hanson, C. E. Ackerman, K. M. Scow. Biodegradation of methyl *tert*-butyl ether by a bacterial pure culture. Appl Environ Microbiol, 65:4788–4792, 1999.
3. J. P. Salanitro, P. C. Johnson, G. E. Spinnler, P. M. Maner, H. L. Wisniewski, C. Bruce. Field scale demonstration of enhanced MTBE bioremediation through aquifer bioaugmentation and oxygenation. Environ Sci Technol, 34:4152–4162, 2000.

4. R. A. Deeb, H. Y. Hu, J. R. Hanson, K. M. Scow, L. Alvarez-Cohen. Substrate interactions in BTEX and MTBE mixtures by an MTBE-degrading isolate. Environ Sci Technol, 35:312–317, 2001.
5. P. M. Bradley, J. E. Landmeyer, F. Chapelle. Widespread Potential for Microbial MTBE degradation in surface-water sediments. Environ Sci Technol, 35:658–662, 2001.
6. K. T. Finneran, D. R. Loveley. Anaerobic degradation of methyl *tert*-butyl ether (MTBE) and *tert*-butyl alcohol (TBA). Environ Sci Technol, 35:1785–1790, 2001.
7. R. A. Deeb, S. Sue, D. Spiers, M. C. Kavanaugh. Field studies to demonstrate the cost and performance of air stripping for the removal of MTBE from groundwater. CD-ROM Proceedings of the 94th Annual Meeting and Exhibition of the Air & Waste Management Association, Orlando, FL, 2001, Paper 1034.
8. W. D. Byers. Control of emissions from an air stripper treating contaminated groundwater. Environ Prog, 7:17–21, 1988.
9. M. Kosusko. Catalytic oxidation of groundwater stripping emissions. Environ Prog, 7:136–142, 1988.
10. A. Stocking, R. Rodriguez, A. Flores, D. Creek, J. Davidson, M. Kavanaugh. Treatment technologies for removal of methyl tertiary butyl ether (MTBE) from drinking water: air stripping, advanced oxidation processes, granular activated carbon, synthetic resin sorbents, 2nd ed. NWRI-99-08, A Report for The California MTBE Research Partnership, Fountain Valley, CA, 2000.
11. A. Buonicore, W. T. Davis, eds. Air Pollution Engineering Manual. Air & Waste Management Association, Van Nostrand Reinhold, New York, 1992.
12. C. D. Cooper, F. C. Alley. Air Pollution Control: A Design Approach, 2nd ed. PWS Engineering, Boston, 1994.
13. D. H. F. Liu, B. G. Liptak, eds. Environmental Engineers' Handbook, 2nd ed. Lewis, New York, 1997.
14. R. H. Perry, D. W. Green, J. O. Maloney, eds. Perry's Chemical Engineers' Handbook, 7th ed. McGraw-Hill, New York, 1997.
15. B. M. Converse, E. D. Schroeder. Estimated cost associated with biodegradation of methyl tertiary-butyl ether (MTBE). In: Health and Environmental Assessment of MTBE Volume V: Risk Assessment, Exposure Assessment, Water Treatment & Cost-Benefit Analysis, A Report to the Governor and Legislature of the State of California, SB 521, November 12, 1998.
16. K. T. Chuang, S. Cheng, S. Tong. Removal and destruction of benzene, toluene, and xylene from wastewater by air stripping and catalytic oxidation. Ind Eng Chem Res, 31:2466–2472, 1992.
17. A. A. Keller, S. Sirivithayapakorn, M. Kram. Field test of treatment process for remediation of soil and water contaminated with MTBE. 93rd Annual Meeting and Exhibition of the Air & Waste Management Association, Salt Lake City, UT, 2000, Paper 1037.

18. J. Scarano. Biological degradation of methyl tertiary butyl ether (MTBE) in a vapor phase biofilter. MS thesis, University of California, Davis, June 2000.
19. J. R. Hart. Dioxin/furan emissions from ultra lean, post-flame, surface-catalyzed combustion of polychlorinated C_2 vapors, presented at the 28th International Symposium on Combustion, 2000, Poster 2-EO1.
20. J. P. Salanitro, L. A. Diaz, M. P. Williams, H. W. Wisnieski. Isolation of a bacterial culture that degrades methyl-t-butyl ether. Appl Environ Microbiol, 60:2593–2596, 1994.
21. B. Eweis, D. P. Y. Chang, E. D. Schroeder. Meeting the Challenge of MTBE Biodegradation. CD-ROM Proceedings of the 90th Annual Meeting and Exhibition of the Air & Waste Management Association, Toronto, Ontario, Canada, 1997, Paper 97-RA133.06.
22. J. B. Eweis, E. D. Schroeder, D. P. Y. Chang, K. M. Scow. Biodegradation of MTBE in a pilot-scale biofilter. In: G. B. Wickramanayake, R. E. Hinchee, eds. Natural Attenuation: Chlorinated and Recalcitrant Compounds. Battelle Press, Columbus, OH, 1998, pp 341–346.
23. J. B. Eweis, N. Watanabe, E. D. Schroeder, D. P. Y. Chang, K. M. Scow. MTBE biodegradation in the presence of other gasoline components. National Ground Water Association Conference, The Southwest Focused Ground Water Conference Discussing the Issue of MTBE and Perchlorate in Ground Water, Anaheim, CA, June 3–4, 1998.
24. K. Romstad, J. Scarano, W. Wright, D. P. Y. Chang, E. D. Schroeder. Performance of a full-scale compost biofilter treating gasoline vapor. Proceedings, 1998 Biofilter Conference, University of Southern California, October 22–23, 1998.
25. D. P. Y. Chang, E. D. Schroeder, K. M. Scow, B. M. Converse, J. Scarano, N. Watanabe, K. Romstad. Experience with laboratory and field-scale ex situ biodegradation of MTBE. Extended abstract. USEPA/API Workshop on Biodegradation of MTBE, Cincinnati, OH, February 1–3, 2000.
26. N. Y. Fortin, M. A. Deshusses. Treatment of methyl *tert*-butyl ether vapors in biotrickling filters. 1. Reactor startup, steady state performance, and culture characteristics. Environ Sci Technol, 33:2980–2986, 1999.
27. N. Y. Fortin, M. A. Deshusses. Treatment of methyl *tert*-butyl ether vapors in biotrickling filters. 2. Analysis of the rate-limiting step and behavior under transient conditions. Environ Sci Technol, 33:2987–2991, 1999.
28. N. Y. Fortin, M. Morales, Y. Nakagawa, D. D. Focht, M. A. Deshusses. Methyl *tert*-butyl ether (MTBE) degradation by a microbial consortium. Environ Microbiol, 3:407–416, 2001.
29. A. Hernández, M. Magaña, B. Cárdenas, S. Hernández, S. Revah, S. Queney, Richard Auria. Methyl *tert*-butyl ether (MTBE) elimination by cometabolism: laboratory and biofilter pilot-scale results. CD-ROM Proceedings of the 94th Annual Meeting and Exhibition of the Air & Waste Management Association, Orlando, FL, 2001, Paper 1037.
30. J. S. Devinny, M. A. Deshusses, T. S. Webster. Biofiltration in Air Pollution Control, Lewis, Boca Raton, FL, 1999.

31. C. Church, P. G. Tratnyek, K. M. Scow. Pathways for the degradation of MTBE and other fuel oxygenates by isolate PM1. Extended abstract. USEPA/API MTBE Biodegradation Workshop, Cincinnati, OH, February 1–3, 2000.
32. M. Fan. Using a sequencing batch reactor to determine monod kinetic coefficients for MTBE bio-degradation. MS thesis, University of California, Davis, June 2000.
33. G. J. Wilson, A. P. Richter, M. T. Suidan, A. D. Venosa. Aerobic biodegradation of gasoline oxygenates MTBE and TBA. Water Sci Technol, 43:277–284, 2001.
34. J. B. Eweis. Biodegradation of MTBE in both liquid and vapor phases. PhD thesis, University of California, Davis, August 2000.

13

MTBE Removal by Air Stripping and Advanced Oxidation Processes

Cris B. Liban
Los Angeles County Metropolitan Transportation Authority, Los Angeles, California, U.S.A.

Sun Liang
Metropolitan Water District of Southern California, La Verne, California, U.S.A.

I. H. (Mel) Suffet
UCLA School of Public Health, Los Angeles, California, U.S.A.

I. INTRODUCTION

Methyl *tertiary*-butyl ether (MTBE) is a fuel additive used as both a replacement for lead in gasoline and an octane booster [1,2]. Most recently, MTBE has been the focus of a significant number of studies, resulting from its detection in and consequent contamination of groundwater and surface drinking water sources [1,3,4]. Removal of MTBE from affected drinking water bodies can be achieved through several water treatment processes, such as air stripping, advanced oxidation, membrane separation, and sorption [5]. This chapter briefly reviews the air stripping and advanced oxidation processes. It also collates cost comparison information for each of these technologies as shown in existing literature. Finally, the chapter also outlines the effectiveness of each individual technology in removing MTBE from water.

II. AIR STRIPPING

Air stripping is a technology that has been proven to productively remove MTBE [6,7]. In air stripping, volatile organics are partitioned from water by greatly increasing the surface area of the contaminated water exposed to air.

Air stripping involves continuously contacting the contaminated water with a large volume of air to transfer a significant fraction of the volatile organic carbon compounds (VOCs) in the air phase [8]. Pollutant removal efficiency is a function of the design of the air stripping tower and the contaminants' Henry's law constant. Since air stripping involves only mass transfer from the water to the air phase, additional treatment of the exiting contaminated air stream may be required. Off-gas treatment is typically required if the effluent vapors contain 1 lb/day (0.45 kg/day) or more [6]. Thresholds can change however, based on local air quality regulations [9,10].

In addition, the air stripping process is temperature sensitive, significantly increasing the Henry's law constant (Fig. 1) [10]. Increasing the operating temperature will consequently improve performance, and reduce tower dimensions, and capital and operations and maintenance cost. Table 1 outlines the advantages and disadvantages of an air stripping system treating MTBE.

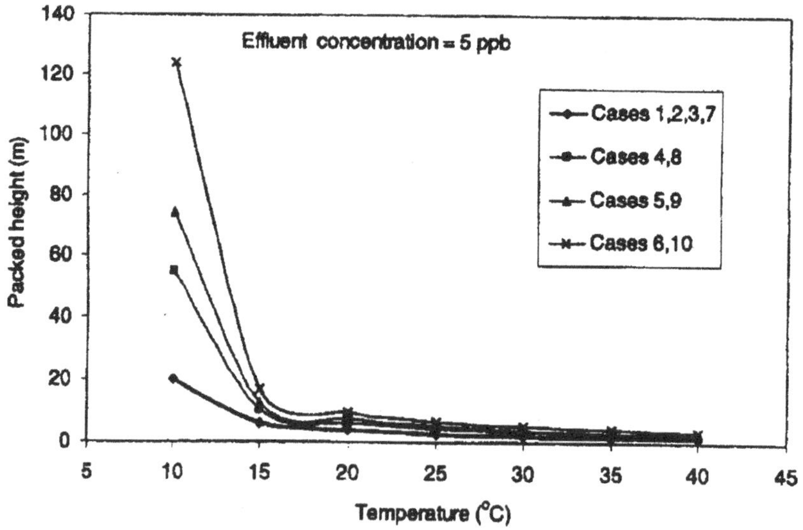

Figure 1 Packing height as a function of temperature, 5 μg/L effluent. (Adapted from Ref. 10.)

Table 1 Advantages and Disadvantages of MTBE Treatment Using Air Stripping Technology

Advantages	Disadvantages
Proven technology that can achieve high removal efficiencies	High concentration variations may require posttreatment polishing using GAC
Mechanically reliable	Flow rates greater than 1000 gal/min may require multiple units
Packed air stripping units can be designed to treat up to 6000 gal/min	Technology does not necessarily eliminate MTBE in the environment, simply transfers MTBE from the water phase to the air phase; vapor treatment necessary
	Carbonate scaling of packing and piping
	Process is temperature-sensitive
	Biological problems can reduce treatment efficiency
	Pretreatment of influent water may be necessary, depending on the water source

There are a number of air stripping methods commercially available to treat MTBE. These include packed towers, diffused aeration, low-profile (tray) aeration, and spray aeration. The cost effectiveness of each of these technologies has been recently reviewed [11]. Among the types of aeration devices, packed-tower and low-profile air strippers are most cost-effective. Between these two technologies, the most efficient device to use depends on the system flow rate applied. The packed-tower air strippers can treat up to 99.98% efficiency with flow rates of as much as 6000 gal/min or higher [12]. Greene and Barnhill [13] show that low-profile air strippers are most efficient, however, at low flow rates (2–10 gal/min), cleaning at a rate of 97.5% removal of MTBE.

Keller et al. [10] and Stocking et al. [12] provide MTBE treatment cost analysis using various comparable technologies that include air stripping. Table 2 summarizes air-stripping cost information from these two references in terms of flow rate, removal efficiency, and technology. Keller et al.'s [10] cost table was modified to be consistent with Stocking et al.'s [12] values.

Packed-tower air strippers provide a more cost-effective way to treat MTBE. Treating off-gasses will add minor additional cost per thousand gallons. Depending on the off-gas treatment option, the price per thousand gallons could be as much as double the cost to air strip without treating the off-gas.

Table 2 Amortized Cost, Dollars per 1000 gal Treated Using Air Stripping[a]

Flow (gal/min)	Influent (μg/L)	Effluent (μg/L)	Efficiency (%)	Packed tower[b]	Low profile[b]	Packed tower[c]
10	100	5	95	$1.54	d	$2.35–$3.51
	500	5	99	$2.30	d	$2.68–$4.60
	1000	5	99.5	$2.65	d	$2.84–$5.14
	5000	5	99.9	$3.22	d	$3.55–$7.45
60	20	0.5	97.5	$1.75	$1.86	e
	200	5	97.5	$1.75	$1.80	e
	200	0.5	99.75	$1.82	$1.89	e
	2000	20	99	$1.79	$1.90	$3.08
	2000	5	99.75	$1.82	$2.02	$3.20
100	100	5	95	$0.40	d	$0.73–$1.08
	500	5	99	$0.59	d	$0.84–$1.58
	1000	5	99.5	$0.68	d	$0.88–$1.86
	5000	5	99.9	$0.88	d	$0.97–$2.81
500	100	5	95	$0.25	d	$0.41–$0.78
600	20	0.5	97.5	$0.34	$0.92	e
	200	5	97.5	$0.34	$0.96	$0.59
	200	0.5	99.75	$0.37	$1.09	$0.62
	2000	20	99	$0.36	$0.96	$0.90
	2000	5	99.75	$0.37	$1.09	$0.91
1000	100	5	95	$0.23	d	$0.33–$0.66
6000	20	0.5	97.5	$0.16	$0.48	$0.39
	200	5	97.5	$0.16	$0.48	$0.39
	200	0.5	99.75	$0.17	$0.64	$0.40
	2000	20	99	$0.17	f	f
	2000	5	99.75	$0.18	f	f

[a]Modified from data collected by Keller et al. [10] and Stocking et al. [12]. Only costs associated with ≥95% removal efficiency are included in this table.
[b]No off-gas treatment.
[c]Off-gas treatment assumed. Keller et al. [10] assumed the following off-gas treatment: thermal oxidation without heat recovery, thermal oxidation with heat recovery, GAC, and gas-phase biofilter. Stocking et al. [12] assumed either GAC or thermal oxidation.
[d]Not reported.
[e]Treatment not required.
[f]Not evaluated.

Keller et al. [10] determined that preheating the influent water could reduce power cost if thermal or catalytic oxidation is used to treat off-gas vapors. Further, at high flow rates (>100 gal/min), gas-phase biofiltration is the most cost-effective air treatment option. The most cost-effective vapor treatment for low flow rates ($\leq$100 gal/min) is thermal oxidation with heat recovery to preheat the influent water. Granular activated carbon (GAC) is not cost-effective for air treatment under any of the cases they studied, given the low MTBE sorption capacity for GAC, even considering virgin coconut-based carbon.

Packed-tower air stripping appears to be the lowest-cost technology compared to other treatment options such as GAC, advance oxidation process, hollow-fiber membrane, and resin sorption, regardless of the flow rate. However, other than the packed-tower air strippers at La Crosse, Kansas, and Rockaway Township, New Jersey, there appears to be a lack of published data for air-stripper application [12,14]. In fact, it appears that GAC is the most preferred method of MTBE treatment [15]. Preferential ratio in favor of GAC among these treatment options ranges from 1 to 6 depending on the treatment option being compared.

III. ADVANCED OXIDATION TREATMENT

Oxidation treatment for MTBE destruction can be categorized into established and emerging technologies based on the existing literature and industry's experience with treatment technologies. The established technologies consist of ozone/H_2O_2, ozone/UV, and UV/H_2O_2. The emerging technologies consist of ultrasonic irradiation, ultrasonic irradiation/ozone, and high energy electron beam (HEEB) irradiation. A cost comparison among vendors for this treatment technology is presented in Table 3.

A. Established Technologies

1. Ozone/H_2O_2

Karpel Vel Leitner [16] studied the reaction of ozonation and ozone/H_2O_2 on oxygenated additives such as MTBE in dilute aqueous solution. Experiments conducted in a semicontinuous reactor with MTBE showed that ozone/H_2O_2 is a more effective treatment than ozone alone. The applied ozone doses of 6 mg/L used in the dilute aqueous solution tested resulted in 80% degradation of MTBE for ozone alone (at pH 8) and 1.7 mg ozone/mg MTBE for ozone/H_2O_2. *tertiary*-butyl formate (TBF), formaldehyde, and *tertiary*-butyl alcohol (TBA) were identified as the

Table 3 Total Amortized Operating Costs for AOPs

		Amortized Operating Costs ($/1,000 gal)		
Flow (gal/min)	Influent MTBE (μg/L)	Calgon Carbon Co.	Applied Process Tech., Inc.	Oxidation Systems, Inc.
60	20	2.18	2.63	2.25
60	200	2.50	2.68	2.61
60	2000	3.47	3.31	3.03
600	20	0.57	0.82	0.62
600	200	0.96	0.90	0.67
600	2000	1.75	1.13	0.86
6000	20	0.32	0.35	0.39
6000	200	0.60	0.43	0.48
6000	2000	1.24	0.59	0.64

Source: Data from Ref. 23.

ozone/H_2O_2 by-products of MTBE. Dyksen et al. [17] performed pilot tests by using an in-line application of ozone and H_2O_2 to evaluate process issues for the removal of organic chemicals such as trichloroethylene (TCE), tetrachloroethylene (PCE), *cis*-1,2-dichloroethylene (*cis*-1,2-DCE), and MTBE. The results indicated that ozone/H_2O_2 is more effective than ozone alone for removals of TCE, PCE, *cis*-1,2-DCE, and MTBE. Nondetectable levels of MTBE were recorded using an ozone dose of 8 mg/L, with a contact time of 3–6 min, an H_2O_2-to-ozone ratio of 0.5, and an initial MTBE level of 2.7 μg/L.

Liang et al. [18] conducted a pilot-scale study to investigate the effectiveness of ozone and ozone/H_2O_2 processes for MTBE removal in surface water. Under the tested conditions, ozone/H_2O_2 was again more effective than ozone alone for MTBE removal. The results indicated that ozone/H_2O_2, with 4 mg/L of ozone and 1.3 mg/L of H_2O_2, could achieve average MTBE removals of approximately 78% for two source water supplies used (CRW and SPW) with initial MTBE concentrations of 25–100 μg/L. Liang et al. [19] further studied treatment of MTBE contaminated groundwater with ozone and ozone/H_2O_2 processes. Applied ozone dose of greater than 10 mg/L was necessary to reduce the MTBE concentrations at approximately 200 and 2000 μg/L to below the California secondary maximum contaminant level (MCL) of 5 μg/L (see Figs. 2 and 3). Again, ozone/H_2O_2 was much more effective than ozone alone for removing MTBE in the water matrix tested. Both ozone and ozone/H_2O_2 processes produced TBF, TBA, acetone, and aldehydes. When ozone dose of 4 mg/L was applied to water with background bromide of 0.3 mg/L, more than

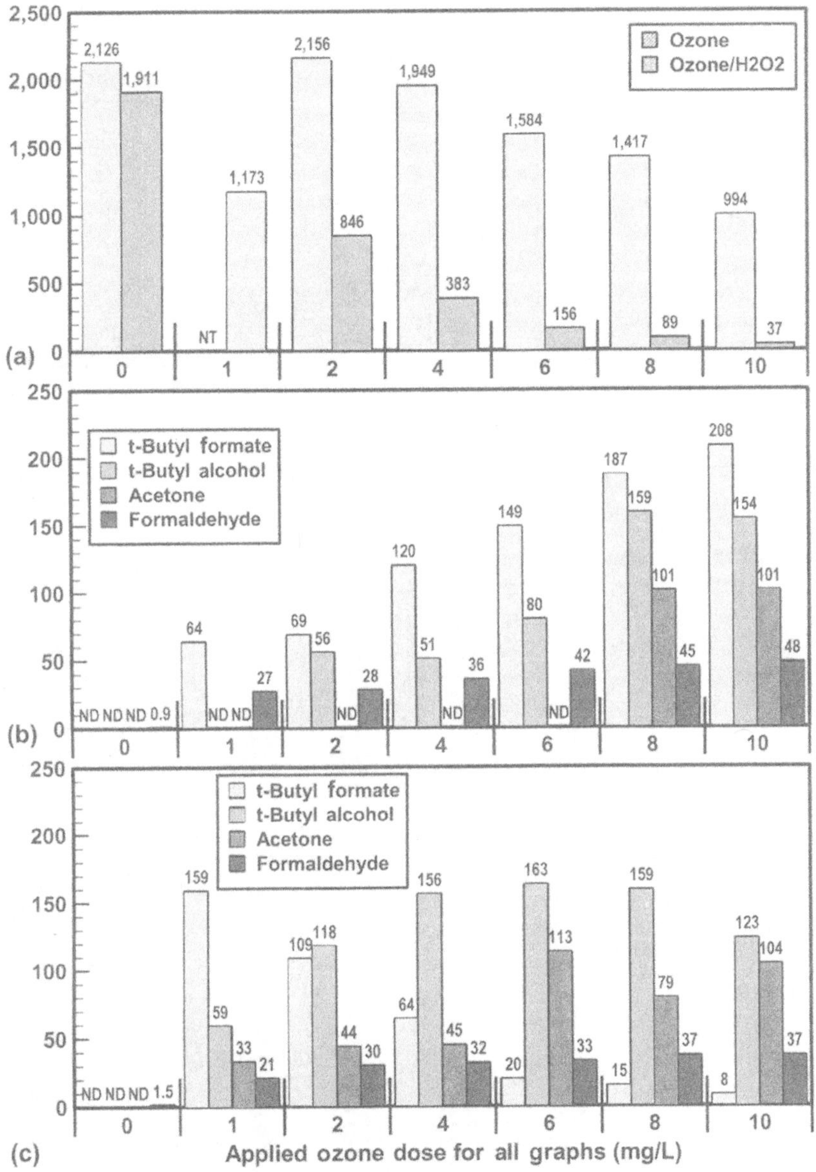

Figure 2 MTBE degradation and formation of MTBE by-products by ozone and ozone/H_2O_2 at high initial MTBE concentration. Read from left to right.

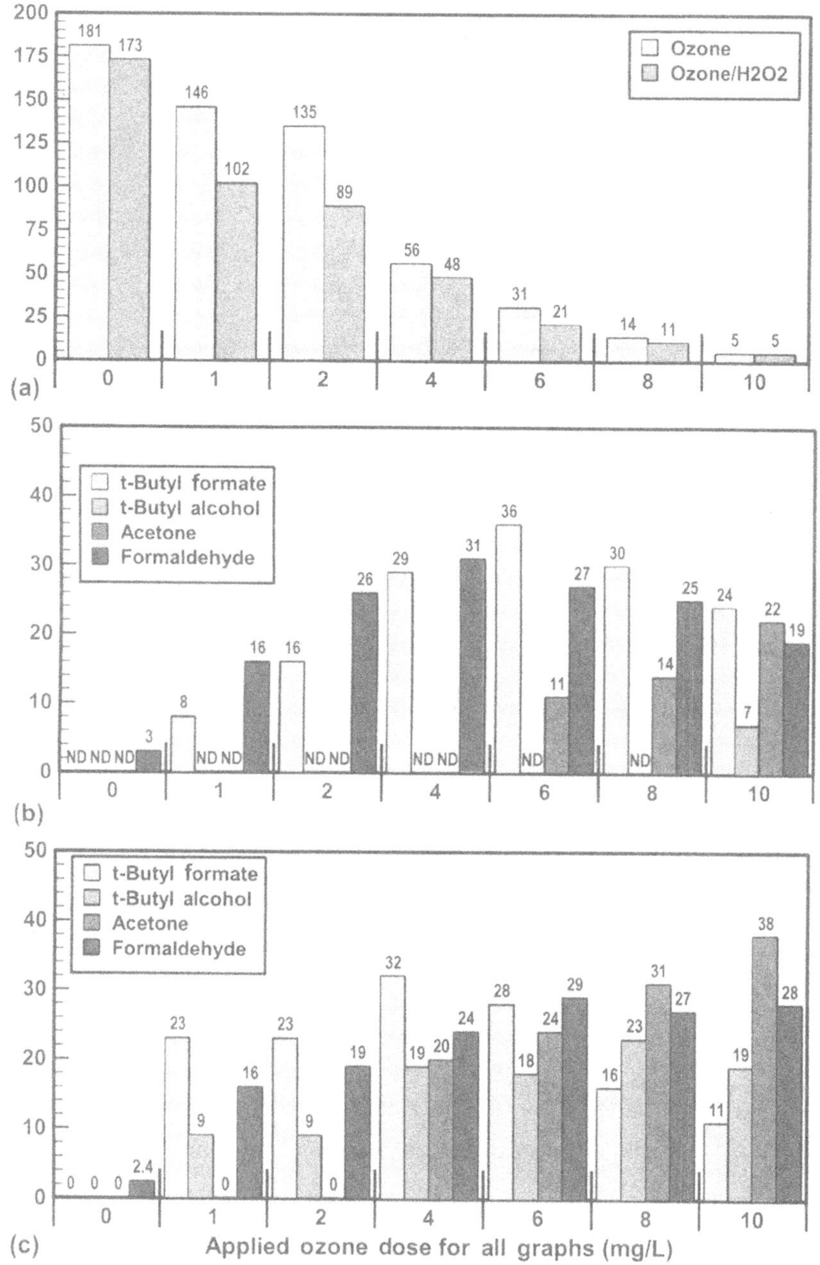

Figure 3 MTBE degradation and formation of MTBE by-products by ozone and ozone/H_2O_2 at low initial MTBE concentration. Read from left to right.

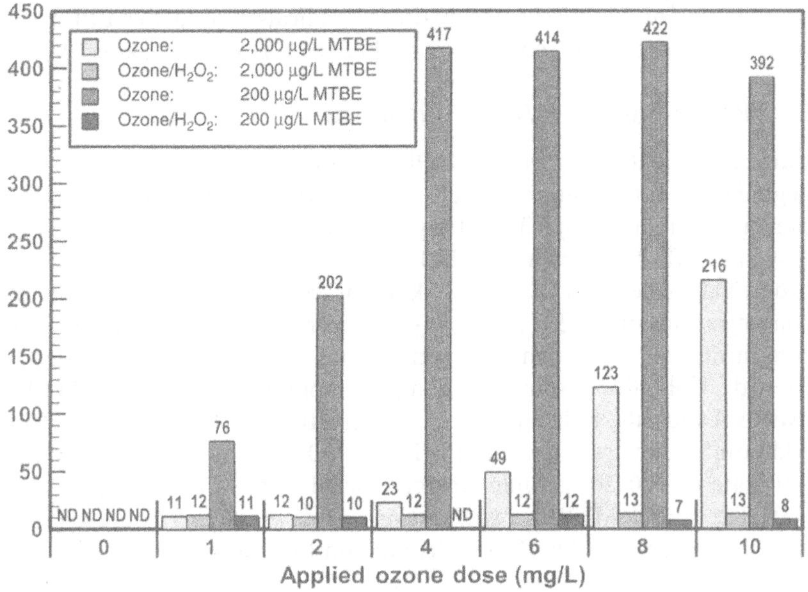

Figure 4 Effect of ozone and ozone/H_2O_2 on bromate formation. Read from left to right.

400 µg/L of bromate was formed, which far exceeds the MCL for bromate at 10 µg/L. Addition of H_2O_2 significantly reduced the bromate formation, but as much as 23 µg/L of bromate was formed under certain circumstances (see Fig. 4).

2. Low-Pressure (LP) UV Irradiation

Wagler and Malley [20] conducted bench-scale studies to determine the effectiveness of LPUV, H_2O_2, and LPUV combined with H_2O_2 to remove MTBE from contaminated groundwater in New Hampshire. In general, a simulated groundwater treated by LPUV or by H_2O_2 alone produced less than 10% removal of MTBE after 2 hr of equilibration time in the pH range 6.5–8.0. The combination of LPUV and H_2O_2 produced more than 95% removal of MTBE after 40 min of exposure time within the pH range 5.5–10.

Chang and Young [21] determined the kinetics of LPUV/H_2O_2 degradation of MTBE using a recirculating batch reactor with a low-pressure mercury lamp. With a spiked MTBE level of 10 mg/L, LPUV/H_2O_2 treatment resulted in 99.9% removal of MTBE in water, with the major by-product identified as TBF. The second-order rate constant for the degradation of MTBE by the hydroxyl radicals during the $LPUV/H_2O_2$ treatment process was reported to be 4.82×10^9 L $\times$ mole^{-1} $\times$ sec^{-1}.

3. Medium-Pressure (MP) UV Irradiation

Cater et al. [22] examined the $MPUV/H_2O_2$ process to determine MTBE degradation rates using a batch reactor with a 1-kW Rayox medium-pressure mercury lamp. The study showed that MTBE could be treated easily and effectively with UV/H_2O_2 process with E_{EO} (electrical energy per order) values between 0.2 and 7.5 kWh/m^3/order, depending on the initial concentrations of MTBE and H_2O_2. In general, at the higher initial MTBE concentrations, the higher concentrations of MTBE by-products are formed. Thus, the by-products compete more effectively with MTBE for hydroxyl radicals, reducing the overall treatment efficiency of MTBE. In addition, a pilot-scale test using $MPUV/H_2O_2$ treatment technology with a flow rate of 100–300 gal/min was conducted at the City of Santa Monica [23]. The initial MTBE concentrations of 50–1500 µg/L were effectively reduced to less than 1 µg/L with irradiation time up to 10 min.

4. Pulsed UV irradiation

A series of bench-scale tests was carried out to investigate the performance of the pulsed-UV and pulsed-UV/H_2O_2 processes for removing MTBE with varying levels of radical scavengers or UV-absorbing species in the background water matrix [24,25]. The water sources tested comprised of various levels of total organic carbon (TOC) and alkalinity, and the effects of these background water quality on the removal of MTBE during the pulsed-UV process was investigated. Some water sources also contained other variables such as NO_3^-, which can affect the pulsed-UV/H_2O_2 process by filtering out some of the available UV light.

For all types of water tested, excellent removal of MTBE was observed. With irradiation time of 12 min at the UV pulse rate of 3 flashes per second (Hz), less than 5–6 µg/L of MTBE was measured in the effluent (greater than 99% removal of MTBE) for all water sources spiked with initial MTBE concentrations of 1000–2000 µg/L and H_2O_2 concentration of 7–20 mg/L. The typical MTBE removal efficiency with 6-min irradiation time at 3 Hz was greater than 94%, which is indicative of the efficiency of the process for removing MTBE. The presence of NO_3^- at the levels of 10–20 mg/L inhibited the rate of the removal of MTBE. In fact, the first-order rate of reduction (k) was 4.3×10^{-3} sec^{-1} for Santa Monica water, with background NO_3^- of 20 mg/L, compared with 8.6×10^{-3} sec^{-1} for Long Beach water and Miami Dade water, where the NO_3^- levels were lower than 2 mg/L (Table 4). With the addition of H_2O_2, however, more

Table 4 Rate Constants for MTBE Degradation by Pulsed-UV Only with Different Water Sources

Conditions	Kinetics rate	Santa Monica groundwater[a]	Suffolk County groundwater	Long Beach groundwater	Miami Dade plant effluent
3-Hz UV frequency	k (sec^{-1})	4.3×10^{-3}	7.0×10^{-3}	8.6×10^{-3}	8.6×10^{-3}
0 mg/L H_2O_2	r^2	0.9759	0.9922	0.9919	0.9822
3-Hz UV frequency	k (sec^{-1})	7.8×10^{-3}[a]	7.6×10^{-3}	8.3×10^{-3}	6.3×10^{-3}
7 or 20 mg/L H_2O_2	r^2	0.9544	0.9857	0.9327	0.9156
Hydroxyl radical controlling factors		Santa Monica ground water	Suffolk County groundwater	Long Beach ground water	Miami Dade plant effluent
TOC (mg/L)		0.5	0.56	3.5	3.86
Total alkalinity (mg/L)		327	13	150	48
NO_3^- (mg/L)		20.5	9.61	ND	1.89

[a]With 20 mg/L H_2O_2; ND, not detected.

consistent removal of MTBE was achieved regardless of the water quality characteristics. For Santa Monica water, the rate of MTBE reduction increased to $7.8 \times 10^{-3}\,sec^{-1}$ with the addition of 20 mg/L of H_2O_2.

The effects of background water characteristics were much more evident in the rate of formation and the fate of the MTBE degradation by-products. For example, with high level of background NO_3^-, contribution of the photolysis of MTBE or water may be small since the UV light is absorbed strongly by NO_3^-. As a result, intermediate by-products may be formed less than water with low level of the background NO_3^-. In addition, the water with a high level of NO_3^- yielded higher levels of TBF and TBA, two primary degradation products of MTBE during the advanced oxidation process. Because of the hydroxyl radical-controlled case, the sequential oxidation of MTBE and MTBE degradation by-products would give rise to each species followed by some lag time for the next by-products to develop peak concentration, similar to the case for ozone/H_2O_2 process.

B. Emerging Technologies

Kang and Hoffmann [26] investigated the kinetics and mechanism of the sonolytic degradation of MTBE at an ultrasonic frequency of 205 kHz and power of 200 W/L. The observed first-order degradation rate constant for the reduction of MTBE increased from $4.1 \times 10^{-4}\,sec^{-1}$ to $8.5 \times 10^{-4}\,sec^{-1}$ as the concentration of MTBE decreased from 1.0 to 0.01 mM. In the presence of ozone, the sonolytic rate of destruction of MTBE was accelerated substantially. The rate of MTBE sonolysis with ozone was enhanced by a factor of 1.5–3.9, depending on the initial concentration of MTBE. TBF, TBA, methyl acetate, and acetone were found to be the primary intermediates and by-products of the degradation reaction with yields of 8, 5, 3, and 12%, respectively.

MTBE pilot studies have shown the ability of HEEB to reduce MTBE [27]. MTBE was reduced to below the detection limit (87 μg/L) after cumulative doses of 665 and 2000 krad were applied to initial MTBE concentrations of 2300 and 31,000 μg/L, respectively. The primary MTBE by-products such as TBF and TBA could also be treated to low residual concentrations. Another pilot study was done in southern California for MTBE destruction with a flow rate of 100 gal/min in a recycle mode [23]. The concentrations of MTBE were reduced from approximately 170–224 μg/L to below 5 μg/L with greater than 97% removal efficiencies. Primary MTBE by-products, such as TBF and TBA, were reduced concurrently with MTBE, suggesting that HEEB could also effectively remove by-products. The treatment efficiency of MTBE was affected by background water quality; e.g., water sources with lower level of TOC had higher MTBE removal efficiency.

IV. SUMMARY

Air stripping is a physicochemical process that involves the phase transfer of MTBE from the liquid phase to the vapor phase. Unlike air stripping, advanced oxidation is a destructive process and converts MTBE through chemical transformation. Each of these processes has its respective advantages and disadvantages, including process inefficiency and formation of unacceptable by-products. Consequently, the process design should address not only achieving the best MTBE reduction, but also minimizing or treating the by-product formation during the MTBE treatment process. In addition, background water quality can have an impact on the effectiveness of MTBE destruction and the formation of MTBE by-products depending on the operating conditions. Regardless of the process used, a pilot-scale study is recommended prior to the implementation of either technology since the effectiveness of MTBE reduction is site-specific.

REFERENCES

1. P. J. Squillace, J. S. Zogorski, W. G. Wilber, C. V. Price (1996). Preliminary assessment of the occurrence and possible sources of MTBE in groundwater in the United States, 1993–1994. Environ Sci Technol, 30(5):1721–1730.
2. P. J. Squillace, M. J. Moran, W. W. Lapham, C. V. Price, R. M. Clawges, J. S. Zogorski (1999). Volatile organic compounds in untreated ambient groundwater of the United States, 1985–1995. Environ Sci Technol, 33(23): 4176–4187.
3. A. M. Happel, E. H. Beckenbach, R. U. Halden (1998). An Evaluation of MTBE Impacts to California Groundwater Resources, UCRL-AR-13089. Environmental Protection Department, Environmental Restoration Division, Lawrence Livermore National Laboratory, University of California, Livermore, CA.
4. Blue Ribbon Panel on Oxygenates in Gasoline (1999). Achieving Clean Air and Clean Water: The Report of the Blue Ribbon Panel on Oxygenates in Gasoline, EPA420-R-99-021. Washington, DC: U.S. Government Printing Office.
5. T. Shih, E. Khan, W. Rong, M. Wangpaichitr, J. Kong, I. H. Suffet (1999). Sorption for removing methyl tertiary butyl ether from drinking water. AWWA Natl Conf Proc, Chicago, IL, June.
6. D. Friday, J. Greene, T. Barnhill. (2001). Effective treatments of MTBE for municipal drinking water systems. Contaminated Soil, Sediment, Water, Spring: 29–31.

7. D. M. Creek, J. M. Davidson (1998). The performance and cost of MTBE remediation technologies. In: Proc Petroleum Hydrocarbons and Organic Chemicals in Groundwater—Prevention, Detection, and Remediation, NGWA/API Conf, by National Ground Water Association, Westerville, OH.
8. Federal Remediation Technologies Roundtable (FRTR). (2001). Remediation Technologies Screening Matrix and Reference Guide, Version 3.0. http://www.frtr.gov/matrix2/top_page.html.
9. G. Tchobanoglous, F. L. Burton (1991). Wastewater Engineering, Treatment, Disposal and Reuse. 3rd ed. New York: McGraw-Hill.
10. A. A. Keller, O. C. Sandall, R. G. Rinker, M. M. Mitani, B. Bierwagen, M. J. Snodgrass (2000). An evaluation of physiochemical treatment technologies for water contaminated with MTBE. Ground Water Monitor Remediation, Fall, 20(4): 114–126.
11. MTBE Research Partnership (1998). Evaluation of Treatment Technologies for Removal of Methyl Tertiary-Butyl Ether (MTBE) from Drinking Water: Air Stripping, Advanced Oxidation Process (AOP), Granular Activated Carbon (GAC). Sacramento, CA: Association of California Water Agencies.
12. A. Stocking, R. Rodriguez, A. Flores, D. Creek, J. Davidson, M. Kavanaugh (2000). Executive summary. In: Treatment Technologies for Removal of Methyl Tertiary-Butyl Ether (MTBE) from Drinking Water: Air Stripping, Advanced Oxidation Process, Granular Activated Carbon, Synthetic Resin Sorbents. 2nd ed. National Water Research Institute, Fountain Valley, CA.
13. J. Greene, T. Barnhill (2001). Proven solutions for MTBE household drinking water. Contaminated Soil, Sediment, Water, Spring: 79–80.
14. R. Cataldo, E. Moyer (2001). Remediation of releases containing MTBE at gas station sites. Contaminated Soil, Sediment, Water, Spring: 87–90.
15. D. Woodward, D. Sloan (2001). Common myths, misconceptions, and assumptions about MTBE: where are we now? Contaminated Soil, Sediment, Water, Spring: 16–19.
16. Karpel Vel Leitner (1994). Oxidation of methyl *tert*-butyl ether (MTBE) and ethyl *tert*-butyl ether (ETBE) by ozone and combined ozone/hydrogen peroxide. Ozone Sci Eng, 16:41.
17. J. E. Dyksen, R. Ford, K. Raman, J. K. Schaefer, L. Fung, B. Schwartz, M. Barnes (1992). In-Line Ozone and Hydrogen Peroxide Treatment for Removal of Organic Chemicals. AWWA Research Foundation, Denver, CO.
18. S. Liang, L. S. Palencia, R. S. Yates, M. K. Davis, J. M. Bruno, R. L. Wolfe (1999). Oxidation of MTBE by ozone and PEROXONE processes. J. Am Water Works Assoc, 91(6):104.
19. S. Liang, R. S. Yates, D. V. Davis, S. J. Pastor, L. S. Palencia, J. M. Bruno (2001). Treatability of groundwater containing methyl *tertiary*-butyl ether (MTBE) by ozone and PEROXONE and identification of by-products. J Am Water Works Assoc, 93(6):110.
20. J. L. Wagler, J. P. Malley Jr (1994). The removal of methyl tertiary-butyl ether from a model groundwater using UV/peroxide oxidation. J NEWWA, 108(3):236.

21. P. Chang, T. Young (1998). Reactivity and By-products of Methyl Tertiary-Butyl Ether Resulting from Water Treatment Processes. Risk Assessment, Exposure Assessment, Water Treatment and Cost-Benefit Analysis (Vol. V). Report to Governor and Legislature of the State of California. University of California, Berkeley.
22. S. R. Cater, M. I. Stefan, J. R. Bolton, S. A. Ali (2000). UV/H_2O_2 treatment of methyl *tert*-butyl ether in contaminated waters. Environ Sci Technol, 34(4):659–662.
23. California MTBE Research Partnership 2000. Treatment Technologies for Removal of Methyl Tertiary Butyl Ether (MTBE) from Drinking Water: Air Stripping, Advanced Oxidation Processes, Granular Activated Carbon, and Synthetic Resin Sorbents. 2nd ed. National Water Research Institute, Fountain Valley, CA.
24. J. H. Min, S. Liang, C. D. Church, C. S. Chou, M. Kavanaugh. Investigation of MTBE removal and fate of MTBE by-products with the pulsed-UV (PUV)/hydrogen peroxide (H_2O_2) process. Proc., 2001 AWWA Annual Conference, Washington DC, June 17–21, 2001.
25. S. Liang, J. H. Min, R. S. Yates, C. S. Chou, M. Kavanaugh. Effects of water quality on destruction of methyl *tert*-butyl ether (MTBE) by the pulsed-UV/H_2O_2 process. Proc., 2001 AWWA WQTC, Nashville, TN, November 11–15, 2001.
26. J. M. Kang, M. R. Hoffman (1998). Kinetics and mechanism of the sonolytic destruction of methyl *tert*-butyl ether by ultrasonic irradiation in the presence of ozone. Environ Sci Technol, 32:3194–3199.
27. W. J. Cooper, G. Leslie, P. M. Tornatore, W. Hardison, P. A. Hajali (2000). MTBE and priority contaminant treatment with high electron beam injection. In: Chemical Oxidation and Reactive Barriers, eds., G. B. Wickramanayake, A. R. Gavaskar, A. S. C. Chen. 2(6):209–216.

A. APPENDIX: SUPPLEMENTAL INFORMATION

A1. Oxidation and Photolysis of MTBE

Ozonation of MTBE can result from direct reaction with molecular ozone or indirect reaction with radical oxidant species (mainly the hydroxyl radicals) [A1]. Of these two reactions, oxidation of ethers by molecular ozone is known to occur very slowly (second-order kinetics constant $<1\ L \times mole^{-1} \times sec^{-1}$). By contrast, oxidation of ethers by radical oxidants is extremely rapid. Hydroxyl radicals react with MTBE at a rate of $1.6 \times 10^9\ L \times mole^{-1} \times sec^{-1}$ (Table A-1) [A2].

UV photolysis of MTBE can occur on direct and indirect photolysis. Direct photolysis involves light absorption by MTBE followed by chemical reaction of MTBE in its electronically excited state. Indirect photoreactions

Table A-1 Hydroxyl Radical Rate Constants for MTBE and By-products

Compound	Reaction	Kinetic Rate ($L \times mole^{-1} \times sec^{-1}$)	Reference
MTBE	$(CH_3)_3COCH_3 + \bullet OH \rightarrow (CH_3)_3COCH_2\bullet + H_2O$	1.6×10^9	A2
		4.82×10^9	21
t-Butyl formate	$(CH_3)_3COCHO + \bullet OH \rightarrow$ products	1.19×10^9	21
t-Butyl alcohol	$(CH_3)_3COH + \bullet OH \rightarrow \bullet CH_2C(CH_3)_2OH + H_2O$	4.8×10^8	A2
Isopropyl alcohol	$(CH_3)_2CHOH + \bullet OH \rightarrow (CH_3)_2COH + H_2O$	1.6×10^9	A2
Acetone	$CH_3COCH_3 + \bullet OH \rightarrow$ products	1.1×10^8	A2
Formaldehyde	$HCHO + \bullet OH \rightarrow$ products	1×10^9	A2
Acetaldehyde	$CH_3CHO + \bullet OH \rightarrow CH_3CO + H_2O$	7.3×10^8	A2
Heptaldehyde	$CH_3(CH_2)_6CHO + \bullet OH \rightarrow$ products	$\sim 5 \times 10^9$	(Estimated)
Glyoxal	$HCOCHO + \bullet OH \rightarrow \bullet COCHO + H_2O$	6.6×10^7	A2
Methyl glyoxal	$CH_3COCHO + \bullet OH \rightarrow$ products	$\sim 1 \times 10^9$	(Estimated)

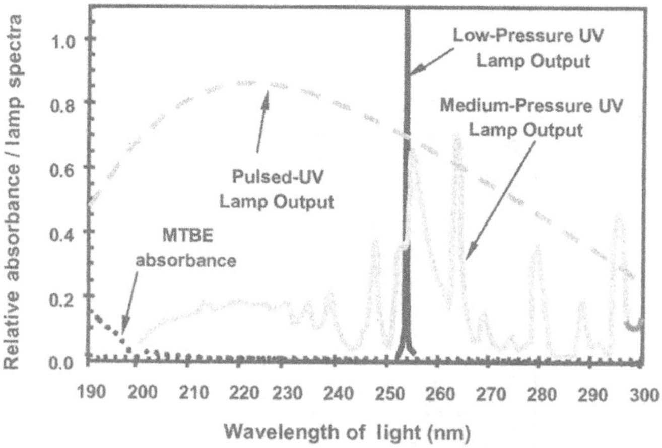

Figure A-1 MTBE absorption spectrum and characteristics of various UV lamps.

of MTBE are mediated by hydroxyl radicals. MTBE does not absorb UV light above the wavelength of 200 nm (Fig. A-1) [A3,A6]. Thus, the MTBE degradation by a monochromatic, low-pressure UV lamp, which emits light at 254 nm, is done primarily via oxidation of MTBE by hydroxyl radicals generated during the UV process (Table A-2). At the wavelength below 200 nm, MTBE weakly absorbs UV light. Thus, with a polychromatic UV lamp, such as a medium-pressure or a pulsed UV lamp, MTBE may be directly photolyzed in the absence of hydroxyl radicals by UV light below 200 nm [A3]. At wavelengths below 200 nm, however, water can also absorb UV light to produce hydroxyl radicals by photolysis of water. The amount of MTBE is typically small compared with the amount of water in the contaminated site. Thus, the reduction of low concentration of MTBE in pure water may be the result of the oxidation by hydroxyl radicals (from photolysis of water) rather than by the direct photolysis of MTBE.

Therefore, the degradation of MTBE by UV processes can be described mainly due to the oxidation reactions initiated by the highly reactive hydroxyl radicals (•OH) generated in water via the following reactions:

$$H_2O_2/\text{UV process}: \quad H_2O_2 \rightarrow 2\,\bullet OH \qquad (\lambda < 300\,\text{nm})$$

$$O_3/\text{UV process}: \quad O_3 + H_2O \rightarrow H_2O_2 + O_2 \qquad (\lambda > 300\,\text{nm})$$

$$2O_3 + H_2O_2 \rightarrow 2\,\bullet OH + 3O_2$$

$$\bullet OH + \text{MTBE} \rightarrow \text{oxidation by-products}$$

$$(k_{\text{OH:MTBE}} = 1.9 \times 10^9\,\text{M}^{-1}\,\text{sec}^{-1})$$

Table A-2 Characteristics of Typical LP-, MP-, and Pulsed-UV Lamps

Characteristics	Low pressure	Medium pressure	Pulsed-UV
Emission	Monochromatic (85–90% at 253.7 nm)	Polychromatic (185–1,367 nm)	Polychromatic (185–1,000 nm)
Mercury vapor pressure (torr)	10^{-3} to 10^{-2}	10^2 to 10^4	N/A
Operating temperature (°C)	40 to 60	500 to 800	15,000 K
Arc length (cm)	40 to 75	5 to 40	15
Lifetime (hr)	8,000 to 10,000	2,000 to 5,000	$>100 \times 10^8$ pulses
Light intensity (relative)	Low	High	High

The chemical effects of sonolysis of MTBE by ultrasonic irradiation are the direct result of the formation of cavitation microbubbles [A5]. The rapid implosion of cavitation bubbles undergoes a thermal dissociation to yield extremely reactive radicals. As a result, near the bubble/water interface there may be thermal decomposition of MTBE, and/or secondary reactions can take place between MTBE and reactive radicals. UV photolysis and ultrasonic irradiation of MTBE are predominantly accomplished by the reactions with hydroxyl radicals.

HEEB irradiation simultaneously produces approximately equal concentrations of oxidizing and reducing species. Irradiation of aqueous solutions results in the formation of the aqueous electron, e^-_{aq}, hydrogen atom, H•, and •OH.

$$H_2O \rightsquigarrow 2.6e^-_{aq} + 0.6H\bullet + 2.7\bullet OH + 0.45H_2 + 0.7H_2O_2 + 2.6H_3O^+$$

The aqueous electron, e^-_{aq}, reacts with MTBE at a rate of $1.75 \times 10^7\,L \times mole^{-1} \times sec^{-1}$, while the hydrogen atoms, H•, react with MTBE at a rate of greater than $3.49 \times 10^6\,L \times mole^{-1} \times sec^{-1}$ [A6]. In comparison, the MTBE reaction rate with hydroxyl radicals is approximately 90 and 460 times faster than with aqueous electrons and hydrogen atoms, respectively.

A2. MTBE By-product Formation

During the oxidation and photolysis treatment of MTBE, certain by-products are formed. These include TBF, TBA, isopropyl alcohol, acetone, formaldehyde, and methyl acetate [A7]. Currently, these contaminants are not regulated for drinking water supplies, but some of these constituents pose health concerns to the public. In fact, the California Department of Health Services (CDHS) has proposed an action level of 12 µg/L for TBA for its cancerous effect in 1999 [A8]. In addition to the organic oxidation by-products, inorganic by-products can also be formed. While ozonation and ozone/H_2O_2 of water containing bromide leads to the formation of bromate, photolysis and UV/H_2O_2 processes do not yield bromate in the effluent [A9].

A3. Possible Transformation Pathways of MTBE

There are a number of pathways by which degradation of MTBE may occur under oxidation processes. However, the major possibilities are hydrolysis and reaction with hydroxyl radical. A possible mechanism for oxidation of MTBE is given in Figure A-2, which shows either direct conversion of MTBE to *t*-butyl alcohol or oxygen addition to MTBE to the methyl group

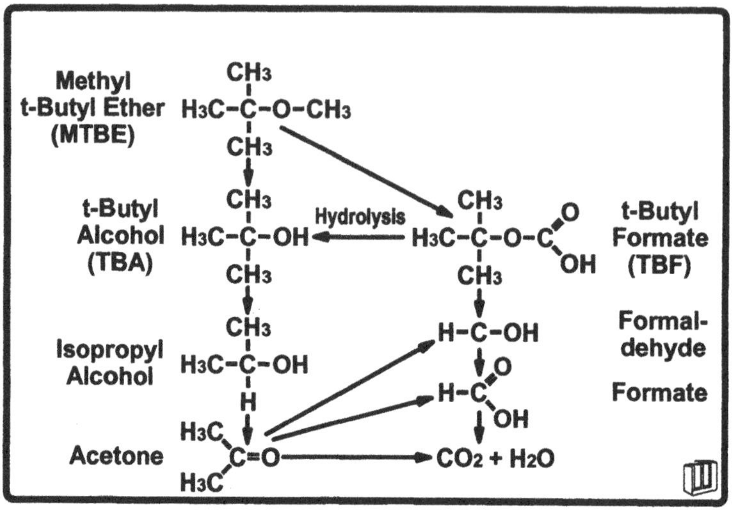

Figure A-2 Pathway of MTBE degradation by-products.

to form *t*-butyl formate [A10]. Then, TBF can either be followed by hydrolysis to TBA or be oxidized to formaldehyde. TBA can then lose a methyl group to form isopropyl alcohol, which in turn can be converted to acetone. The acetone may then go on to formaldehyde or formate and ultimately to carbon dioxide (CO_2) and water.

A4. Water Quality Impacts

The specific water quality matrix of the source water largely determines the effectiveness of oxidation of MTBE. Water quality parameters such as alkalinity, total organic carbon (TOC) or natural organic matter (NOM), nitrate (NO_3^-), iron or manganese, turbidity, pH, and other chemicals can affect the MTBE reduction by scavenging hydroxyl radicals or competing for UV light to hinder the effectiveness of the process.

In general, high concentration of alkalinity serves as a hydroxyl radical scavenger by reaction of bicarbonate HCO_3^- and carbonate (CO_3^{2-}) with hydroxyl radicals. Background NOM could also serve as an effective hydroxyl radical scavenger, but depending on the characteristics of NOM, a photochemical process, such as a pulsed-UV process, could generate small amount of hydroxyl radicals [A11,A12] and organic peroxy radicals [A13]. The background organic matter, however, is generally a more important sink than a source of hydroxyl radicals.

Peyton and co-workers [A14] demonstrated the effect of alkalinity and dissolved organic carbon (DOC) concentration on the efficiency of hydroxyl radical attack on a target compound. As the DOC and alkalinity concentrations increase, hydroxyl radicals produced during the advanced oxidation processes are scavenged by DOC and carbonate, and consequently the relative efficiency of hydroxyl radicals decreases significantly compared to the efficiency without DOC and alkalinity.

The other parameter that plays a significant role in the UV treatment process is NO_3^-. Water with an elevated level of NO_3^- can produce hydroxyl radicals by a photochemical reaction of NO_3^- without the addition of H_2O_2. The irradiation of NO_3^- results in the primary photochemical processes as described below [A15].

$$NO_3^- + UV \rightarrow NO_3^{-*} \rightarrow NO_2^- + O(^3P) \tag{1}$$

$$\rightarrow NO_2^- + \bullet O^- \tag{2}$$

$$\bullet O^- + H_2O \rightarrow \bullet OH + OH^- \tag{3}$$

Overall, the steady-state concentration of hydroxyl radicals, which are responsible for MTBE degradation during an advanced oxidation process, can be correlated by the simplified general equation shown below [A12].

$$\frac{[\bullet OH]_{ss} \propto [NO_3^-]}{[HCO_3^-] + [DOC]} \tag{4}$$

Photolysis of Fe(III) is the other possible source of hydroxyl radical from natural water during a UV process [16]. Depending on the concentration of these hydroxyl radical-generating species in a specific water matrix, the addition of H_2O_2 may not contribute significantly to improving the MTBE degradation, since large amounts of hydroxyl radicals may be generated from the background constituents.

In addition, systems relying on UV irradiation for the dissociation of H_2O_2 or O_3 exhibit a decrease in efficiency as turbidity increases. Turbidity lowers the transmittance of the source water and thus lowers the penetration of the UV light into the source water.

REFERENCES

A1. American Water Works Association Research Foundation (1991). Ozone in Water Treatment: Application and Engineering (B. Langlais, D. A. Reckhow, D. R. Brink, eds.). Cooperative Research Report, Lewis Publishers, Chelsea, MI.

A2. G. V. Buxton, C. L. Greenstock, W. P. Helman, A. B. Ross (1988). Critical review of rate constants for reactions of hydrated electrons, hydrogen atoms and hydroxyl radicals (•OH/•O^-) in aqueous solution. J Phys Chem Ref Data, 17(2): 513–886.

A3. H. P. Schuchmann, C. V. Sonntag (1973). The UV photolysis (=185 nm) of liquid *t*-butyl methyl ether. Tetrahedron, 29:1811–1818.

A4. J. P. Malley, Jr., R. R. Locandro, J. L. Wagler (1993). Innovative Point-of-Entry (POE) Treatment for Petroleum Contaminated Water Supply Wells. Final Report. U.S.G.S. New Hampshire Water Research Center, Durham, NH, September.

A5. T. Mason, J. P. Lorimer (1998). Sonochemistry: Theory, Applications, and Uses of Ultrasound in Chemistry. Ellis Norwood, Chichester, England.

A6. S. P. Mezyk, W. J. Cooper, D. M. Bartels, T. Tobien, K. E. O'Shea. Radiation chemistry of alternative fuel oxygenates—substituted ethers. Am Chem Soc Div Environ Chem Preprints Ext Abstr, 40:1.

A7. S. Liang, R. S. Yates, D. V. Davis, S. J. Pastor, L. S. Palencia, J. M. Bruno (2001). Treatability of groundwater containing methyl *tertiary*-butyl ether (MTBE) by ozone and PEROXONE and identification of by-products. J AWWA, 93(6):110.

A8. California Department of Health Services (CDHS) (2000). Drinking Water Standards: Unregulated Chemicals Requiring Monitoring. http://www.dhs.ca.gov/ps/ddwem/chemicals /MCL/unregulated.htm.

A9. J. M. Symons, M. C. H. Zheng (1997). Technical note: does hydroxyl radical oxidize bromide to bromate? J AWWA, 89(6):106.

A10. Church, et al. (1997). Assessing the in-situ degradation of methyl *tert*-butyl ether (MTBE) by-product identification at the sub-ppb level using direct aqueous injection GC/MS. Natl Meet Am Chem Soc Div Environ Chem Repr Ext Abstr, 37(1):411.

A11. P. P. Vaughan, T. E. Thomas-Smith, N. V. Blough. (1998). Photochemical production of the hydroxyl radical and the hydrated electron from colored dissolved organic matter. Am Chem Soc Div Environ Chem Preprints Ext Abstr, 38:2.

A12. R. P. Schwarzenbach, P. M. Gschwend, D. M. Imboden (1993). Environmental Organic Chemistry. New York: Wiley.

A13. J. Hoigné, B. C. Faust, W. R. Haag, F. E. Scully, R. G. Zepp (1989). Aquatic Humic Substances as Sources and Sinks of Photochemically Produced Transient Reactants Advances in Chemistry Series (I. H. Suffet and P. MacCarthy, eds.). Washington, DC: American Chemical Society.

A14. G. R. Peyton, O. J. Bell, E. Girin, M. H. LeFaivre, J. Sanders (1998). Effect of Bicarbonate Alkalinity on Performance of Advanced Oxidation Process. Denver, Co: AWWA Research Foundation and American Water Works Association.

A15. R. G. Zepp, J. Hoigne, H. Bader (1987). Nitrate-induced photooxidiation of trace organic chemicals in water. Environ Sci Technol, 21(5):443–450.

A16. W. Stumm, J. J. Morgan (1996). Aquatic Chemistry, Chemical Equilibria and Rates in Natural Waters. 3rd ed. New York: Wiley.

14

Adsorption Process for the Removal of MTBE from Drinking Water

Tom C. Shih
State of California Environmental Protection Agency, Los Angeles, California, U.S.A.

I. H. (Mel) Suffet
UCLA School of Public Health, Los Angeles, California, U.S.A.

I. INTRODUCTION

Adsorption of a substance involves its accumulation at the interface between two phases. The molecule that adsorbs at the interface is called an adsorbate, and the solid on which adsorption occurs is the adsorbent [1]. Adsorbents of interest in water treatment include activated carbon; ion-exchange resins; adsorbent resins; metal oxides, hydroxides, and carbonates; activated alumina; clays; and other solids that are suspended in or in contact with water [1].

Adsorption is a viable and important process in the removal of organic compounds from water. The primary advantages of adsorption over other treatment technologies are that adsorption is an established technology that requires no off-gas treatment and generates few (if any) by-products. Further, the physical and chemical characteristics of methyl *tertiary*-butyl ether (MTBE) are such that the compound will preferentially remain in solution, rendering most traditional treatment technologies ineffective. Adsorption, in particular by granular activated carbon (GAC), has been shown to be an efficient, cost-effective, and proven technology for treating organic pollutants [2–5], including MTBE, from contaminated water [6–8]. The cost of GAC treatment, however, is now escalating because of the increasing costs of handling, transporting, and disposing of spent GAC [2].

Synthetic resins, both carbonaceous and polymeric, have been developed for the treatment of potable water. Since the late 1970s, synthetic carbonaceous resins, in particular, have been extensively tested and found to be effective in removing organic chemicals from contaminated groundwater [9]. However, the high unit cost price of the synthetic resins has until now prevented their widespread application. Recent improvements in synthetic resins are making them attractive and viable adsorbents, and their capability to be regenerated on-site through steam stripping, solvent extraction, or microwave regeneration are making the life-cycle cost of a resin adsorption system competitive relative to other treatment technologies, including GAC. Studies have shown their superiority compared to GACs for removing a number of organic chemicals [2,10,11]. Recent work [8,11–13] has also demonstrated, relative to GAC, their greater capacities for MTBE, their resistance to fouling by natural organic matter (NOM) present in the natural water sources, their resistance to interference from other organics, and their greater affinity to *tertiary*-butyl alcohol (TBA). TBA, a gasoline additive, impurity, and oxidation by-product of MTBE, frequently co-occurs with MTBE in MTBE-contaminated groundwater.

The cost effectiveness of various adsorption systems is evaluated traditionally through pilot columns. Pilot columns have been shown to be accurate and reliable predictors of breakthrough behavior in full-scale systems. However, pilot columns have equivalent operation time as full-scale systems, and coupled with the high capital investments required for the pilot system and the necessity to screen all types of adsorbents, it becomes clear that the operation of pilot columns as a screening and optimization tool is not feasible [14]. Isotherms and rapid small-scale column tests (RSSCTs) are two state-of-the-art bench-scale tests that have been shown to be accurate and reliable predictors of breakthrough behavior of full-scale adsorber systems [3,15–17].

Important differences exist between isotherm tests and dynamic tests such as RSSCT. The isotherm is a static method that is used to assess the equilibrium adsorption capacity of the adsorbents for the solute of interest, and is usually run in batch reactors with different ratios of carbon dose to initial component concentration. Because equilibrium adsorption capacity for an adsorbent is a major determinant of the adsorbent's performance in columns, the isotherm can be used to compare and to screen candidate adsorbents. However, the static nature of the isotherm tests limits the extension of its predictive capability to full-scale operational parameters [14].

The RSSCT procedures developed by Crittenden et al. [3,15,16] used dimensional analysis to elucidate the relationship between the breakthrough curves of full-scale and small-scale columns. The RSSCTs have been shown [3,14–17] to yield small-scale column breakthrough curves that are equivalent to those of a full- or pilot-scale breakthrough curves in a much shorter time period, thus allowing the initial optimization of design or operational parameters. Comparatively, dynamic tests such as the RSSCTs are a more sophisticated and time-consuming testing procedure than the isotherms.

A large number of adsorbents under a variety of conditions (e.g., different water sources, different competitive adsorbate composition and concentrations) can be tested simultaneous through the isotherm testing procedure. However, it is by definition a static test that yields adsorption capacity under equilibrium conditions. Consequently, the isotherm tests may be used as a preliminary screening tool for comparing the relative effectiveness of different adsorbents and the magnitude of competitive effects arising from competing chemical species. The specific or operational adsorption capacities may then be derived from the performance of RSSCTs using the most promising adsorbents selected from the isotherm tests [6,7].

This chapter evaluates the performance and cost effectiveness of different adsorbents in removing MTBE from contaminated water, with emphasis on GAC and synthetic carbonaceous resins. Research results from isotherms and RSSCTs will be used to assess the performance of adsorbents in removing MTBE in various conditions that may affect adsorbent performance, such as low to high MTBE influent concentrations, the presence of competitors such as NOM, TBA, and volatile organics present in gasoline (e.g., benzene, toluene, and *p*-xylene [BTX], and different process and flow configurations. This chapter will also present a detailed cost analysis based on adsorbent usage rates predicted by either RSSCTs [7,13] or the AdDesignS model [18].

II. PRELIMINARY PERFORMANCE COMPARISON OF ADSORBENTS

Studies have demonstrated the coconut GAC and synthetic carbonaceous resin to have the highest equilibrium adsorption capacities for MTBE [7,8,13,19]. Shih et al. [19] investigated the equilibrium adsorption capacities of 11 different adsorbents for MTBE. Their findings are summarized in Figure 1, which depicts the results of linear regressions of the isotherm data after log-transformation for 11 adsorbents conducted with 1000 μg/L of influent MTBE in organic-free water. The results indicate the coconut GAC (Calgon GRC-22) had the highest equilibrium adsorption capacity for

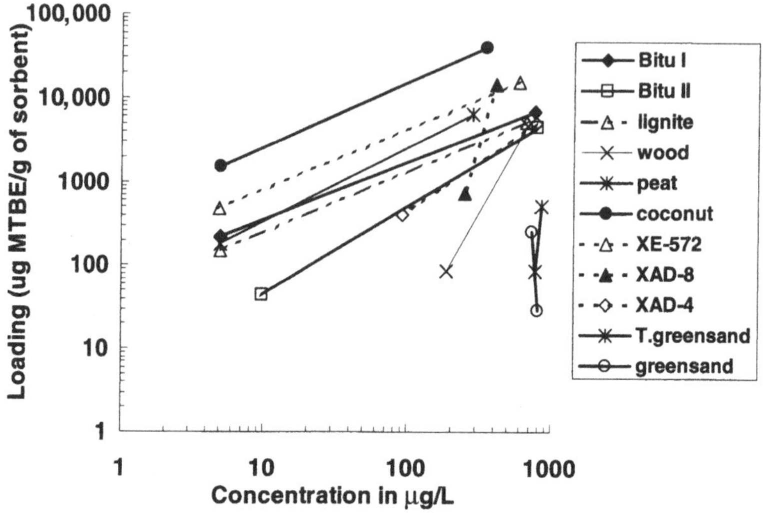

Figure 1 Isotherm of 11 GACs with 1000 µg/L influent MTBE in organic-free water. Best-fit line is shown for clarity. Actual data points are not shown. The two points on each line are not actual data points. Lines were obtained by linear regression of a series of data. (Adapted with permission from Ref. 19. Copyright American Water Works Association.)

MTBE, followed by the synthetic carbonaceous resin Ambersorb 572. The data for these two adsorbents were also well represented by the Freundlich model. Other investigators have arrived at similar results. Both Davis and Powers [11] and Malley et al. [12] found the equilibrium adsorption capacities of the Ambersorb 572 and 563 carbonaceous resins to be among the highest, with Ambersorb 563 greater than 572.

As coconut GAC and synthetic carbonaceous resin have been shown to be the most promising candidates for treating MTBE-contaminated water, the rest of the chapter will be devoted to assessing the performance of synthetic carbonaceous resin Ambersorb 563 (manufactured by Rohm and Haas) and two coconut GACs (PCB and CC-602 GAC, manufactured by Calgon Carbon and U.S. Filter, respectively) in removing MTBE from contaminated water under conditions that may adversely affect adsorbent performance.

III. IMPACT OF NOM ON ADSORPTION PERFORMANCE

NOM is a complex mixture of high-molecular-weight compounds, such as fulvic and humic acids, hydrophilic acids, and carbohydrates, which varies with location [1]. NOM can affect the adsorption of trace organics to GACs in three ways: (1) NOM can reduce the number of adsorption sites available to target organics by competition for sites [1]; (2) by pore blockage [1]; and (3) irreversible adsorption by NOM to GAC adsorption sites may permanently remove those sites from adsorption to trace organics [20].

The competitive effects attributed to NOM on GAC performance have been well documented. Studies have shown that the presence of NOM can cause significant reduction in the adsorption capacity of GACs for target organics [1,8,21–23]. Shih et al. [7] investigated the performance of two coconut GACs (PCB and CC-602) in removing MTBE from three natural water sources and attempted to correlate the total organic carbon (TOC) content to GAC performance parameters (e.g., carbon usage rate [CUR] and adsorption capacity). Studies have used TOC, dissolved organic carbon (DOC), UV absorbance at 254 nm (UV_{254}), or specific UV absorbance (SUVA) as surrogates for quantification of NOM [3,8,24].

The water quality parameters for the three natural water sources are shown in Table 1. Each of these water sources has different sources of NOM, i.e., South Lake Tahoe Utility District (SLTUD) (pine forest), Arcadia Well Field (ARWF) (groundwater from desert flora and urban runoff), and Lake Perris (LP) (Colorado River water stored in a lake environment which changes with the season and nutrient levels). The TOC concentration of the water sources ranged from 0.2 mg/L (low) to 3.2 mg/L (medium) for groundwater from SLTUD and for surface water from LP, respectively.

Figure 2 is a plot of isotherms of the two coconut GACs in both organic-free water and in the three natural water sources [7]. The isotherms were conducted with 1000 μg/L of influent MTBE. The Freundlich parameters are presented in Table 2. The Freundlich K is related primarily to the capacity of the adsorbent, while the coefficient $1/n$ is a function of the strength of adsorption [1].

Table 1 Analysis of Water Quality Parameters

Water source	pH	Conductivity (μmhos/cm)	TOC (ppm)	UV_{254} Abs	SUVA	MTBE (μg/L)	TBA (μg/L)
ARWF	7.8	1130	1.0	0.008	0.8	<1	<1
SLTUD	7.9	77	0.2	0.004	2.0	<1	<1
LP	8.5	640	3.2	0.068	2.1	5–10	<1

SUVA = specific UV absorbance (UV_{254} absorbance 100/TOC).
Source: Data from Ref. 7.

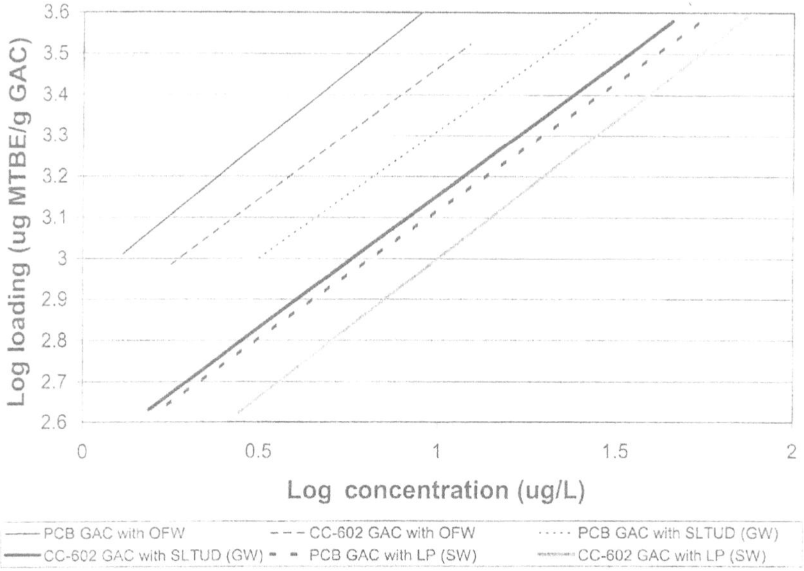

Figure 2 Isotherm of two coconut GACs in organic-free water and in three natural waters. Best-fit line is shown for clarity. Actual data points are not shown. (From Ref. 7.)

Both Figure 2 and Table 2 show the PCB GAC to have higher equilibrium adsorption capacity than the CC-602 GAC in all three water sources investigated. More important, Figure 2 and Table 2 indicate the magnitude of reduction in equilibrium adsorption capacity of the GAC

Table 2 Freundlich Parameters for PCB and CC-602 GAC

Water source	Coconut GAC	Freundlich constant K $(\mu g/g)(L/\mu g)^{1/n}$	Freundlich coefficient $1/n$
Organic-free water	PCB	853.89	0.700
	CC-602	665.27	0.645
SLTUD (groundwater)	PCB	489.22	0.623
	CC-602	323.52	0.645
ARWF (groundwater)	PCB	416.39	0.623
	CC-602	251.88	0.684
LP (surface water)	PCB	308.60	0.629
	CC-602	207.87	0.682

Source: Data from Ref. 7.

for MTBE due to competitive adsorption of NOM present in the natural water sources (as characterized by the TOC concentration and SUVA), with higher reductions resulting from higher NOM of the water sources.

While studies have shown the presence of NOM can cause significant reduction in the adsorption capacity of GACs for target organics including MTBE, isotherm studies [2,8,19] conducted thus far have found NOM to have limited impact on the performance of carbonaceous resins (e.g., Ambersorb 563). The inaccessibility of the narrow internal pores in carbonaceous resins to NOM may be the primary reason the carbonaceous resins as a group are resistant to NOM fouling as compared to GACs [8]. Hand et al. [2] conducted TCE isotherm studies on Ambersorb 563 carbonaceous resin and Filtrasorb 400 GAC that had been preexposed to groundwater containing NOM (1–3 mg/L). They found that after 10 weeks of NOM exposure, the equilibrium capacity of carbonaceous resin for TCE was still unaffected, whereas there was an average decrease of 35% for the GAC. Shih et al. [19] conducted isotherm studies using GRC-22 coconut GAC and Ambersorb 572 carbonaceous resin in groundwater from Santa Monica, California ([TOC] = 0.5 mg/L). The effects of NOM on the equilibrium capacity of coconut GAC and Ambersorb 572 for MTBE are shown in Figures 3 and 4. Significant reduction in the equilibrium adsorption capacity of the coconut GAC was observed by the addition of NOM, while the Ambersorb 572 resin was relatively unaffected by the presence of NOM. These findings, in particular the reduction in adsorption capacity of coconut GACs due to competitive effect from NOM, is further supported by results obtained from a dynamic column study applying the RSSCT methods by Shih et al. [7], in which GAC performance (e.g., liters of water treated per gram of GAC) was demonstrated to decreased (in some cases as much as fourfold) as the measured TOC content of the water source used increased.

IV. IMPACT OF OTHER COMPOUNDS ON MTBE ADSORPTION IN MULTISOLUTE SYSTEMS

Competitive adsorption of MTBE with aqueous soluble fuel components is more relevant than single-solute or pure adsorption in water treatment applications, since MTBE occurs with these fuel components. Competitive adsorption due to the presence of other organic compounds in addition to NOM can cause further reductions in adsorbent performance for target organics as a greater and greater percentage of the available sites on the adsorbent become utilized by competing chemicals. The magnitude of competitive effect in a multisolute system is complicated by factors such as

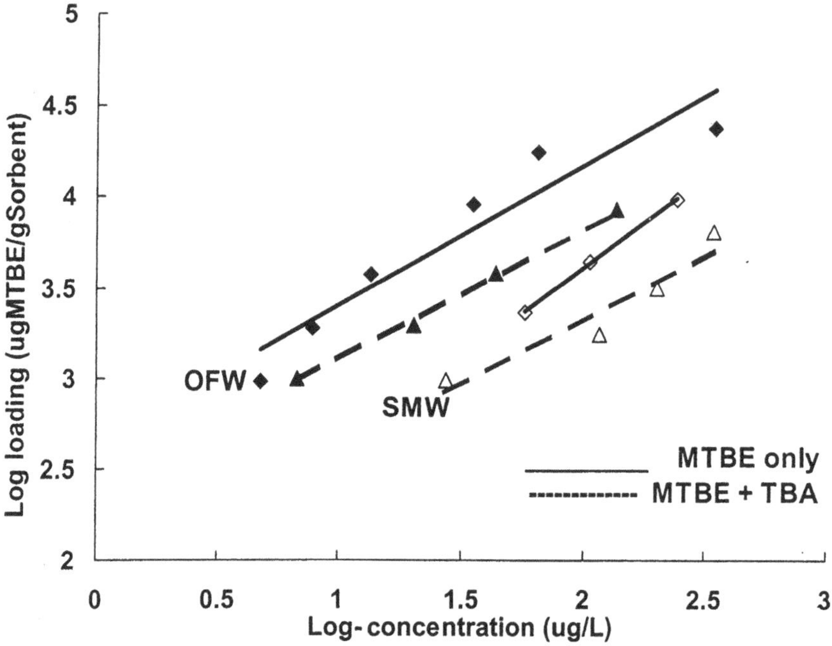

Figure 3 Isotherm of coconut GAC (GRC-22) in organic-free water (OFW) or in Santa Monica water (SMW) with MTBE (1000 μg/L) or MTBE and *t*-butyl alcohol (100 μg/L). (Reproduced with permission from Ref. 19. Copyright American Water Works Association.)

the type of adsorbent and the variety of competing species each having different adsorbabilities and concentrations [1,5,21].

Studies have shown significant reduction in both resin's and GAC's capacity for MTBE in the presence of BTEX fuel components. However, even in the presence of BTEX, the resin's performance has shown to remain superior to GACs. Davis and Powers [11] conducted bi-solute MTBE isotherm studies on carbonaceous resins (Ambersorb 563 and Ambersorb 572) and GAC (Fitrasorb 400) in the presence of 43.2 mg/L of *m*-xylene as a representative BTEX compound. They found the resins' and GAC's equilibrium capacities (at 1 mg/L MTBE influent concentration) decreased approximately 10–20% and 33%, to 8–12 mg/g and 2 mg/g, respectively. Shih et al. [7] conducted dynamic column tests (RSSCTs) on coconut GACs (PCB and CC-602) with MTBE and benzene, toluene, and *p*-xylene (BTX) in one groundwater (SLTUD) and one surface water (LP) source. Their findings indicate increases in carbon usage rates (CURs) in the range 40–50%. Recent RSSCTs conducted with Ambersorb 563 resin in LP surface water by Shih et al. [13] indicate a greater than fivefold reduction in

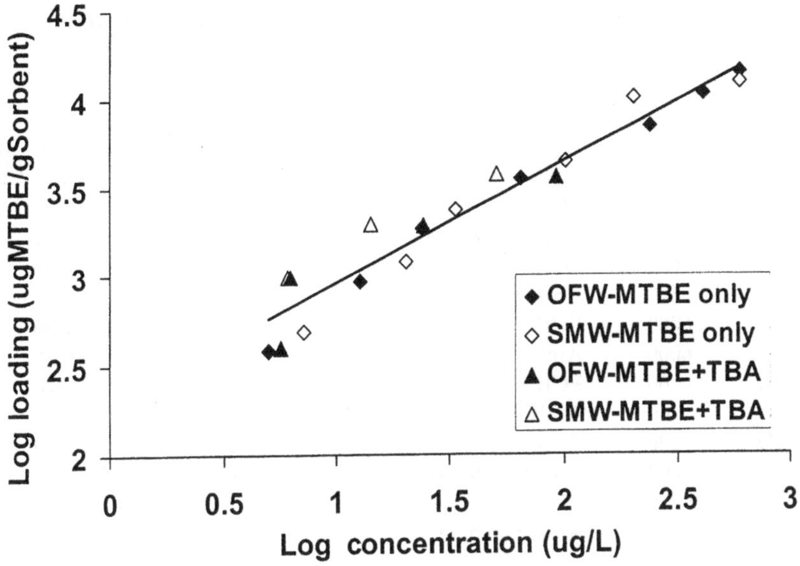

Figure 4 Isotherm of synthetic carbonaceous resin (Ambersorb 572) in organic-free water (OFW) or in Santa Monica water (SMW) with MTBE (1000 μg/L) or MTBE and *t*-butyl alcohol (100 μg/L). (Reproduced with permission from Ref. 19. Copyright American Water Works Association.)

CUR for Ambersorb 563 resin as compared to coconut GAC (PCB) in the presence of MTBE and BTX (Table 3).

Similarly, in the presence of TBA and MTBE, the resin's performance has shown to remain superior to GACs. Studies have demonstrated no observed reduction in adsorption capacity for MTBE in the presence of TBA for synthetic carbonaceous resins. Shih et al. [19] conducted isotherms on Ambersorb 572 resin and GRC-22 coconut GAC with organic-free water at 100 μg/L of TBA and 1000 μg/L of MTBE. The equilibrium adsorption capacity of the Ambersorb 572 resin for MTBE was unaffected by competition from TBA, while significant reduction in adsorption capacity of the GRC-22 GAC was observed (Figs. 3 and 4). In addition, recent RSSCTs conducted with Ambersorb 563 resin in ARWF groundwater by Shih et al. [13] indicate an almost sixfold reduction in CUR for Ambersorb 563 resin as compared to coconut GAC (PCB) in the presence of MTBE and TBA (Table 3).

These findings are not surprising given the hydrophobic nature of these resins and GACs, and the greater affinity of the hydrophobic BTEX fuel components to these adsorbents. The physical properties (e.g., high water solubility, low K_{ow}, etc.) of MTBE make its affinity to GAC and resin

Table 3 Comparison of GAC Versus Resin Cost for Treating MTBE in the Presence of SOCs

Water source	Flow rate[a] (gpm)	Adsorbent[b]	Influent MTBE (μg/L)	SV CUR at breakthrough[f] (lb/1000 gal)	Cost[e] ($/1000 gal)
LP	60	PCB	20[c]	1.95	4.76
LP	60	Ambersorb 563	20[c]	0.38	2.51
LP	600	PCB	20[c]	1.95	2.35
LP	600	Ambersorb 563	20[c]	0.38	1.02
LP	6000	PCB	20[c]	1.95	1.36
LP	6000	Ambersorb 563	20[c]	0.38	0.36
SLTUD	60	PCB	1,000[d]	2.25	5.11
SLTUD	60	Ambersorb 563	1,000[d]	0.38	4.53
SLTUD	600	PCB	1,000[d]	2.25	2.60
SLTUD	600	Ambersorb 563	1,000[d]	0.38	1.32
SLTUD	6000	PCB	1,000[d]	2.25	1.52
SLTUD	6000	Ambersorb 563	1,000[d]	0.38	0.59

[a]GAC system flow rate configurations: 60 gal/min (5000-lb beds, 3 in series), 600 gal/min (20,000-lb beds, 2 parallel lines, 3 in series), and 6000 gal/min (20,000-lb beds, 12 parallel lines, 3 in series). GAC modeled based on EBCT of 20 and 10 min for LP and SLTUD water, respectively. Resin modeled based on EBCT of 2.5 min.
[b]Ambersorb 563 and PCB are manufactured by Rohm and Haas and Calgon Co., respectively.
[c]Influent MTBE of 20 μg/L and 303 μg/L of BTX (98 μg/L benzene, 104 μg/L toluene, 98 μg/L *p*-xylene).
[d]Influent MTBE of 1000 μg/L and 100 μg/L of TBA.
[e]Carbon cost estimated using assumed capacity increase of 50% due to in-series operation.
[f]Breakthrough is calculated at effluent goal of 5 μg/L.
SOCs, synthetic organic chemicals.
Source: GAC and resin SV CURs from Refs. 7 and 13, respectively. GAC and resin cost assumptions from Ref. 8.

low compared to other organics such as BTEX that are commonly associated with MTBE in gasoline-contaminated waters. Competitive displacement of MTBE from adsorption can occur by those organics with greater affinity to GAC, resulting in column effluent MTBE concentrations exceeding those of the influent. Figure 5 shows an example of a chromatographic effect [5,20,25] in which the effluent is greater than the influent concentration up to 30% at 10 L per gram of GAC, and remains present until the conclusion of the RSSCT run (at approximately 26.5 L per gram of GAC). This suggests that the more strongly adsorbed organics such as BTX are displacing the less strongly adsorbed ones such as MTBE.

Competitive displacement and the resulting chromatographic effect are critical factors to consider in the design of full-scale GAC operations [5,21].

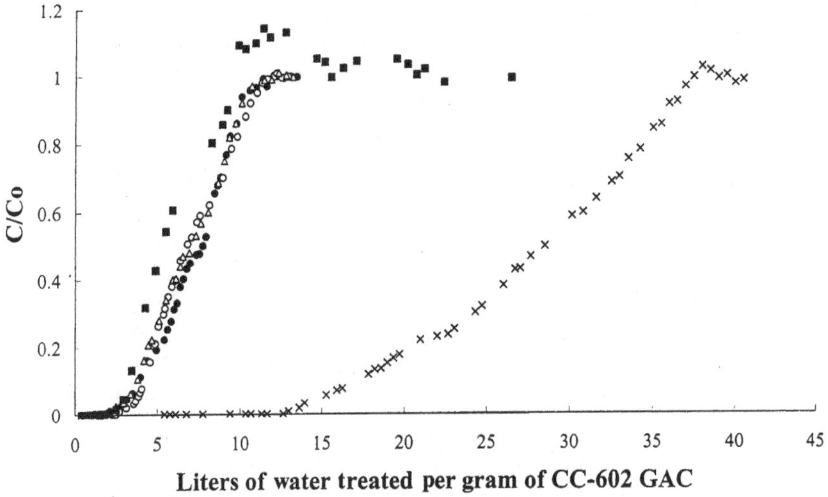

Figure 5 Breakthrough curves for the RSSCTs conducted with groundwater from South Lake Tahoe Utility District. • 1950 ppb MTBE at 10 min, EBCT; ○ 2052 ppb MTBE at 20 min, EBCT; △ 1964 ppb MTBE at 10 min, EBCT; × 20 ppb MTBE at 10 min, EBCT; ■ 1958 ppb MTBE and 1977 ppb BTX at 10 min, EBCT. (From Ref. 7.)

Ignorance of these factors can result in the appearance of a MTBE pulse in the effluent that is greater than the influent concentration (chromatographic effect). Competitive displacement including the chromatographic effect create the need for careful monitoring of GAC operations and earlier replacement of the GAC bed, before it reaches complete saturation [1,26], in order to maintain stable operation and to prevent the occurrence of a MTBE pulse. In addition, as shown in Figure 5, competitive displacement of MTBE by BTX can significantly increase the adsorbent usage rate and reduce even further the removal efficiency and adsorbent capacity.

V. SELECTION OF DESIGN PARAMETERS

One of the most important design considerations is the selection of the proper EBCT to fully utilize the adsorbent capacity [16]. The EBCT is defined as the bed volume (BV) divided by the flow rate Q, and is typically used to estimate the number and volume of adsorbent vessels required. Optimum EBCT is related to the length of the mass transfer zone, which is primarily affected by the hydraulic loading, the background water matrix, and the physical characteristics of the adsorbent and the organic compound (e.g., adsorption strength) to be removed [5,8]. As the EBCT increases, the adsorbent usage rate should decrease until a minimum value is reached. Similarly, the corresponding bed life should increase to a maximum value. Studies seem to indicate that an EBCT of 10 min should be sufficient for normal GAC operations involving organics with moderate affinity to GAC [5,27,28]. However, since MTBE has comparatively high water solubility and low GAC affinity, longer EBCTs may be required.

Shih et al. [7] investigated the dependence of GAC performance on EBCT for three natural water sources. They concluded that the background matrix (e.g., NOM) contributed significantly to the observed pattern. As the water source investigated changed from very low TOC (0.2 mg/L for SLTUD), to low TOC (1.0 mg/L for ARWF), to medium TOC (3.2 mg/L for LP), an increasing dependence of GAC performance on EBCT was observed. There was little or no difference in GAC capacity and bed life as the EBCT was altered for RSSCTs conducted in groundwater from SLTUD and ARWF. In contrast, significant decrease in GAC usage rate (30–40%) was found for RSSCTs conducted at 20- rather than at 10-min EBCT in surface water from LP. It is suspected that the higher NOM content of the surface water over the groundwater sources caused a greater competitive adsorption effect that made more sites on the GAC unavailable to MTBE, thus decreasing its rate of adsorption. Combined with the increased flow rate for the 10- versus the 20-min EBCT, this could have resulted in decreases in GAC capacity and bed life as one changed the EBCT from 10 to 20 min.

In contrast, synthetic resin adsorption systems have been shown to require a much shorter EBCT for MTBE applications than GAC systems. Preliminary results from the Charnock well field in Santa Monica, California, indicate that resins, specifically Ambersorb 563, require an EBCT of approximately 5 min [8], and recent RSSCTs (see Table 3) conducted with Ambersorb 563 resin by Shih et al. [13] have shown a five- to six-fold superior performance in MTBE applications with resin compared to GACs at EBCT as short as 2.5 min.

The shorter EBCT and superior adsorption kinetics of the carbonaceous resin as compared to GACs are thought to be derived from the resin's greater percentage of mesopores and macropores (relative to GAC), which facilitate the transport of adsorbate molecules to adsorption sites [8].

VI. COST ANALYSIS FOR GAC AND RESIN SYSTEMS

Tables 4 and 5 present summaries of GAC and resin cost estimates, respectively. As coconut GAC and synthetic carbonaceous resin (e.g., Ambersorb 563) have been shown to be the most promising candidates

Table 4 Summary of GAC Cost Estimates Derived from RSSCTs

Water source	Flow rate (gpm)[a]	Influent MTBE (μg/L)	Capital cost ($)[b]	Annual O&M ($)[c]	Total annual cost ($)	Unit cost ($/1000 gal)
LP	60	20	233,533	106,669	125,489	3.98
LP	600	20	1,018,673	474,440	556,531	1.76
LP	6,000	20	5,978,730	2,709,600	3,191,406	1.01
SLTUD	60	20	233,533	68,244	87,064	2.76
SLTUD	600	20	1,018,673	188,240	270,331	0.86
SLTUD	6,000	20	5,978,730	992,440	1474206	0.47
ARWF	60	200	233,533	86,369	105,189	3.34
ARWF	600	200	1,018,673	323,240	405,331	1.29
ARWF	6,000	200	5,978,730	1,802,400	2,284,206	0.72
SLTUD	60	2,000	233,533	200,919	219,739	6.97
SLTUD	600	2,000	1,018,673	1,176,440	1,258,531	3.99
SLTUD	6,000	2,000	5,978,730	6,921,600	7,403,406	2.39

[a]System flow rate configurations: 60 gal/min (5000-lb beds, 3 in series), 600 gal/min (20,000-lb beds, 2 parallel lines, 3 in series), and 6000 gal/min (20,000-lb beds, 12 parallel lines, 3 in series).
[b]Capital cost is based on detailed cost estimates and assumptions presented in Appendix 4A in Ref. 8, and includes cost of carbon adsorption vessels, piping, valves, electrical, site work, contractor O&P, engineering, and contingency. Amortization is based on 30-year period at 7% discount rate.
[c]Annual O&M cost based on CURs shown in the Appendix and detailed cost estimates and assumptions presented in Appendix 4A in Ref. 8.
Source: RSSCT data from Ref. 7.

Table 5 Summary of Resin Cost Estimates (Series Operation)

Flow rate (gal/min)	Influent MTBE (μg/L)	Resin + steamcapital cost ($)	Resin + steam O&M ($)	Annual cost of regeneration ($)	Total (resin, steam, and regeneration) ($/1000 gal)
60	20	16,307	61,928	825	2.51
60	200	26,710	103,662	1,086	4.17
60	2000	26,710	113,212	2,804	4.53
600	20	212,454	105.359	2,807	1.02
600	200	212,454	152,598	5,411	1.17
600	2000	212,454	180,435	22,800	1.32
6,000	20	804,448	308,685	8,295	0.36
6,000	200	1,007,690	461,320	34,482	0.48
6,000	2,000	1,007,690	819,231	41,520	0.59

Source: Data from Ref. 8.

for treating MTBE-contaminated waters, their performance under various treatment scenarios will be used as the basis for their cost evaluation. Treatment scenarios considered for both GAC and resin systems include the following: flow rate (60, 600, and 6000 gal/min), influent MTBE (20, 200, and 2000 μg/L), presence of other synthetic organic chemicals (SOCs) in addition to MTBE (e.g., BTEX and TBA), and competitive adsorption due to NOM from both surface water and groundwater sources.

For GAC systems, data obtained through RSSCTs [7] conducted in various combinations of water sources, EBCTs, and influent MTBE and SOCs concentrations were utilized to generate single-vessel CURs and breakthrough times (see Tables A-1, A-3, and A-5 in the Appendix). The predicted CURs, vessel life, and number of change-outs per year under in-series operations (see Tables A-2, A-4, and A-6 in the Appendix) were then derived by conservatively assuming a 50% increase in single-vessel breakthrough times to account for the increased specific throughput due to in-series operations. These assumptions are supported by field and industry experience [8]. The number of change-outs per year was used as the basis for deriving annual operation and maintenance (O&M) costs, which include the cost of the replacement GAC per change-out, cost of change-out labor/transport, O&M, power, and analytical testing. Similarly, the capital costs were developed using standardized assumptions for design/engineering, contractor overhead and profit (O&P), and contingency, and these costs were amortized based on 30-year period at 7% discount rate [8].

In contrast, unlike the GAC system in which the adsorbent usage rates determine the vessel life, the number of change-outs, and contribute a significant percentage to the total cost of the system, the resin usage rates have negligible impact on the final cost of the system. This is due to the fact that synthetic carbonaceous resins (e.g., Ambersorb 563) can be regenerated on-site with almost no loss of capacity, and the cost of regeneration is low compared to the total cost of the system (see Table 5). The California MTBE Research Partnership [8] conducted detailed economic analyses of resin systems. The cost analyses evaluated roughly 30 scenarios, and included operating design scenarios such as series versus carousel operation, and postregeneration options (e.g., hazardous waste disposal, air stripping, resin, GAC, or microbial degradation). Steam was the only regeneration option evaluated, since there are still unresolved uncertainties in the application effectiveness and approval for use of microwave and solvent regeneration techniques [8]. Recent ongoing field applications at Charnock Well Field in Santa Monica, California, however, suggest that solvent regeneration may prove to be more cost effective than steam regeneration [29].

Key findings on the treatment cost of synthetic resin by California MTBE Research Partnership with respect to series operation are shown in Table 5. Series rather than carousel operation is shown here, since series operation is much easier to operate (lower monitoring, sampling, and O&M costs), and the total cost difference between series and carousel operations is small (less than 10%). To determine time to breakthrough and the number of bed volumes treated, the AdDesignS model by Mertz et al. [18] was used to predict resin adsorption capacity and kinetics. Other assumptions mirror those performed for the GAC system, including those for design/engineering, contractor O&P, contingency, and discount rate and period of amortization (see Ref. 8).

Recent improvements in resin technology has increased the cost-effectiveness of the resin system such that it has been demonstrated to be competitive in cost to the most cost-effective GAC systems [8,13]. In fact, as shown in Tables 4 and 5, the cost of the GAC system is, for most of the scenarios evaluated, more than for the complementary resin system. This difference becomes even more pronounced at higher influent MTBE concentrations and flow rates, where the high unit resin cost ceases to be a factor in computing total treatment cost. In fact, the differential MTBE treatment cost between resin and GAC systems balloons from ~10% (at 60 gal/min and 20 μg/L and 600 gal/min and 200 μg/L) to ~50% (at 6000 gal/min and 200 μg/L), and finally to over fourfold difference (at 6000 gal/min and 2000 μg/L).

As shown in Section III, in contrast to resin systems, GAC performance in MTBE applications can be severely affected by the type and concentration of NOM present in the water source. RSSCTs conducted by Shih et al. [7] demonstrated that, depending on the effluent goal, an increase in CURs of over fourfold could occur with even a modest increase in NOM concentration. Table 4 illustrates the substantial difference in GAC treatment cost that can result from alteration in the background water matrix. The increase in estimated GAC treatment costs for operations conducted with LP surface water rather than with SLTUD groundwater ranged from 44% to 115% depending on the flow rate applied (Table 4). Clearly, the higher NOM concentration (characterized by the TOC concentration) of the LP surface water as compared to SLTUD groundwater caused a greater "fouling" of the GAC and resulted in significantly reduced GAC performance in terms of CURs, which in turn is translated to the observed higher treatment costs for removing MTBE from LP water.

In addition, even though the performance of both GAC and resin operations are dependent on the presence of other SOCs, GAC operation has been shown to be much more sensitive [7,19] and more susceptible to competitive displacement and the resulting chromatographic effect [7,13] than the complementary resin system. Section IV presents a brief review of the present knowledge on this topic. Table 3 presents a comparison of

GAC and resin performance and cost for treating MTBE in the presence of other SOCs such as BTEX and TBA. As shown in Table 3, when the SOCs present are BTEX compounds, the SV CURs for PCB coconut GAC are over fivefold the values observed for the Ambersorb 563 resin, and the higher CURs for the PCB coconut GAC are in turn translated into higher treatment costs in the range of approximately two- to fourfold relative to resin unit treatment costs, depending on the applied flow rate. Similarly, RSSCTs conducted in relatively "clean" SLTUD groundwater demonstrate that when TBA is present with MTBE, the SV CURs for PCB coconut GAC are over six times the values observed for the Ambersorb 563 resin, resulting in GAC unit treatment costs relative to resin in the range of 1.1- to 2.6-fold (Table 3).

VII. SUMMARY

Adsorption is a viable and important process in the removal of organic compounds from water. The physical and chemical characteristics of MTBE cause the compound to preferentially remain in solution, rendering most traditional treatment technologies ineffective. Careful evaluation is thus needed to define the optimum conditions under which the adsorption process would be cost-effective.

Isotherms and RSSCTs are two bench-scale tests that have been shown to be accurate and reliable predictors of breakthrough behavior of full-scale adsorber systems. Studies have demonstrated through the application of isotherms and RSSCTs that the performance of the Ambersorb 563 synthetic carbonaceous resin (e.g., Ambersorb 563) is superior to that of the GAC (e.g., coconut GAC) in removing MTBE from contaminated water, in particular under conditions of competitive adsorption. The significantly greater adsorption capacity and superior kinetics of the synthetic resin for MTBE, its resistance to competitive effects arising from NOM and other SOCs, the unique capability of the resin for on-site regeneration without loss of capacity, combined with reductions in resin unit costs, are contributing to making the life-cycle cost of the resin adsorption system competitive relative to other treatment technologies including GAC. As was demonstrated in Section VI, MTBE cost analyses performed have in general indicated lower costs for resin systems compared to GACs. In addition, as the degree of competition and/or the rate of influent flow increased, the discrepancy in cost between the two systems will be magnified further.

Future research should concentrate on investigating promising adsorbents identified through screening processes such as the isotherms.

The adsorption properties of the adsorbents may then be quantified by site-specific dynamic column testing such as the RSSCTs. Full-scale systems performance evaluations under actual field conditions may then serve to provide validate comparisons to dynamic column testing results.

REFERENCES

1. V. L. Snoeyink, S. R. Summers. (1999). Adsorption of organic compounds. In: Water Quality & Treatment, A Handbook of Community Water Supplies. 5th ed. Washington, DC: McGraw-Hill, pp 13.1–13.83.
2. D. W. Hand, J. A. Herlevich Jr., D. L. Perram, J.C. Crittenden. (1994). Synthetic adsorbent versus GAC for TCE removal. J Am Water Works Assoc, 86(8), 64–72.
3. J. C. Crittenden, et al. (1989). Prediction of GAC Performance Using Rapid Small-Scale Column Tests. American Water Works Association Research Foundation and American Water Works Association, Denver, CO.
4. I. H. Suffet, J. Mallevialle, E. Kawczyski, eds. (1995). Advances in Taste-and-Odor Treatment and Control, AWWARF: Denver, CO; pp 151–200.
5. I. H. Suffet, M. C. McGuire eds. (1980). Activated Carbon Adsorption of Organics from the Aqueous Phase, Vol. 1. Ann Arbor Science: Ann Arbor, MI.
6. T. Shih, M. Wangpaichitr, I. H. Suffet. (2001). Evaluation of the adsorption process for the removal of methyl tertiary butyl ether (MTBE) from drinking water. In: A. Diaz, D. Drogos, eds. Oxygenates in Gasoline: Environmental Aspects. American Chemical Society Symposium Series No. 799.
7. T. Shih, M. Wangpaichitr, I. H. Suffet. (2001). Evaluation of granular activated carbon technology for the removal of methyl tertiary butyl ether (MTBE) from drinking water. Submitted.
8. The California MTBE Research Partnership. (2000). Treatment Technologies for Removal of Methyl Tertiary Butyl Ether (MTBE) from Drinking Water. 2nd ed. National Water Research Institute, Fountain Valley, CA.
9. J. W. Neely, E. G. Isacoff. (1982). Carbonaceous Adsorbents for the Treatment of Ground and Surface Waters. Pollution Engineering and Technology, Vol. 21. Marcel Dekker, New York.
10. D. L. Gallup, E. G. Isacoff, D. N. Smith III. (1996). Use of Ambersorb carbonaceous adsorbent for removal of BTEX compounds from oil-field produced water. Environ Prog, 15(3), 197–203.
11. S. W. Davis, S. E. Powers. (1999). Alternative sorbents for removing MTBE from gasoline-contaminated groundwater. J Environ Eng, 126(4), 354–360.
12. J. P. Malley, R. R. Locandro, J. L. Wagler. (1993). Innovative point-of-entry (POE) treatment for petroleum contaminated water supply wells. Final Report, U.S.G.S. New Hampshire Water Resources Research Center, Durham, NH, September 1993.

13. T. Shih, M. Wangpaichitr, I. H. Suffet. (2003). Evaluation of granular activated carbon technology for the removal of methyl tertiary butyl ether (MTBE) from drinking water. Water Research, 37:275–385.
14. R. S. Summers, et al. (1992). Standardized Protocol for the Evaluation of GAC. American Water Works Association Research Foundation and American Water Works Association, Denver, CO.
15. J. C. Crittenden, J. K. Berrigan, D. W. Hand. (1986). Design of rapid small-scale adsorption tests for a constant surface diffusivity. J Water Pollution Control Fed, 58(4):312–319.
16. J. C. Crittenden, et al. (1987). Design consideration for GAC treatment of organic chemicals. J Am Water Works Assoc, 79(1):74–81.
17. D. W. Hand, et al. (1989). Design of fixed bed adsorbers to remove multicomponent mixtures of volatile and synthetic organic chemicals. J Am Water Works Assoc, 81(1):67–77.
18. K. A. Mertz, F. Gobin, D. W. Hand, D. R. Kokanson, J. C. Crittenden. (1994). Adsorption Design Software for Windows (AdDesignS). Houghton, MI: Michigan Technical University.
19. T. Shih, E. Khan, W. Rong, M. Wangpaichitr, J. Kong, I. H. Suffet. (1999). Sorption for removing methyl tertiary butyl ether from drinking water. American Water Works Association Natl Conf Proc, Chicago, June 1999.
20. R. J. Baker, I. H. Suffet. (1989). Frontal chromatographic concepts to study competetive adsorption, humic substances and halogenated organic substances in drinking water. In: I. H. Suffet, P. MacCarthy, eds. Aquatic Humic Substances: Influence on Fate and Treatment of Pollutants. Advances in Chemistry, Vol. 219. American Chemical Society, Washington, DC.
21. I. H. Suffet. (1980). An evaluation of activated carbon for drinking water treatment: National Academy of Science report. J Am Water Works Assoc, 72:41.
22. G. Chen, B. W. Dussert, I. H. Suffet. (1997). Evaluation of granular activated carbon for removal of methyl isoborneol to below odor threshold concentrations in drinking water. Water Res, 31(5):1155–1163.
23. G. Zimmer, H. J. Brauch, H. Sontheimer. (1989). Activated carbon adsorption of organic pollutants. In: I. H. Suffet, P. McCarthy, eds. Aquatic Humic Substances: Influence on Fate and Treatment of Pollutants. Advances in Chemistry, Vol. 219. American Chemical Society, Washington, DC.
24. C. Munz, J. L. Walter, G. Baldauf, M. Boller, R. Bland. (1990). Evaluating layered upflow carbon adsorption for the removal of trace organic contaminants. J Am Water Works Assoc, 82:63–76.
25. T. Yohe, I. H. Suffet, P. R. Cairo. (1981). Specific organic removal by granular activated carbon pilot contactors. J Am Water Works Assoc, 73: 402–410.
26. A. A. Keller, O. C. Sandall, R. G. Rinker, et al. (1998). Health and Environmental Assessment of MTBE, Vol. 5, Risk Assessment, Exposure Assessment, Water Treatment and Cost Benefit Analysis. Report to the

Governor and Legislature of the State of California as sponsored by SB521, Vol. 5, pp 1–35.
27. O. T. Love, R. G. Eilers. (1982). Treatment of volatile organic compound in drinking water. J Am Water Works Assoc, 74(8):413.
28. R. S. Summer, et al. (1997). DBP Precursor Control with GAC Adsorption. American Water Works Research Foundation, Denver, CO.
29. Kennedy Jenks. (2001). Tom Shih, Personal Communication with Larry Leong, Kennedy Jenks, Inc., October 2001.

APPENDIX

Table A-1 SV CURs and Breakthrough Times for ARWF Groundwater

Influent MTBE (μg/L)	GAC vendor/ product	EBCT (min)	SV CUR at breakthrough[b,c] (lb/1000 gal)	SV breakthrough[c] (days)
50	Calgon/PCB	10	0.72	62.0
200	Calgon/PCB	10	1.20	37.4
200	Calgon/PCB	20	0.82	54.4
1000[a]	Calgon/PCB	10	2.25	19.8

[a]*tert*-Butyl alcohol at 100 μg/L.
[b]Breakthrough is calculated at effluent goal of 5 μg/L.
[c]Breakthrough is calculated assuming GAC single-vessel EBCT of 20 min.
SV, Single vessel; CURs, carbon usage rates.
Source: RSSCT data from Ref. 7.

Table A-2 Predicted CURs and Vessel Life Using In-Series Operation for ARWF Groundwater

Influent MTBE (μg/L)	GAC vendor/ product	EBCT (min)	Predicted usage rate[b] (lb/1000 gal)	Predicted vessel life[b] (days)	Predicted change-outs per year[b]
50	Calgon/PCB	10	0.48	93.0	3.9
200	Calgon/PCB	10	0.80	56.1	6.5
200	Calgon/PCB	20	0.55	81.6	4.5
1000[a]	Calgon/PCB	10	1.50	29.7	12.2

[a]*tert*-Butyl alcohol at 100 μg/L.
[b]Values estimated using assumed capacity increase of 50% due to in-series operation.
CURs, carbon usage rates.
Source: RSSCT data from Ref. 7.

Table A-3 SV CURs and Breakthrough Times for SLTUD Groundwater

Influent MTBE (μg/L)	GAC vendor/ product	EBCT (min)	SV CUR at breakthrough[b,c] (lb/1000 gal)	SV breakthrough[c] (days)
20	Calgon/PCB	10	0.36	123.7
2000	US Filter/CC-602	20	4.07	12.0
2000[a]	US Filter/CC-602	20	6.56	7.4

[a]VOC = 1977 μg/L BTX (512 μg/L benzene, 540 μg/L toluene, and 925 μg/L *p*-xylene).
[b]Breakthrough is calculated at effluent goal of 5 μg/L.
[c]Breakthrough is calculated assuming GAC single-vessel EBCT of 20 min.
SV, single vessel; CURs, carbon usage rates.
Source: RSSCT data from Ref. 7.

Table A-4 Predicted CURs and Vessel Life Using In-Series Operation for SLTUD Groundwater

Influent MTBE (μg/L)	GAC vendor/ product	EBCT (min)	Predicted usage rate[b] (lb/1000 gal)	Predicted vessel life[b] (days)	Predicted change-outs per year[b]
20	Calgon/PCB	10	0.24	185.5	2.0
2000	US Filter/CC602	20	2.71	18.0	20.3
2000[a]	US Filter/CC602	20	4.37	11.0	33.2

[a]VOC = 1977 μg/L BTX (512 μg/L benzene, 540 μg/L toluene, and 925 μg/L *p*-xylene).
[b]Values estimated using assumed capacity increase of 50% due to in-series operation.
CURs, carbon usage rate.
Source: RSSCT data from Ref. 7.

Table A-5 SV CURs and Breakthrough Times for LP Surface Water

Influent MTBE (μg/L)	GAC vendor/ product	EBCT (min)	SV CUR at breakthrough[b,c] (lb/1000 gal)	SV breakthrough[c] (days)
20	Calgon/PCB	20	1.36	33.1
20	US Filter/CC-602	20	1.67	29.2
20[a]	Calgon/PCB	20	1.95	22.9

[a]VOC = 303 μg/L BTX (108 μg/L benzene, 98 μg/L toluene, and 97 μg/L *p*-xylene).
[b]Breakthrough is calculated at effluent goal of 5 μg/L.
[c]Breakthrough is calculated assuming GAC single-vessel EBCT of 20 min.
SV, single vessel; CURs, carbon usage rates.
Source: RSSCT data from Ref. 7.

Table A-6 Predicted CURs and Vessel Life Using In-Series Operation for LP Surface Water

Influent MTBE (μg/L)	GAC vendor/ product	EBCT (min)	Predicted usage rate[b] (lb/1000 gal)	Predicted vessel life[b] (days)	Predicted change-outs per year[b]
20	Calgon/PCB	20	0.91	50.0	7.3
20	US Filter/CC602	20	1.11	44.0	8.3
20[a]	Calgon/PCB	20	1.30	34.0	10.7

[a]VOC = 303 μg/L BTX (108 μg/L benzene, 98 μg/L toluene, and 97 μg/L *p*-xylene).
[b]Values estimated using assumed capacity increase of 50% due to in-series operation.
CURs, carbon usage rates.
Source: RSSCT data from Ref. 7.

15

Impact of MTBE Phaseout on the Petroleum and Petrochemical Industries

Blake T. Eskew
Purvin & Gertz, Inc., Houston, Texas, U.S.A.

Christopher L. Geisler
Chemical Market Associates, Inc., Houston, Texas, U.S.A.

I. INTRODUCTION

Methyl *tertiary*-butyl ether (MTBE) has played a key role in the development of today's regulatory approach to controlling vehicle emissions and meeting air quality improvement targets in the United States and throughout the world. However, the discovery of MTBE contamination of surface and groundwater caused by gasoline leaks from underground storage tanks, spills from gasoline transfers, and releases into lakes and reservoirs by marine craft has raised concerns over MTBE-contaminated drinking water. MTBE, even at very low concentrations, causes taste and odor problems, and very small amounts can potentially make large volumes of water unfit as a source of drinking water. This has prompted state and federal governments in the United States to review the environmental drawbacks associated with the use of MTBE as a gasoline blending component, and to initiate programs that would either ban or restrict its use. If MTBE use in gasoline is eliminated, the phaseout will have far-reaching impacts on not only the MTBE industry but also the refining and petrochemical industries worldwide.

In this chapter, Purvin & Gertz and CMAI have not attempted to determine whether MTBE is "good" or "bad," nor have we attempted to determine whether the use of MTBE provides benefits that outweigh the risks inherent in its use. These questions can only be resolved through the political process. Instead, we have attempted to assess the impacts of reduced MTBE use to improve understanding of the implications of future political decisions.

II. BACKGROUND AND ASSUMPTIONS

A. Oxygenates and Vehicle Emissions

Research in the early 1980s demonstrated the capability of fuel oxygenates to reduce wintertime carbon monoxide (CO) emissions. The impact of fuel oxygenates on hydrocarbon (HC) emissions was established in the mid-to-late 1980s, culminating in ARCO's EC-1 emissions-controlling gasoline introduced in the California market in 1989 and in the oxygen requirement in reformulated gasoline (RFG) adopted in the Clean Air Act Amendments of 1990 (CAAA). However, advanced automotive technology appears to have substantially reduced the contribution of fuel oxygen content to emissions control, particularly with respect to HC emissions.

The presence of oxygen in gasoline, whether from MTBE, ethanol, or other compounds, has the result of producing a shift toward leaner operation (i.e., increasing the air:fuel ratio) for any given volumetric air:fuel combination. This change in engine stoichiometry is largely responsible for the observed reductions in CO and HC emissions. Oxygenates tend to provide other benefits due to their low sulfur, lack of aromatics, lack of olefins, and high octane, but compounds without oxygen can provide these benefits as well.

The benefits of leaner operation are particularly significant in vehicles which use carburetors rather than fuel injection, and that lack oxygen-sensing electronic feedback fuel control systems. The U.S. automobile fleet of the 1980s was dominated by such vehicles, and these vehicles are still a major element of the fleet in many areas of the world. In North America, Western Europe, Japan, and other more developed markets, however, vehicles with advanced fuel control systems now comprise the largest element of the automobile fleet. Only 6% of the U.S. new vehicle fleet had fuel injection in 1980, growing to over 90% by 1990. In 1999, essentially all new gasoline-powered vehicles employed port fuel injection. In these vehicles, the electronic feedback control systems automatically compensate for the fuel oxygen content by decreasing the air:fuel ratio, with the exceptions of cold operation or periods of rapid acceleration. In addition, fuel oxygen can even result in increased emissions depending on the leanness of the control system's base calibration. Conclusions about the impact of oxygenates on HC emissions typically depend on the study's assumptions regarding the composition of oxygen-free gasoline. However, studies have concluded that improvements in ambient CO levels are linked to the presence of wintertime oxygenate programs. The persistence of the CO impact even as automotive fuel control technology has improved may be due to the cold-start efficacy of oxygenates.

While oxygenates can be viewed as having a beneficial, or at worst neutral impact on CO and HC emissions, the effect of oxygen on nitrogen oxides (NOX) emissions has long been controversial. Vehicles studies have typically shown a small increase in NOX, with the impact increasing with oxygen content. Some tests have shown that ethanol has a particular impact on NOX. As HC emissions from vehicles and other sources have been reduced, appreciation of the role of NOX in ozone formation has increased, and NOX control is likely to be the most difficult task facing U.S. automobile manufacturers over the next decade.

The emissions control strategy embodied in the 1990 CAAA and the California Air Resources Board (CARB) programs is generally viewed as successful in improving urban air quality. However, the contribution of oxygenates to this strategy's continued success in the United States is now diminishing as automotive technology improves. In the current public debate on MTBE and oxygenate use, arguments seem to focus more and more on the "other" benefits and properties of oxygenates. For example, ethanol use is favored due to its energy security benefits, while MTBE's use is defended due its indirect impact of lowering aromatics, olefins, and sulfur. The debate over the need for oxygen in RFG may soon be over, although its role in CO-control regions may continue. In the United States and other more developed markets, fuel improvement efforts are shifting to the "other" benefits, targeting sulfur, aromatics, olefins, and distillation.

Even while oxygenate use is deemphasized in the U.S. market, oxygenate strategies may continue to expand in other regions. Markets with a high population of older vehicles with less advanced fuel control systems can still reap significant air quality benefits from oxygenate programs. Manila, Caracas, Beijing, and many other urban regions would be appropriate candidates, while Mexico City has benefited from an oxygenate program for many years. The increased availability of MTBE from supplies no longer needed in the United States could act to expand oxygenated gasoline programs around the world.

In this context, MTBE and other oxygenates can be viewed as a "transitional" strategy for air quality improvement. In markets with older, less advanced vehicle fleets, oxygenate strategies can provide immediate, widespread benefits. As these areas move toward higher environmental standards and their vehicle fleets improve, oxygenate use can decline without affecting air quality. The U.S. market may now be entering this latter phase of oxygenate use.

B. Future MTBE Use

Efforts to phase out MTBE use began in California in 1997–1998, and culminated in California Governor Gray Davis's March 1999 Executive Order stating that MTBE shall be removed from all gasoline marketed in California at the earliest possible date, but not later than December 31, 2002. Gasoline supply considerations resulted in a one-year delay in implementation of the MTBE ban.

In addition, it appears probable that MTBE use in the remainder of the United States will be reduced or eliminated. A number of states, as well as the United States Environmental Protection Agency (USEPA), have announced plans to eliminate the future use of MTBE. In March 2000, the USEPA and the U.S. Department of Agriculture jointly announced actions to significantly reduce or eliminate the use of MTBE and boost the use of renewable fuels such as ethanol. However, due to the complexities of the 1990 CAAA, it is considered more likely that the issue will be resolved through action of the U.S. Congress. A number of bills have been introduced into the U.S. Congress to address MTBE use, most of which call for a ban on its use in the 2007–2008 time frame.

While the political pressure to ban MTBE use in the United States seems to have advanced rapidly, markets outside the United States have generally shown much less concern over MTBE use. Environmental and industry authorities in Europe, Latin America, and Asia have typically considered the water contamination problem something to be addressed through prevention and maintenance, not through the ban of a useful and valuable gasoline component.

Some countries, including Mexico, have slowed programs to expand oxygenate use and deferred investments in MTBE capacity, but have not acted to reduce current consumption. In Europe, MTBE consumption is generally anticipated to expand over the next few years, due to the scheduled changes in gasoline specifications, particularly the reductions in aromatics, benzene, and sulfur. These countervailing influences on MTBE regulation make its future in international markets extremely uncertain at this time. In the near term, its use is likely to be influenced as much by individual company positions and risk profiles as by national policies. Longer term, the experience of the U.S. market should provide some direction. If contamination problems continue to spread, MTBE may rapidly become politically unpalatable almost everywhere. However, if improved maintenance and handling practices act to reduce or eliminate contamination, international markets are more likely to maintain and expand their use of MTBE. However, with the United States representing almost 60% of the global MTBE demand and 43% of global MTBE production capacity, whatever action is taken in the United States will undoubtedly have a major impact on the state of the industry worldwide.

C. MTBE Phaseout Scenarios

Given the level of uncertainty still associated with the use of MTBE in U.S. gasoline, there are many possible scenarios for future MTBE regulation. This analysis focuses on the results of a scenario incorporating a total phaseout for MTBE in the U.S. gasoline pool.

III. PETROLEUM INDUSTRY IMPACTS

A. Impact on Refining Operations and Investment Patterns

The U.S. refining industry's substantial reliance on MTBE has developed in less than two decades, illustrating the power of regulatory action to reshape the business environment. The elimination of MTBE from California gasoline and potential phaseout throughout the U.S. market will have a significant impact on refining operations and gasoline prices throughout the United States, with further effects in other world markets.

U.S. MTBE consumption reached roughly 300,000 bbl/day, less than 4% of total gasoline production, with a contribution twice that level in the U.S. East Coast and West Coast markets. MTBE's high octane makes its contribution to those regions' gasoline supply very significant, providing about two octane numbers to the regional pool. The removal of MTBE will strain the refining industry's octane capacity at a time when more stringent gasoline sulfur requirements are also reducing octane capacity.

1. United States West Coast

The highly desirable gasoline blending properties of MTBE make it particularly difficult to replace in California due to the stringent California Air Resources Board (CARB) gasoline specifications. Imports (domestic or foreign) of iso-octane, alkylates, and heavy isomerates will be required to allow the region to blend its refinery-produced components and satisfy demand. Ethanol will be required to meet the federal RFG oxygen mandate if it is not lifted in California. While ethanol blending can help recover MTBE's lost octane, its high blending vapor pressure creates difficulties in producing gasoline meeting the stringent CARB Phase 3 specifications, and results in rejection of lighter materials from the gasoline pool. Refining capacity expansion will be necessary and will focus on gasoline desulfurization and octane enhancement.

2. United States East Coast

The U.S. East Coast consumes a lower proportion of MTBE than the West Coast, but faces a similar octane deficit if MTBE is phased out. However, the less stringent federal RFG specifications allow the region to meet future demand with only low levels of imports of additional components. Gasoline desulfurization and octane enhancement are again the major focus of anticipated refinery investment.

3. United States Gulf Coast

The replacement of MTBE is a less acute problem in the U.S. Gulf Coast, due to its relatively low octane and volume contribution in the total pool. Refinery investment will be driven by the tighter sulfur specifications, with the MTBE phaseout resulting in significant investment in octane enhancement.

B. Impact on Butane Supply and Demand

Butane blending by U.S. refineries was dealt another blow by the Phase II federal RFG requirements and the subsequent reductions in gasoline Reid vapor pressure (RVP) in 2000. If MTBE is phased out, alkylation demand for isobutane will increase with refinery alkylation operations. On the other hand, butane feedstock requirements for MTBE producers are certainly at risk with reductions in MTBE demand. A portion of the standalone, butane dehydrogenation-based MTBE capacity in the United States is expected to convert to isooctane production. However, some capacity may shut down, offsetting the increases in consumption of isobutane for the production of alkylate.

C. Impact on Refined Product Prices and Refinery Margins

Reductions in MTBE use will tighten refined product balances, with the greatest impact on the U.S. East and West coasts. Due to the product supply links with the U.S. Gulf Coast, East Coast disruptions will have a significant impact on the Gulf Coast market.

1. Impact on Gasoline Prices and Refining Margins

The complete loss of MTBE would result in an increase in U.S. refinery capacity utilization of about 1.5%. This seemingly minor change will be focused on the East and West Coasts, which are already running near capacity and are quite sensitive to supply disruptions.

The tighter supply/demand balance is anticipated to result in small increases in U.S. Gulf Coast regular-grade gasoline prices in the year of the MTBE phaseout. RFG and premium gasoline prices are expected to show somewhat larger price increases. As capacity responds to higher prices, prices and refining margins would be expected to move back toward traditional levels. However, the potential for supply disruptions to create upward price volatility will be increased, and price spikes could be much higher.

Pricing impacts are expected to be somewhat more pronounced on the West Coast, due to the stringent CARB Phase 3 regulations. The cost of producing CARB Phase 3 gasoline without MTBE is estimated to be several cents per gallon higher than CARB Phase 2. If ethanol is required in CARB Phase 3, prices would likely increase further.

2. Impact on Octane Values

In addition to the tighter gasoline supply/demand balance caused by MTBE phaseout, the loss of MTBE's octane contribution will result in substantial changes in refinery operations. As refiners move toward higher-severity reforming and higher alkylation throughput, octane production costs will increase, and the market value of octane will follow. Octane capacity utilization is projected to increase strongly. Octane values are expected to increase by roughly 20% due to the MTBE phaseout. Octane values can be measured in terms of the differential between premium and regular gasoline, which is expected to increase by about 1.0 cent per gallon (Fig. 1).

D. Impact on International Markets

The proposed phase out of MTBE in the U.S. gasoline pool will undoubtedly have further impact on international markets. Developments on the U.S. East Coast, which is highly dependent on imports from Europe and other exporters to balance demand, will affect prices in those supplier markets.

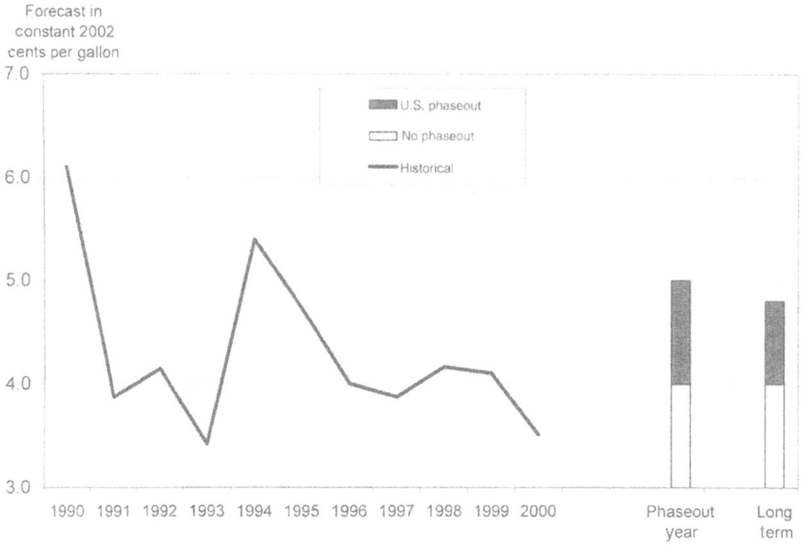

Figure 1 MTBE phaseout impact on USGC octane values (premium versus regular price differential).

1. European Markets

The increased demand for gasoline and gasoline components on the U.S. East Coast is expected to raise margins slightly in Europe. European prices are projected to spike in 2005 with the introduction of new European Union (EU) product specifications. As in the past, the anticipated spike is expected to be short-lived, as constraints in the supply chain due to the introduction of new specifications are resolved through refinery investment and operating changes.

2. Asian Markets

Even as the MTBE phaseout occurs in the United States, economic expansion and associated product demand growth in Asia will again be filling Asian refining capacity. The impact of changes in the U.S. product values will be overshadowed by improvements in local refining profitability, limiting the impact of a U.S. MTBE ban on prices in the region. A short-lived spike in response to California import needs is possible, but will quickly be offset by increased imports of MTBE into the Asian market.

IV. PETROCHEMICAL INDUSTRY IMPACTS

Significant impacts on the petrochemical industry are expected if MTBE use is phased out in U.S. gasoline. Price and supply/demand impacts from the MTBE phaseout will be felt globally in MTBE and methanol markets. Octane value changes resulting from the removal of MTBE will affect propylene, butylenes, and aromatics prices, especially in the United States.

A. Impact on MTBE Production, Trade, and Pricing

World MTBE capacity grew from 8 million tons in 1990 to 25 million tons by 2000. The capacity increase has been largely in response to gasoline oxygenate requirements regulated by the 1990 Clean Air Act Amendments (CAAA) in the United States and also due to MTBE's favorable gasoline blending properties of high octane, no sulfur, aromatics or olefins, and neutral vapor pressure. Increases in MTBE capacity over the past 10 years have been focused on production from isobutane in areas of low-cost gas availability such as the Middle East, Venezuela, and Malaysia. Refinery-based MTBE production has also grown significantly due to the flexibility MTBE provides as a motor gasoline blendstock and the alkylation unit debottlenecking that results when isobutylene is removed for MTBE production.

On-purpose MTBE production from isobutene, along with refinery-based production, account for about 70% of the world's MTBE supply. Production from the raffinate of butadiene extraction accounts for approximately 20% of MTBE production. Nearly the entire balance of global MTBE production is from *tertiary*-butyl alcohol (TBA) dehydration, a co-product of propylene oxide (PO) production. Growth in this route of MTBE production has been stalled because additions of propylene oxide/styrene monomer (PO/SM) units have been the preferred route to supply PO demand growth.

Over the period 1990–2000, global MTBE demand grew at an average annual rate of more than 13%, with total world MTBE demand approaching 22 million tons in 2000. The United States accounted for almost 60% of the total world MTBE demand in 2000. Within the United States, the West Coast market MTBE demand was roughly 35% of the total U.S. MTBE demand in 2002. Due to the importance of the U.S. market, the California MTBE ban and potential restrictions on MTBE use in the rest of the United States have significant impacts on the global MTBE industry.

If MTBE use is banned in U.S. gasoline, world MTBE demand will drop to less than 11 million tons. Our analysis assumes increased consumption in other regions of the world as MTBE use is reduced in the United States. However, gasoline markets in other areas of the world are not large enough to assimilate all of the current MTBE production.

The direction West Europe takes is an important factor in the global MTBE balances. In 2005, new gasoline specifications will be implemented that reduce the amount of allowable aromatics in gasoline from 42% to 35%. MTBE use is expected to increase to help replace the octane lost by these reductions in gasoline aromatics.

We also assume that consumption of MTBE continues to grow with gasoline demand in the current MTBE-consuming countries of Southeast Asia, Northeast Asia, the Middle East, and the Indian Subcontinent. In the Middle East, MTBE use as a percentage of the gasoline consumed in the region is expected to increase. In Japan, a decline in MTBE use is expected, as the climate for MTBE use is more negative.

Some of the current MTBE consumption in the rest of North America and in South America is actually MTBE consumed for production of reformulated gasoline destined for the United States. In the case of a phaseout of MTBE use in the United States, this consumption will be terminated. Figure 2 shows the impact of a MTBE ban in California by the end of 2003 and of a total ban in the United States by 2007.

MTBE plant rationalizations began in 2002 and have continued in 2003 in preparation for the California ban. In addition, on-purpose MTBE economics have been affected by high natural gas prices in the United States. Initially, MTBE plant rationalizations and conversions have been concentrated in North America and include West Coast refinery-based production and some U.S. Gulf Coast isobutane dehydrogenation facilities.

In the case of a complete phaseout of MTBE in the United States, North American MTBE capacity will be seriously affected. We believe nearly all isobutane dehydrogenation production will either convert to isooctane production or shut down. Units based on raffinate from butadiene extraction will also convert, or the producers will sell the raffinate for conventional alkylate production. In fact, it is likely that only the North American MTBE producers that are integrated into other chemical derivatives or TBA dehydration units will continue to produce MTBE. As shown in Figure 3, if MTBE is banned in U.S. gasoline, North American MTBE production capacity will drop from nearly 12 million tons to 2 million tons or below.

Other global MTBE production will be affected if MTBE use is phased out in the United States. With the United States currently importing nearly 30% of the rest of the world's production, it is impossible to assume otherwise. The United States imported more than 4 million tons of MTBE annually prior to the California ban.

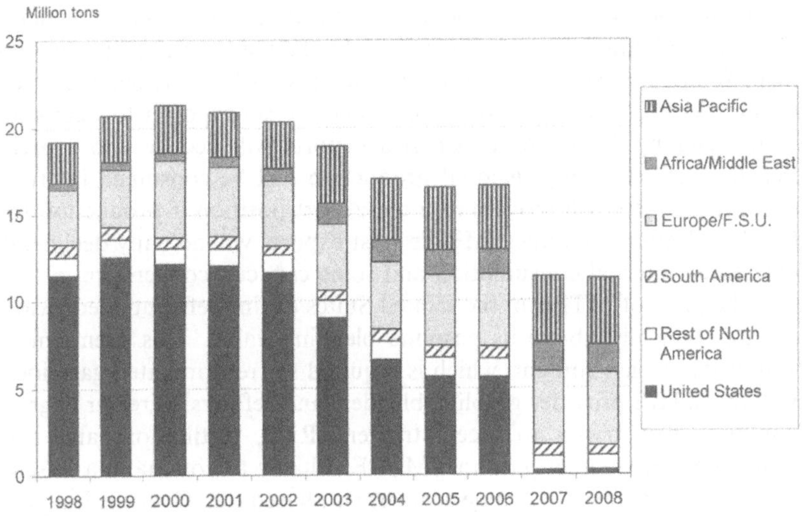

Figure 2 World MTBE demand (phaseout case).

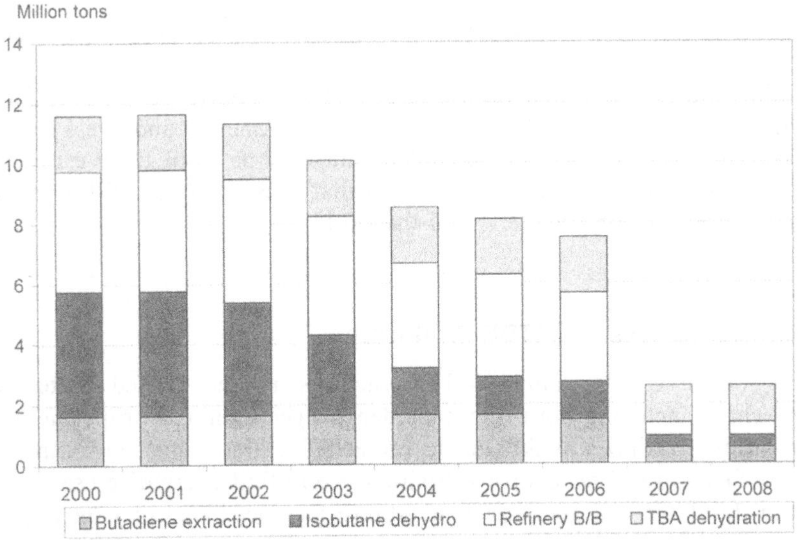

Figure 3 North America MTBE capacity (phaseout case).

In addition to Saudi Arabia and Venezuela, other major suppliers of MTBE to the United States include Canada, the United Arab Emirates, and Brazil. World MTBE trade flows will shift if MTBE use is banned in U.S. gasoline. West Europe will move from a historic net exporter to a consistent net importer by 2004. South America exports will decline and a greater percentage of remaining regional production will be consumed internally. The United States will move into a net export position if production from TBA dehydration continues. Middle East exports will actually decline, due to increased internal consumption and some capacity conversion.

The price of MTBE in the United States during reformulated gasoline season is typically above its gasoline blending value. This premium is a result of its oxygen content, which is required for reformulated gasoline. In addition, MTBE provides gasoline blenders and refiners a greater degree of flexibility when trying to meet stringent RVP, distillation, and sulfur requirements in summer gasoline. MTBE is likely to command a lessening premium over its blending value as its use is phased out. Over the long term, global MTBE prices are expected to erode as global MTBE operating rates decline due to reduced U.S. demand. This price erosion will be tempered by MTBE capacity rationalizations, with some units converting to isooctane production or shutting down in favor of conventional alkylate production.

The U.S. Gulf Coast MTBE pricing mechanism is expected to change in the case of a complete phaseout of MTBE in the United States. The United States will switch from a large MTBE importer to a net exporter, with any excess production moving to South America and West Europe. Therefore, MTBE prices in the United States will be set by the West Europe price less freight. In fact, it is expected that West Europe will become the price-setting region for the rest of the world.

B. Alternatives for MTBE Producers

The future of standalone MTBE facilities in the United States and elsewhere, including both isobutane dehydrogenation and propylene oxide co-product facilities, is difficult to project. The amount of sunk capital in these facilities creates a strong driving force to find alternative uses, rather than shutdown and abandonment. MTBE facilities based on captive olefin streams produced by refineries, or closely linked petrochemical plants, are likely to have alternative disposition for isobutylene in existing alkylation facilities. However, technologies for isooctane production or polymerization of mixed butylene streams may be chosen by some refiners, particularly those with large volumes of olefins or limited alkylation capacity.

If MTBE is phased out in the United States, most standalone MTBE facilities will need to choose between production of alkylate or isooctane or face shutting down due to their inability to compete in global MTBE markets. In a base standalone MTBE configuration, normal butane is isomerized to isobutane and the isobutane is then dehydrogenated to isobutylene. The isobutylene is then reacted with methanol to produce MTBE. Alternately, isobutylene can be dimerized and then hydrogenated to isooctane or can be reacted with isobutane to produce conventional alkylate. Figure 4 compares the process flow of the alternatives with the MTBE process.

Conversion from MTBE production to alkylate production would require the isobutylene from the butane dehydrogenation process to be routed to a new alklylation unit, where it would be reacted with isobutane to produce alkylate. Two competing technologies are generally available, one using hydrofluoric acid catalyst and another using sulfuric acid catalyst. For purposes of this analysis, sulfuric acid alkylation was used to evaluate the conversion of existing MTBE facilities to alkylate production.

The alkylation reaction produces a mixture of various branched paraffins. Isooctane is one of these components. Pure isooctane can be produced by dimerizing isobutylene, which selectively produces the olefin isooctene. Isooctene can then be hydrogenated to produce isooctane. Several licensors are currently marketing isooctane production technology. Process information typical of selective conversion of isobutylene to isooctane was used for this analysis.

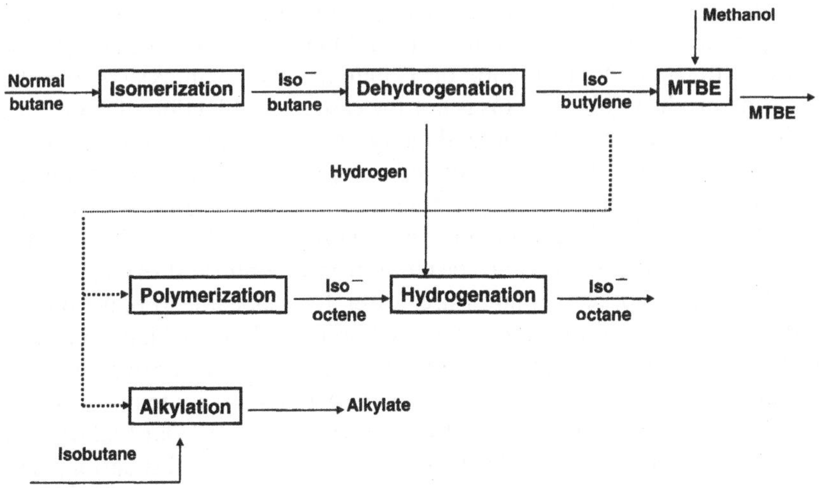

Figure 4 MTBE plant conversion alternatives.

Table 1 Process Charge and Yield Comparison (Percent of Butane Feed)

	MTBE	Alkylation	Isooctane
Charge			
Normal butane	100	100	100
Isobutane	—	91	—
Methanol	35	—	—
Total feed	135	191	100
Yield			
MTBE	104	—	—
Alkylate	—	145	—
Isooctane	—	—	70

Both potential alternative uses for an MTBE facility fully utilize the butane isomerization and dehydrogenation sections of the existing plant, but require new process units to convert the isobutylene to a saleable gasoline component. The differences between alternatives lie in the product yield, product qualities and value, and capital and operating costs. The charge and yield for the alternative processes are compared to MTBE production in Table 1.

The three cases each assume that the butane dehydrogenation section operates at the same isobutylene production rate. Data are presented on a volume basis to highlight the differences in product volumes for the various alternatives. Product yields as a percentage of total feed are comparable for the alternatives. However, production of MTBE and alkylate enable the producer to combine methanol or isobutane with isobutylene, thereby increasing product volume. This is not the case for production of isooctane. In the case of alkylation, more than twice the product volume can be produced compared to production of isooctane given the same normal butane feed rate. This is a major consideration for producers evaluating alternatives to MTBE production, especially when faced with volume reductions in their gasoline pool resulting from the removal of MTBE.

Although product volumes are lower for isooctane production, its octane is higher and vapor pressure is lower, leading to higher product values than butylene alkylate. Isooctane also has a more favorable midpoint distillation, as shown by the T-50. T-50 is defined as the temperature at which 50% of the material has distilled overhead in a standard laboratory distillation. Table 2 shows the vapor pressure, octane, distillation, and 2007 blending value ratio to reformulated gasoline of MTBE compared to alkylate and isooctane. For reference, we have included similar data

Table 2 Alternative Products for MTBE Producers

	Vapor pressure (RVP, psia)	Octane (R+M)/2	T-50 (°F)	2005 Price ratio to USGC premium RFG
Component				
MTBE (isobutylene + methanol)	8.0	110	131	1.22
Alkylate (isobutylene + isobutane)	4.0	93	231	1.04
Isooctane (isobutylene + isobutylene)	2.0	99	211	1.12
References				
Premium reformulated gasoline	6.8	93	<220	1.00
Ethanol	18.0	113	173	1.23

for summer reformulated premium gasoline and ethanol. Ethanol is the other most prevalent oxygenate used in U.S. motor gasoline.

The data show the value of MTBE is not only its oxygen content but also its high octane, low T-50, and neutral vapor pressure relative to reformulated gasoline. Although ethanol has higher octane than MTBE, its high blending vapor pressure is a major drawback and one of the primary reasons that MTBE has become the preferred oxygenate used in reformulated gasoline in the United States. Conventional alkylate and isooctane are less valuable blending components for premium gasoline due to their lower octane. However, these components have very advantageous vapor pressures and are expected to become even more integral components of reformulated gasoline blends if MTBE use is phased out. In addition, if oxygen content continues to be mandated in reformulated gasoline, isooctane and alkylate values are likely to increase, as these products will be required to offset the higher vapor pressure of ethanol in reformulated blends.

Capital costs for conversion to isooctane can be significantly lower than grass-roots sulfuric acid alkylation. It may also be possible to convert some existing MTBE reactors to isooctane production, which would tend to lower the capital cost required for some isooctane conversions.

Overall economics for converting MTBE production to an alternative product will vary depending on a producer's specific configuration. Alkylation is likely to be favored by refiners that can accommodate additional butylenes in their existing alkylation unit without significant capital outlay. Refiners with limited alkylation capacity may choose to convert their MTBE production to isooctane to partially replace the octane and volume contribution of MTBE. We believe standalone MTBE producers are more likely to convert their production to iso-octane rather than alkylation, due to the lower capital cost. These producers will have to adjust their fixed cost structures to account for the lower product volume and revenue generated by isooctane relative to MTBE.

C. Impact on Methanol Production, Trade, and Pricing

World methanol demand reached 31 million tons in 2000, having grown at an average rate of nearly 6% in the 1990s. MTBE has been methanol's fastest growing derivative, with an average growth rate over the decade of more than 13%. North America accounts for about 30% of world methanol demand, with the United States alone representing one-fourth of total world demand. In the United States, MTBE accounted for over 40% of methanol demand in 2000. A phaseout of MTBE use in the United States will have a significant impact on world methanol demand and an even greater impact on U.S. methanol demand.

World methanol capacity growth has more than kept pace with world methanol demand over the 1990s. The most recent capacity additions have been focused on methanol production in areas with access to low-cost or stranded gas. In fact, during the period 1995–2000, more than 5 million tons of new methanol capacity were added in Chile, Trinidad, and Saudi Arabia. Significant expansions also occurred in the countries of Norway, New Zealand, and Venezuela during this same period.

After capacity increases through the early 1990s, significant rationalization and consolidation is now occurring in the high-cost-gas region of North America. Methanol, like MTBE, is a highly traded commodity. In 2000, almost 50% of total world demand was traded between the 10 major regions. Similar to MTBE, the United States is a major importer of methanol. The United States imports methanol primarily from the South American countries of Trinidad, Venezuela, and Chile, as well as from Canada. U.S. methanol imports have increased in recent years as domestic capacity has been shut down in favor of lower-cost imports. Asia and Europe join the United States in importing methanol from the Middle East and South America.

In the event of a 2007 MTBE phaseout in the United States, world methanol demand for MTBE would drop from a high of more than 8 million tons in 2000 to 4.4 million tons in 2007. In North America, the loss of more than 3 million tons of methanol demand represents nearly one-third of the region's total demand. Figure 5 shows the significant impact an MTBE phaseout will have on total methanol demand in North America.

With such a significant reduction in U.S. methanol demand, continued rationalization in the North American region is likely. Global methanol

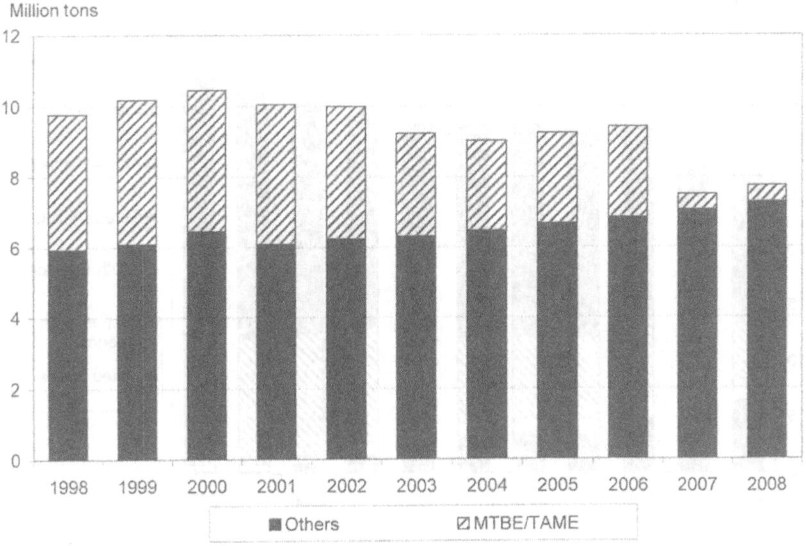

Figure 5 North America methanol demand (phaseout case).

production costs are primarily a function of the cost of gas as feedstock. North American methanol producers are at a significant production cost disadvantage due to the high price they must pay for gas.

High-valued electrical power generation and home heating alternative markets elevate the gas price and increase volatility in this region. In comparison, remote areas with no alternative markets for gas routinely set gas prices at a level that provides economic incentive for derivative production.

We estimate an additional 2.5–3.0 million tons of disadvantaged methanol capacity will need to be shed if MTBE use is banned in U.S. gasoline. The bulk of these capacity rationalizations will occur in North America, with lesser amounts in Europe and Southeast Asia.

North America will continue to lose its world capacity share as a result of significant increases in South American and Middle East methanol production. This new global production, along with the MTBE phaseout in the United States, will result in further declines in North America methanol production.

The impact of the MTBE phaseout on world methanol supply/demand is evident, although somewhat masked by the larger market developments (Fig. 6) such as the potential new supply for methanol-to-olefins technology.

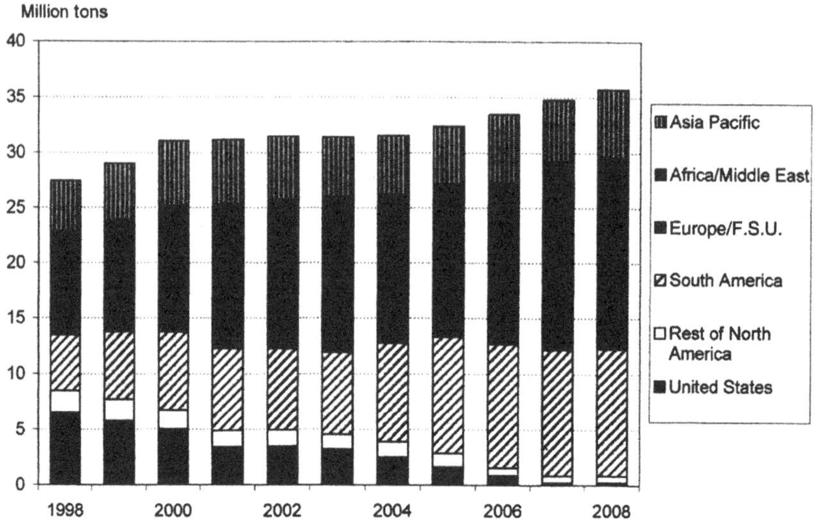

Figure 6 World methanol production (phaseout case).

Trade flows of methanol are expected to be similar to today even if MTBE use is reduced or banned in the United States. This is different than the projections for MTBE for two reasons.

The first major difference is the significant use of methanol in North American for formaldehyde and other derivative production. The second difference is our belief that the bulk of the methanol capacity closures will occur in North America, thereby allowing a still significant level of methanol imports even though demand for MTBE production is reduced.

The mechanisms for setting regional methanol prices are not expected to change even if MTBE use in gasoline is banned in the United States. North American demand for other derivative production remains significant. World trade flows are not expected to be materially affected due to the forecast rationalization of methanol capacity in North America. Expectations are for margins to erode as global methanol operating rates decline due to reduced North American demand. This margin erosion will be less pronounced due to capacity rationalizations over the period.

D. Impact on Other Petrochemical Markets

1. Propylene

With the impending decline in MTBE usage in the U.S. motor gasoline pool, propylene alkylate and propylene polygas/dimate become possible candidates to replace the "clean" volume and octane that will be lost by the removal of MTBE. However, for most refineries, the high values of propylene that can be realized by sales to petrochemicals will be well above the netbacks that propylene will attain in the alkylation or dimersol/polygas processes.

Therefore, propylene alkylation in refineries is expected to increase, but only slightly as compared to the increase in propylene demand by petrochemical derivatives. Refiners are expected to continue to increase average propylene yields from FCC units in order to meet rapidly growing petrochemical demand (along with the relatively small incremental alkylation demand that could occur with a phaseout of MTBE).

Since incremental demand for propylene alkylation in refineries will compete with incremental demand from petrochemical derivatives, refinery propylene market prices will move slightly higher as MTBE is phased out of gasoline. The slightly higher future market prices expected for propylene as a result of the impending MTBE phaseout are not expected to materially affect demand for propylene's end-use derivatives.

2. Butylenes (Raffinate-1 and Raffinate-2)

Before 1990, prices for butylene raffinate from butadiene extraction (raffinate-1) and isobutylene removal (raffinate-2) were based on their value in the production of butylene alkylate for the gasoline pool. As the oxygenate programs in the Clean Air Act Amendments of 1990 were implemented, values for raffinate-1 came to be based on its use in the production of MTBE, providing higher values to the raffinate-1 producer than sales to the refining industry for butylene alkylate production. With the phaseout of MTBE, valuations for raffinate-1 streams will once again be set by use in the production of butylene alkylate or other high-octane components.

The raffinate producer will find that its product value will be highly dependent on the end-use gasoline market that its customer will be targeting. Conventional gasoline prices will provide the lowest returns, reformulated gasoline somewhat higher, and gasoline for the California region—CARB gasoline—will provide the highest value by far.

If MTBE is banned, most raffinate-2 produced from MTBE plants will be lost. Some new and existing raffinate-1 producers may be forced to produce isooctane in order to produce the volume of raffinate-2 that is needed for feedstock to chemical plants that cannot tolerate significant concentrations of isobutylene.

3. Aromatics

Aromatics are major components of gasoline in the United States, and are also major contributors to the chemical industry. There are two main drivers for aromatics demand: chemical demand and the demand by the gasoline pool for low-vapor-pressure/high-octane blending components.

Refineries must maintain adequate octane values in the gasoline pool. The removal of aromatics for chemical use depletes the octane value and volume of the gasoline pool. In the past, lead alkyls supplemented the gasoline pool to raise the octane rating. More recently, MTBE has filled this role, and to some extent, MTBE has supported the availability of aromatics for chemical use.

If MTBE is phased out, it is expected that refiners will respond by increasing the severity of reforming operations to replace some of the lost octane contribution of MTBE. Given the prediction of higher reformer severity and also reformer capacity utilization, one would expect an increase in the concentration of aromatics in the gasoline pool. If a U.S. refiner increases aromatics in reformulated gasoline, then it will have to offset this increase with other nonaromatic blending components in order to avoid an increase in toxic emissions.

The aromatic components of reformate that are the most desirable for chemical use are benzene, toluene, and xylenes. Table 3 shows their properties and impact on toxic emissions.

Benzene concentration in U.S. gasoline is currently limited to 1 vol% in reformulated gasoline and is the variable with the greatest leverage on toxic emissions. A phaseout of MTBE use is not likely to have a significant impact on benzene supply for chemicals, as refiners in general have already removed the majority of contained benzene in reformulated gasoline.

Of the aromatics being discussed, toluene is the most desirable for gasoline blending, and is an excellent component for premium gasoline or to correct a gasoline blend that is off on vapor pressure or octane. Increases in toluene have a lesser impact on toxic emissions than does benzene,

Table 3 Selected Aromatics Properties

Property	Units	Reformate	Benzene	Toluene	Mixed xylenes
Reid vapor pressure	psia	6	3.2	0.8	1.2
Road octane	(R+M/2)	95	96	104	104
T50/boiling point	°F	248	176	231	282
Emissions impact	Percent[a]	N/A	7.4	0.5	0.5

[a]Percent decrease in toxic emissions resulting from a 0.5 vol% reduction in component.

and toluene's boiling point is not as negative as mixed xylenes' on the T-50 of finished gasoline.

If MTBE is phased out, will U.S. refiners be less likely to part with this valuable gasoline component? We believe the answer to this question is that the toluene will be made available, but the transaction will need to occur at a higher price.

We expect that toluene prices will increase for two reasons if MTBE is phased out. First, the underlying value of octane increases, which drives up the blending value of toluene. In fact, our assumption is that the blending value of toluene remains closely linked to reformulated gasoline pricing in the summer and then falls back to its value in conventional gasoline in the winter. Second, in the near term, the greater blending value of toluene will decrease refiners' incentive to spend capital to recover toluene for chemical use. Chemical toluene prices and margins will need to rise in order to justify additional toluene recovery.

Although octane values in Asia are currently at historic lows, we expect that over the coming years they will equilibrate more closely with values in West Europe and the United States. A phaseout of MTBE use in the United States will impact this improving regional parity. The MTBE phaseout results in a greater increase in octane values in the United States. As a result, regional blending values will diverge. This will have the effect of further disadvantaging U.S. producers of mixed xylenes and paraxylene relative to West Europe and Asia and could affect future xylene trade flows.

REFERENCE

1. MTBE Phaseout—Global Impact on the Petroleum & Petrochemical Industry. Copyright 2000 by Chemical Market Associates, Inc., and Purvin & Gertz, Inc.

Index

T - #0028 - 111024 - C0 - 229/152/23 - PB - 9780367394486 - Gloss Lamination